Handbook of Applied Physics

Handbook of Applied Physics

Edited by
Faith Sherman

Willford Press
www.willfordpress.com

Published by Willford Press,
118-35 Queens Blvd., Suite 400,
Forest Hills, NY 11375, USA

ISBN: 978-1-64728-356-8

Cataloging-in-Publication Data

Handbook of applied physics / edited by Faith Sherman.
p. cm.
Includes bibliographical references and index.
ISBN 978-1-64728-356-8
1. Physics. 2. Physical sciences. 3. Dynamics. I. Sherman, Faith.
QC21.3 .H36 2022
530--dc23

For information on all Willford Press publications
visit our website at www.willfordpress.com

Willford Press

Contents

Preface

The purpose of the book is to provide a glimpse into the dynamics and to present opinions and studies of some of the scientists engaged in the development of new ideas in the field from very different standpoints. This book will prove useful to students and researchers owing to its high content quality.

Applied physics deals with the application of the principles of physics for solving practical and technological problems. It is also involved in the application of physics to other areas of science. Applied physics seeks to develop new technologies and solve engineering problems. Some of its branches are accelerator physics, biophysics, communication physics, experimental physics and microfluidics. There are numerous fields where applied physics is used for developing technologies such as stealth technology, sonar, lidar, computational physics, quantum technology and semiconductors. It also plays a key role in improving practical investigations such as in experimental nuclear physics and experimental particle physics. This book discusses the fundamentals as well as modern approaches of applied physics. It explores all the important aspects of this field in the present day scenario. This book is a vital tool for all researching and studying this field.

At the end, I would like to appreciate all the efforts made by the authors in completing their chapters professionally. I express my deepest gratitude to all of them for contributing to this book by sharing their valuable works. A special thanks to my family and friends for their constant support in this journey.

Editor

1

Plasmon-enhanced second harmonic generation in semiconductor quantum dots close to metal nanoparticles

Pablo M. Jais,[1,2*] Catalina von Bilderling,[1,3†] Andrea V. Bragas[1,2‡]

We report the enhancement of the optical second harmonic signal in non-centrosymmetric semiconductor CdS quantum dots, when they are placed in close contact with isolated silver nanoparticles. The intensity enhancement is about 1000. We also show that the enhancement increases when the incoming laser frequency ω is tuned toward the spectral position of the silver plasmon at 2ω, proving that the silver nanoparticle modifies the nonlinear emission.

I. Introduction

Second Harmonic Generation (SHG) is the second order nonlinear process for which two photons with the same frequency ω interact simultaneously with matter to generate a photon of frequency 2ω. The nonlinear response of a material to an applied field becomes evident in the presence of intense fields, as the one given by a focused pulsed laser.

Due to symmetry requirements for second order nonlinear processes, the leading dipolar term of the SHG is forbidden for centrosymmetric materials. However, this condition is relaxed at surfaces, where the symmetry is broken. On the other hand, for a small and perfectly spherical particle of any type of material under homogeneous illumination, the SHG from the surface is also zero since the geometrical shape recovers the symmetry [1].

The presence of inhomogeneities in the electromagnetic field recovers the SHG even for symmetric materials, both from bulk and surface, as pointed out by Brudny *et al.* [1]. Those inhomogeneities may be, for instance, the consequence of a strong focusing of the field, the proximity of a surface or the presence of another particle. Besides, electronic resonances and enhancement of the field around the particle will increase the SHG and, in many cases, would make it detectable.

It is very well-known, as one of the main results of nano-optics, that the field around a metal nanoparticle is confined and enhanced when the incoming wavelength is resonant with the surface plasmons sustained by the nanoparticle. This effect is exploited in many different applications of metal nanoparticles: volume reducers for diffusion measurements [2], optical probes for high resolution imaging [3–5], biosensors [6], nanoheaters [7] and surface enhanced Raman scattering (SERS) [8, 9], among others.

In the present paper, we show the enhancement

*E-mail: jaisp@df.uba.ar

†E-mail: catalina@df.uba.ar

‡E-mail: bragas@df.uba.ar

[1] Instituto de Física de Buenos Aires, CONICET, Argentina.

[2] Laboratorio de Electrónica Cuántica, Departamento de Física, Facultad de Ciencias Exactas y Naturales, Universidad de Buenos Aires, Intendente Güiraldes 2160, Pabellón I - Ciudad Universitaria, Buenos Aires, C1428EHA, Argentina.

[3] Centro de Microscopías Avanzadas, Facultad de Ciencias Exactas y Naturales, Universidad de Buenos Aires, Argentina.

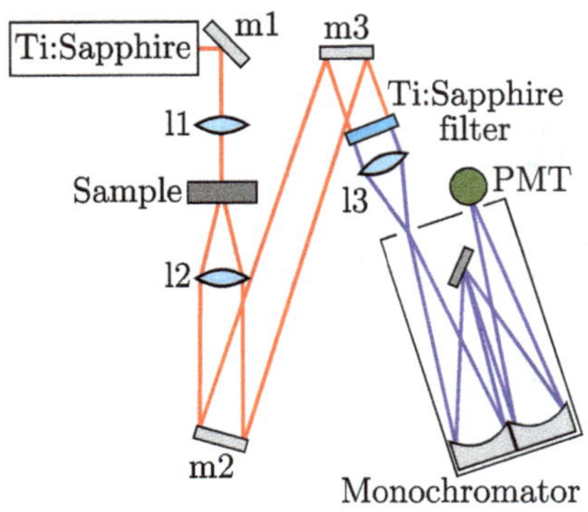

Figure 1: Experimental setup. PMT: photomultiplier tube; m1 to m3: mirrors, l1 to l3: lenses.

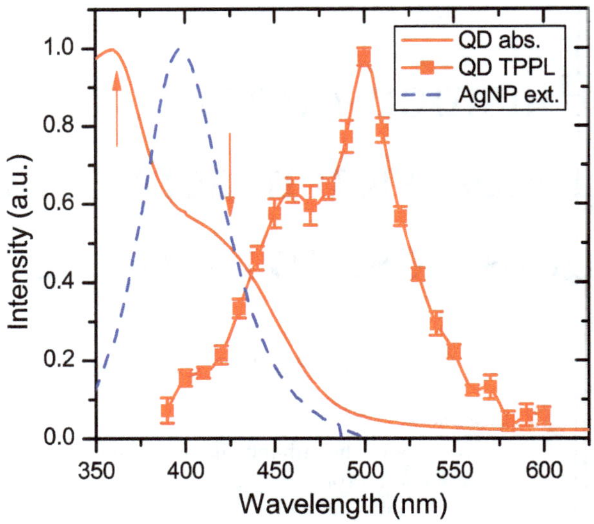

Figure 2: Absorption (red line) and two-photon photoluminescence, TPPL (red squares), of the CdS QDs in ethanol solution. The dotted blue line is the extinction of the Ag NP in aqueous solution.

of the SHG when non-centrosymmetric CdS quantum dots are placed in close contact with silver nanoparticles. The plasmon of the nanoparticles is resonant at the second harmonic (SH) frequency and enhances the emission. As far as we know, there are few works on the enhancement of SHG (either in semiconductor QDs or in molecules) produced by the resonant excitation of metal nanoparticles [10–13].

II. Materials and methods

The experiment was performed in transmission, by performing spectrally-resolved photon counting (Fig. 1). The laser was a KMLabs tunable mode-locked Ti:Sapphire with 50 fs pulse width, 400 mW average power, 80 MHz repetition rate and tunable in the range 770–805 nm. The resolution of the monochromator was 2 nm in all experiments. Silver nanoparticles (NPs), of average diameter 20 nm, were synthesized as in Marchi *et al.* [14], while the CdS quantum dots (QDs), 3 nm in diameter, as in Frattini *et al.* [15].

The absorption spectra for both NPs and QDs in solution are shown in Fig. 2. The arrows mark the first and second excitonic transitions in the QDs, at (362 ± 4) nm and (425 ± 15) nm, obtained by the second derivative method [16]. Figure 2 also shows the two-photon photoluminescence (TPPL) spectrum of the QDs in solution, excited at $\lambda = 780$ nm and measured with the setup shown in Fig. 1. A Stokes shift was observed in the TPPL, in accordance with the same effect reported for the (one photon) photoluminescence of CdS QDs [17].

NP samples were prepared by placing a drop of the NP solution on a glass substrate until the solvent evaporated. The average NP concentration was ≈ 10 particles per μm^2. However, optical and SEM images revealed that the metal NPs are inhomogeneously distributed on the substrate, with a high concentration of particles in the fringe (drop boundary).

A CdS QD sample was prepared according to the same protocol. However, the QD solution has a very high concentration of aminosilanes. They form a thick layer when dried, so the QDs are immersed in an aminosilane matrix. The estimated height of this layer was $\approx 5\,\mu m$, and the average QD concentration was $\approx 10^5$ particles per μm^2 (inferred from AFM images taken at a 1:1000 diluted concentration).

For the mixed NP-QD sample, we first dried a drop of the metal NP solution, and then we deposited a drop of CdS QD solution on top of it. The resulting structure is schematically shown in Fig. 3.

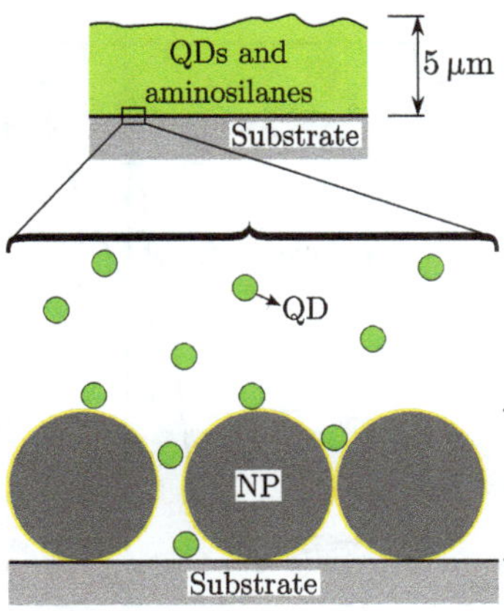

Figure 3: Diagram of the NP-QD sample.

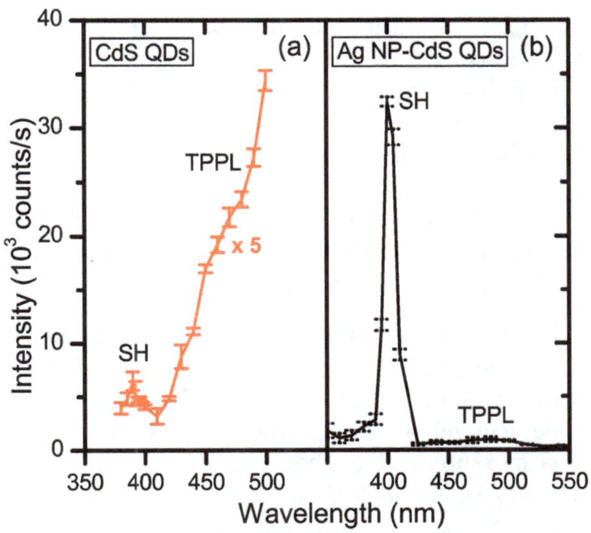

Figure 4: SH and TPPL spectra for (a) the CdS QD sample (multiplied by 5) and (b) the mixed Ag NP-CdS QD sample. Note that the SH peak was 20 times higher in the mixed sample, and the ratio between the SH and the TPPL changes drastically.

III. Results and discussion

For the silver NP samples we did not detect SH signal above the noise level of the experiment (of about 200 counts per second). This result agrees with the fact that silver is a centrosymmetric material and the NPs are almost spherical, even considering that their interaction should produce a weak SH signal. On the other hand, Fig. 4a shows that the CdS QD sample presents a small SH peak (as expected for a non-centrosymmetric material) and a strong TPPL signal. The central wavelength of the excitation laser for this measurement was 784 nm, and its FWHM was 26 nm.

However, a strong SH signal was measured in the mixed NP-QD sample, as shown in Fig. 4b. The SHG was 20 times larger in this sample, and the SH/TPPL ratio increased significantly. In addition, in the mixed sample, the TPPL decayed after a few seconds of laser irradiation while the SHG remained stable over time. In this case, the central wavelength was 799 nm, and the FWHM was also 26 nm.

The observations can be attributed to the enhancement of the nonlinear field around the QDs due to silver NP plasmon resonance at 2ω.The incoming light at ω is not enhanced by the silver NP, but creates a weak nonlinear signal at 2ω in the QDs. This outgoing signal is enhanced by the field enhancement factor (η), so that the SHG intensity is enhanced by η^2. Note that in the analogous case of SERS, the incoming and outgoing signals are enhanced giving an intensity enhancement of η^4 [18].

The magnitude of this enhancement is not easy to calculate since the mixed NP-QD sample contains several hot-spots (usually in the concentrated NP fringe), and their intensities were wildly different. Nevertheless, the highest intensities recorded for the NP-QD sample were $\approx 10^4$ cts/s, while the highest ones for the QD sample were ≈ 1700 cts/s. To estimate the enhancement, we used the analytical enhancement factor (AEF), defined as [19]

$$\mathrm{AEF} = \frac{I_{\mathrm{NP-QD}}/\rho_{\mathrm{NP-QD}}}{I_{\mathrm{QD}}/\rho_{\mathrm{QD}}}, \tag{1}$$

where I and ρ are the SHG intensity and the surface concentration of QDs in the corresponding samples.

We must take into consideration that there is a thick layer of QDs that are not in the enhancement region of the NPs, i.e., the measured signal in the NP-QD sample is the sum of the intensity coming from QDs close to the NPs that are enhanced, and the intensity from QDs far from the NPs that are not enhanced. If we keep only the enhanced fraction,

$$\mathrm{AEF} \approx \frac{I_{\mathrm{NP-QD}} - I_{\mathrm{QD}}}{I_{\mathrm{QD}}} \frac{\rho_{\mathrm{QD}}}{\tilde{\rho}_{\mathrm{NP-QD}}}, \tag{2}$$

where $\tilde{\rho}_{\mathrm{NP-QD}}$ is the concentration of QDs near the NPs. Since the NP enhancement decays one order of magnitude in 5 nm [20], we can take $\tilde{\rho}_{\mathrm{NP-QD}} \approx \rho_{\mathrm{QD}} \frac{5\,\mathrm{nm}}{5\,\mu\mathrm{m}}$. This gives AEF $\approx 5\,10^3$. This is just an estimation of the order of magnitude of the enhancement, since the uncertainties in the thicknesses and maximum intensities are very high, but this estimation is consistent with the enhancements usually found in SERS experiments [18].

To study the spectral dependence of the enhancement, and gain a deeper insight into the behavior of the system, the incoming laser light was tuned while keeping the laser always on the same spot. This measurement is shown in Fig. 5, where the intensity is the photon flux at half the laser wavelength, normalized to the signal provided by a 40 fs pulse with 275 mW average power, according to the following formula:

$$I_{\mathrm{norm}} = I \frac{\delta t}{40\,\mathrm{fs}} \left(\frac{275\,\mathrm{mW}}{P} \right)^2 . \quad (3)$$

The signal at the SH wavelength was, for each point of Fig. 5, a local maximum in the spectrum.

Figure 5 shows that the SH from the QD sample increases toward the excitonic resonance in about a factor of 2 for the whole tuning range, a value similar to the one reported by Baranov *et al.* [21]. It is worth noting that the contribution of the TPPL to this increase is much smaller than the SH signal throughout the measured range (see Fig. 4). Showing a very different spectral response, the SH for the mixed NP-QD system sharply increases in more than a factor of 7 close to the resonance of the silver plasmon.

This measurement reinforces the hypothesis that the silver NP is modifying the nonlinear response of the QDs thanks to the resonant excitation of the silver NP plasmons. However, it must be noted that the enhancement does not exactly follow the plasmon resonance spectrum shown in Fig. 2. The reason for this is still unknown, but it is possible that the plasmon resonance was shifted due to the interaction among NPs. Unfortunately, the tuning range of the laser was much smaller than the spectral width of the plasmon resonance.

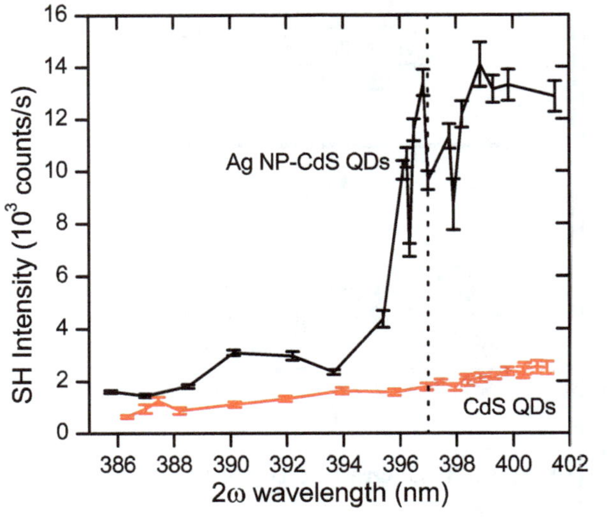

Figure 5: SHG as a function of the laser excitation wavelength. The horizontal scale is the second harmonic of the incident photon. Red symbols show the SHG for the QD sample, while the black symbols show the SHG for the mixed NP-QD sample. The dotted line marks the spectral position of the maximum of the plasmon resonance.

IV. Conclusions

We have observed a strong enhancement ($\approx 10^3$) of the SHG when CdS QDs are mixed with silver NPs, compared with a CdS QD sample. The spectral dependence of the enhancement shows that the NP plasmons are resonantly excited by the SH emission of the QDs. These measurements are evidence that the SH enhancement is mediated by nanoparticle plasmons. This effect can be applied to significantly improve traditional applications of SH measurements such as the study of surface deposition and orientation of molecules, among others.

Acknowledgements - We thank Claudia Marchi for her help with the NP synthesis and Nora Pellegri for providing the QDs. We also thank Alejandro Fainstein for the insightful discussions. This work was supported by the University of Buenos Aires under Grant X010 and by ANPCYT under Grant PICT 14209.

---

[1] V L Brudny, B S Mendoza, W L Mochan, *Second-harmonic generation from spherical particles*, Phys. Rev. B **62**, 11152 (2000).

[2] L C Estrada, P F Aramendía, O E Martínez, *10000 times volume reduction for fluorescence correlation spectroscopy using nano-antennas*, Opt. Express **16**, 20597 (2008).

[3] A F Scarpettini, N Pellegri, A V Bragas, *Optical imaging with subnanometric vertical resolution using nanoparticle-based plasmonic probes*, Opt. Commun. **282**, 1032 (2009).

[4] T Kalkbrenner, M Ramstein, J Mlynek, V Sandoghdar, *A single gold particle as a probe for apertureless scanning near-field optical microscopy*, J. Microsc. **202**, 72 (2001).

[5] P Anger, P Bharadwaj, L Novotny, *Enhancement and quenching of single-molecule fluorescence*, Phys. Rev. Lett. **96**, 113002 (2006).

[6] A J Haes, R P Van Duyne, *A nanoscale optical biosensor: Sensitivity and selectivity of an approach based on the localized surface plasmon resonance spectroscopy of triangular silver nanoparticles*, J. Am. Chem. Soc. **124**, 10596 (2002).

[7] A G Skirtach, *et al.*, *Laser-induced release of encapsulated materials inside living cells*, Angew. Chem. Int. Ed. **45**, 4612 (2006).

[8] P Etchegoin, *et al.*, *New limits in ultrasensitive trace detection by surface enhanced Raman scattering (SERS)*, Chem. Phys. Lett. **375**, 84 (2003).

[9] P G Etchegoin, P D Lacharmoise, E C Le Ru, *Influence of photostability on single-molecule surface enhanced Raman scattering enhancement factors*, Anal. Chem. **81**, 682 (2009).

[10] M Ishifuji, M Mitsuishi, T Miyashita, *Enhanced optical second harmonic generation in hybrid polymer nanoassemblies based on coupled surface plasmon resonance of a gold nanoparticle array*, Appl. Phys. Lett. **89**, 011903 (2006).

[11] H A Clark, *et al.*, *Second harmonic generation properties of fluorescent polymer-encapsulated gold nanoparticles*, J. Am. Chem. Soc. **122**, 10234 (2000).

[12] S Baldelli, *et al.*, *Surface enhanced sum frequency generation of carbon monoxide adsorbed on platinum nanoparticle arrays*, J. Chem. Phys **113**, 5432 (2000).

[13] I Barsegova, *et al.*, *Controlled fabrication of silver or gold nanoparticle near-field optical atomic force probes: Enhancement of second-harmonic generation*, Appl. Phys. Lett. **81**, 3461 (2002).

[14] M C Marchi, S A Bilmes, G M Bilmes, *Photophysics of Rhodamine B interacting with silver spheroids*, J. Colloid Interface Sci. **218**, 112 (1999).

[15] A Frattini, N Pellegri, D Nicastro, O de Sanctis, *Effect of amine groups in the synthesis of Ag nanoparticles using aminosilanes*, Mater. Chem. Phys. **94**, 148 (2005).

[16] A I Ekimov, *et al.*, *Absorption and intensity-dependent photoluminescence measurements on CdSe quantum dots: Assignment of the first electronic transitions*, J. Opt. Soc. Am. B **10**, 100 (1993).

[17] Z Yu, *et al.*, *Large resonant Stokes shift in CdS nanocrystals*, J. Phys. Chem. B **107**, 5670 (2003).

[18] E C Le Ru, P G Etchegoin, *Principles of Surface-Enhanced Raman Spectroscopy*, Elsevier Science Ltd, Amsterdam (2008).

[19] E C Le Ru, E Blackie, M Meyer, P G Etchegoin, *Surface enhanced raman scattering enhancement factors: A comprehensive study*, J. Phys. Chem. C **111**, 13794 (2007).

[20] S A Maier, *Plasmonics: Fundamentals and Applications*, Chap. 5.1, Springer Science, New York (2007).

[21] A V Baranov, *et al.*, *Resonant hyper-Raman and second-harmonic scattering in a CdS quantum-dot system*, Phys. Rev. B **53**, R1721 (1996).

2

Experimental determination of L X-ray fluorescence cross sections for elements with $45 \leq Z \leq 50$ at 10 keV

E. V. Bonzi,[1,2*] G. B. Grad,[1†] R. A. Barrea[3‡]

Synchrotron radiation at 10 keV was used to experimentally determine the Ll, $L\alpha$, $L\beta_I$, $L\beta_{II}$, $L\gamma_I$ and $L\gamma_{II}$ fluorescence cross sections for elements with $45 \leq Z \leq 50$, as part of an ongoing investigation at low energies. The measured data were compared with calculated values obtained using coefficients from Scofield, Krause and Puri *et al.*

I. Introduction

This work is part of a systematic investigation on elements with $45 \leq Z \leq 50$, which has been carried out at different energies [1–3]. The L X-ray cross sections were measured with monoenergetic excitation beam at 10 keV.

We report cross sections for each spectral line, according to the resolution of the Si(Li) solid state detector used to resolve individual component lines of the spectral emission. The experimental cross sections were grouped considering the transitions scheme, the energy of the emission lines and the detector resolution.

In general, the fluorescence cross sections obtained in this work show the same trend with Z and broad agreement with the data published by Puri *et al.* [4,5] and Krause [6,7], calculated using Scofield's coefficients [8,9].

II. Experimental Condition

The measurements were carried out at the X-ray Fluorescence beam line at the National Synchrotron Light Laboratory (LNLS), Campinas, Brazil [10]. The components of the experimental setup were:

- Silicon (111) channel cut double crystal monochromator, which can tune energies between 3 and 30 keV. The energy resolution is $3{\cdot}10^{-4}$ to $4{\cdot}10^{-4}$ between 7 and 10 keV.
- A Si(Li) solid state detector, 5 mm thick and 5 mm in diameter, with a resolution of 170 eV at 5.9 keV and a 0.0127 cm thick beryllium window. The model introduced by Jaklevic and Giauque [11] was used to obtain the detector efficiency.
- The whole setup is mounted on a motorized lift table, which allows the vertical positioning of the instruments within the linearly polarized part of the beam.
- To limit the beam size, a motorized computer controlled set of vertical and horizontal

*E-mail: bonzie@famaf.unc.edu.ar
†E-mail: grad@famaf.unc.edu.ar
‡E-mail: rbarrea@depaul.edu

[1] Facultad de Matemática, Astronomía y Física, Universidad Nacional de Córdoba. Ciudad Universitaria. 5000 Córdoba, Argentina.

[2] Instituto de Física Enrique Gaviola (CONICET), 5000 Córdoba, Argentina.

[3] Physics Department, DePaul University, Chicago, IL 60614, USA.

slits (located upstream and downstream of the monochromator) was used.

A set of foil samples (rhodium, palladium, silver, cadmium, indium and tin) was used to determine the L fluorescence cross sections of these elements. The foil samples were provided by Alfa products Inc., with a certified purity of over 99%. The foils thicknesses are shown in Bonzi *et al.* (see Table I) [2].

K emission lines of chlorine, calcium, titanium and iron were measured to determine the geometrical and the detector efficiency factors.

The $K\alpha$ and $L\alpha$ fluorescent spectra were measured by collecting $2{\cdot}10^5$ net counts for each element in order to have the same statistical counting error in all measured spectra.

A system dead time, lower than 1%, was established measuring the fluorescence emission of a Ti sample, adjusting the slit at the exit of the monochromator. All samples were measured with the same slit aperture. Unwanted effects, such as piling up, were avoided using this configuration and the geometric factors were ensured to be the same for all samples. This configuration made it unnecessary to carry out corrections for count losses, spectra distortions or modification of the geometrical arrangement.

III. Spectra analysis

The energy of the emission lines tabulated by Scofield [8,9] and the detector resolution were considered to group the L X-ray fluorescence lines. This line arrangement was used to fit the L spectrum, where the $L\beta$ and $L\gamma$ compound lines have been noted with a Roman subscript according to the most intense contribution line, with its corresponding atomic transition:

- $Ll = L_3 - M_1$,
- $L\alpha = L_3 - M_5 + L_3 - M_4$,
- $L\beta_I = L_2 - M_4 + L_1 - M_2 + L_1 - M_3 + L_3 - N_1$,
- $L\beta_{II} = L_3 - N_5 + L_3 - O_4 + L_3 - O_5 + L_3 - O_1 + L_1 - M_5 + L_1 - M_4 + L_3 - N_4$,
- $L\gamma_I = L_2 - N_4$,
- $L\gamma_{II} = L_1 - N_2 + L_1 - N_3 + L_1 - O_2 + L_1 - O_3$.

The background radiation was fitted using a linear second order polynomial.

The area of the fluorescence peaks was determined as the average of the areas obtained by the adjustment using Hypermet and Gaussian functions. The escape peaks were fitted using a Gaussian function.

As a consequence of the excitation with a linearly polarized photon beam, the contribution to the background was very low. The linear polarization of the incident beam produces negligible scattered radiation at 90^0 with respect to the incident beam direction. The detector position is localized at the same height of the storage ring.

IV. Data Analysis

The expression for the L experimental fluorescence cross sections is [13]

$$\sigma^e_{Li}(Eo) = \frac{I_{Li}}{Io.G.\epsilon(E_{Li}).T(Eo, E_{Li})} \tag{1}$$

where $\sigma^e_{Li}(Eo)$ = experimental Li fluorescence cross sections of the element observed at the energy Eo, with $Li = Ll$, $L\alpha$, $L\beta_I$, $L\beta_{II}$, $L\gamma_I$ or $L\gamma_{II}$; I_{Li} = measured intensity of the Li spectral line; $Io.G.\epsilon(E_{Li})$ = factor comprising the intensity of the excitation beam Io; the geometry of the experimental arrangement G and the detector efficiency $\epsilon(E_{Li})$; Eo = energy of the incident beam, in this case 10 keV; E_{Li} = energy of the Li spectral line; the data was obtained from Scofield [8]; and $T(Eo, E_{Li})$ = correction factor for self absorption in an infinitely thick sample, which is

$$T(Eo, E_{Li}) = \left(\frac{\mu(Eo)}{\sin(\theta_1)} + \frac{\mu(E_{Li})}{\sin(\theta_2)}\right)^{-1} \tag{2}$$

where

$\mu(E)$ = mass absorption coefficient of the sample at energy E from Hubbell and Seltzer [14] and

θ_1 and θ_2 = incidence and take off angles, equal to 45° in the current setup.

In these measurements, all the samples were considered as infinitely thick for X-ray fluorescence.

The factor $Io.G.\epsilon(E)$ was calculated using the following expression

$$Io.G.\epsilon(E_{Ki}) = \frac{I_{Ki}}{\sigma^{\omega F}_{Ki}(Eo).C_i.T(Eo, E_{Ki})} \qquad (3)$$

where

I_{Ki} = measured intensity of the K spectral line,

Ci = the weight concentration of the element of interest in the sample,

$\sigma^{\omega F}_{Ki}(E)$ = K fluorescence cross sections of the element observed at energy E, defined as $\sigma^{\omega F}_{Ki}(Eo) = \sigma_{Ki}(Eo).\omega_K.F_K$, with $\sigma_{Ki}(E)$ = K shell photoionization cross section for the given element at the excitation energy E, from Scofield [8],

ω_K = K shell fluorescence yield, from Krause [6,7] and

F_K = fractional emission rate for $K\alpha$ or $K\beta$ X-rays, from Khan and Karimi [15], defined as

$$F_{K\alpha} = \left[1 + \frac{I_{K\beta}}{I_{K\alpha}}\right]^{-1}; F_{K\beta} = \left[1 + \frac{I_{K\alpha}}{I_{K\beta}}\right]^{-1} \qquad (4)$$

$T(Eo, E_{Ki})$ = correction factor for self absorption in the sample, Eo = energy of the incident beam and E_{Ki} = energy of the K spectral line for a given element, from Scofield [8].

The factor $Io.G.\epsilon(E)$ was previously determined in Ref. [2], where the same geometry and detector were used. Because of this, the $Io.G.\epsilon(E)$ energy dependence is already known and only a scale factor is needed to obtain the correct beam intensity.

Four targets: Cl (NaCl), Ca ($CaHPO_4 \cdot 2H_2O$), Ti (Ti foil) and Fe (Fe foil) emitting fluorescent X-rays in the range from 2.4 keV to 7.0 keV were used to determine the scale factor in this work. Four $K\alpha$ and four $K\beta$ lines were used to fit the scale factor. Jaklevic and Giauque's [11] model was used to fit the detector efficiency.

V. Results and Discussion

L X-ray cross section values obtained in our fluorescence experiment and the theoretical values calculated by using coefficients given by Scofield [8,9], Puri *et al.* [4] and Krause [7] are shown in Table 1 and Figs. 1 to 6.

Puri *et al.* predicted theoretical Coster Kronig and fluorescence values using *ab initio* relativistic calculations, while Krause's values of ω_K, ω_{Li} and f_{ij} were obtained by fitting experimental and theoretical compiled data. In Krause's tables, the theoretical data were calculated for singly ionized free atoms while the experimental data contain contributions from solid state, chemical and multiple ionization effects.

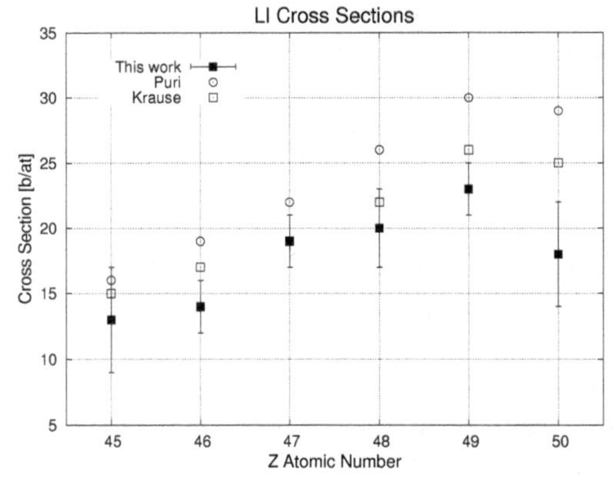

Figure 1: Comparison of Ll cross sections.

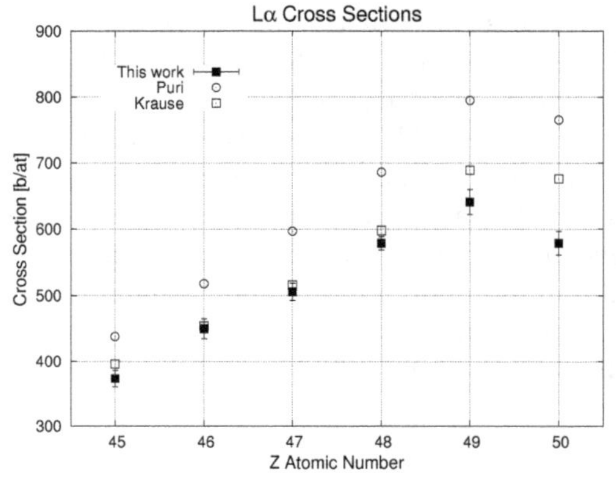

Figure 2: Comparison of $L\alpha$ cross sections.

The $L\alpha$ cross section values have a better agreement with the theoretical values when the intensity peaks are fitted with an Hipermet function instead of a Gaussian function. This happens because the Hipermet function has a tail on the left side that increases the fitted area.

Moreover, the tail of the Hipermet function used to fit the $L\alpha$ peaks diminishes the area and the cross sections of the Ll peaks, accordingly.

The experimental Ll cross sections show a sim-

Element		Ll	$L\alpha$	$L\beta_I$	$L\beta_{II}$	γ_I	$L\gamma_{II}$
	This work	13 ± 4	373 ± 13	163 ± 9	50 ± 5	22 ± 4	7 ± 2
Rh 45	Puri	16	438	194	30	11	7
	Krause	15	396	212	27	11	10
	This work	14 ± 2	450 ± 15	233 ± 10	43 ± 7	25 ± 3	11 ± 2
Pd 46	Puri	19	518	234	43	17	8
	Krause	17	454	249	38	16	11
	This work	19 ± 2	506 ± 13	280 ± 16	59 ± 6	28 ± 3	14 ± 3
Ag 47	Puri	22	597	277	55	22	10
	Krause	19	516	295	48	21	14
	This work	20 ± 3	579 ± 10	371 ± 17	70 ± 6	32 ± 2	15 ± 2
Cd 48	Puri	26	686	328	70	29	11
	Krause	22	598	351	62	28	17
	This work	23 ± 2	641 ± 19	447 ± 17	83 ± 5	37 ± 2	23 ± 2
In 49	Puri	30	795	386	89	37	13
	Krause	26	689	413	77	36	20
	This work	18 ± 4	579 ± 18	454 ± 17	164 ± 5	45 ± 2	33 ± 3
Sn 50	Puri	29	765	599	94	51	38
	Krause	25	676	582	83	48	39

Table 1: Experimental and theoretical L X-ray fluorescence cross sections in Barns/atom at 10 keV. Experimental data (This work), theoretical values calculated using Scofield [8] and Puri [4] and semi-empirical coefficients obtained from Scofield [8] and Krause [6].

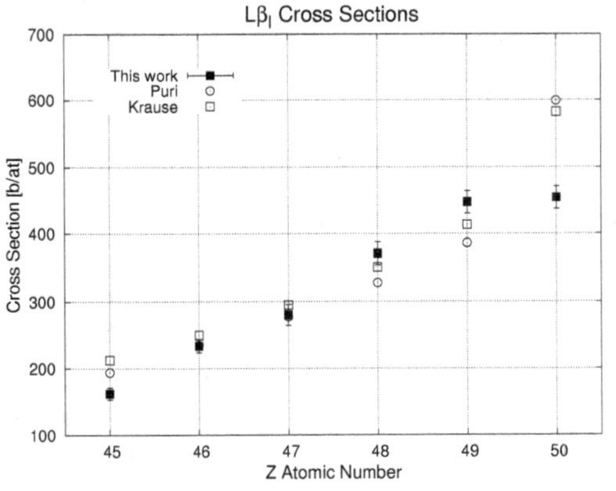

Figure 3: Comparison of $L\beta_I$ cross sections.

ilar Z trend, compared to the data obtained using Krause and Puri *et al.* values. Nevertheless,in general, our results are lower than those.

The $L\alpha$ experimental fluorescence cross section data, Fig. 2, agree well with Krause's values although for elements with higher Z, the experimental values are slightly lower than those from Krause. They are even lower than Puri's *et al.* values, but still showing the same trend with Z.

The $L\beta_I$ experimental fluorescence cross sections show a very good agreement with the theoretical values when the Hipermet function is used to fit the area (see Fig. 3).

The $L\beta_{II}$ measured cross sections show a similar dependence on Z as both theoretical assemblies (Fig. 4).

The $L\beta_I$ Sn experimental value is lower than the data presented by either Puri *et al.* or Krause and the $L\beta_{II}$ Sn experimental value is much higher than both theoretical values. This behavior might be due to the fitting process as both spectra lines are too close in energy; the $L\beta_{II}$ intensity seems to be overestimated while the $L\beta_I$ intensity seems to be underestimated.

A similar behavior is observed for rhodium experimental data although the differences with the theoretical values are much smaller than those for tin.

The experimental $L\gamma_I$ fluorescence cross sections show some differences with the Z trend of the theoretical data: in the lower Z range, the experimental values are higher than the theoretical ones while for higher Z, this difference becomes smaller. Sn values, Z = 50, show a different behavior, being lower than both calculated values (see Fig. 5).

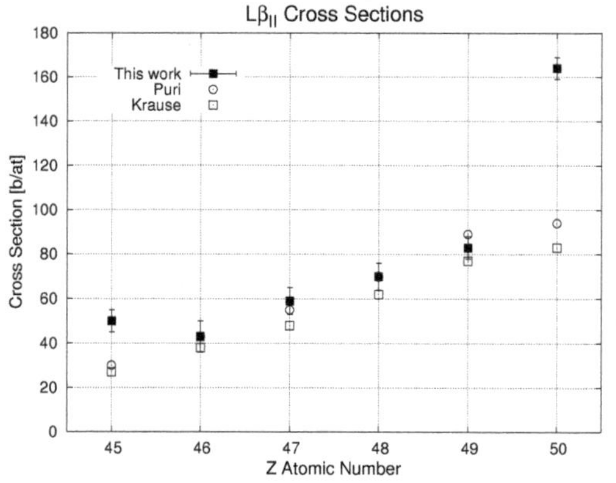

Figure 4: Comparison of $L\beta_{II}$ cross sections.

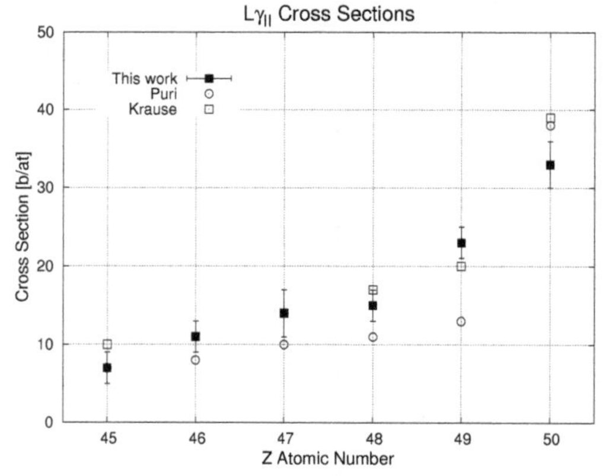

Figure 6: Comparison of $L\gamma_{II}$ cross sections.

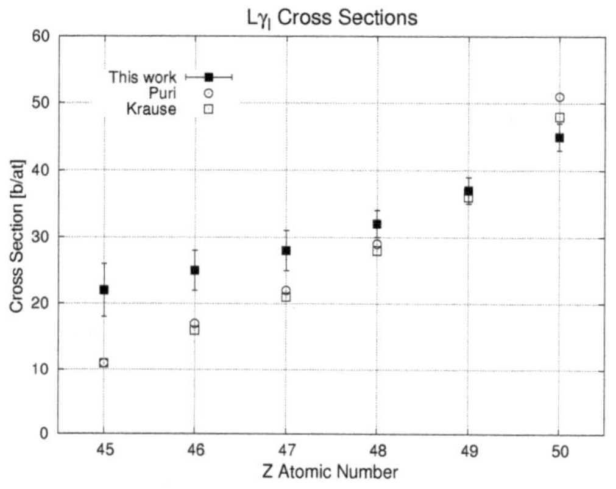

Figure 5: Comparison of $L\gamma_I$ cross sections.

The $L\gamma_{II}$ experimental values show the general Z trend of the values presented by Krause and Puri *et al.*. The experimental values are sometimes higher or lower than the theoretical ones but the range of values is similar to them (see Fig. 6).

To determine the uncertainties of the experimental cross sections, the propagation of errors was carried out in Eq. (1). The uncertainty values are in general around 6-10%, and less than 40% in case of the Ll line.

The uncertainty associated to the $Io.G.\epsilon(E)$ factor was estimated as the mean quadratic deviation of the experimental values ($\leq$ 2%). For the factor $T(Eo, E_{Li})$, a propagation of errors was carried out assuming a 3% error in the values of the mass absorption coefficients, and a 2% error in the sine of the angles due to the sample positioning errors. Krause's ω_K values for elements with $45 \leq Z \leq 50$ have an estimated error of 1%.

The uncertainties of the peak areas were established as half the difference between the areas obtained using Gaussian and Hypermet functions to fit. These uncertainties were the main contribution to the experimental errors of the cross section.

VI. Conclusions

In this investigation, the L X-ray fluorescence cross sections of a group of elements with $45 \leq Z \leq 50$ were measured using a synchrotron radiation source for monoenergetic beams at 10 keV. The polarization properties of the monoenergetic excitation beam and the high resolution of the detector system allowed to reduce the scattered radiation thus obtaining a better signal to noise ratio and a better accuracy for the experimental cross sections.

The cross sections of Ll, $L\alpha$, $L\beta_I$, $L\beta_{II}$, $L\gamma_I$ and $L\gamma_{II}$ lines were measured considering a more detailed group than the usual sets. In Table 1, the comparison between the experimental fluorescence cross section values with the theoretical values calculated using coefficients from Scofield [8, 9], Puri *et al.* [4] and Krause [6] are shown.

Our experimental values are in general in good agreement with the calculated data using Scofield's [8, 9] and Krause's [6] coefficients.

The L cross sections present uncertainties around 6-10% and the less intensive Ll peaks show uncertainties that in some cases come close to 40%, being the fitting uncertainty the most important error source.

The use of the Hypermet function is very convenient to fit the $L\alpha$ and $L\beta$ peaks (see Table 1).

The solid state detector used in our experiments does not have enough energy resolution to resolve each spectral line. A higher resolution detection system would be desirable in order to analyze each spectral line separately.

The Coster Kronig coefficients present large fluctuations in this atomic range and that is the cause of the observed discrepancies.

Acknowledgements - This work was carried out under grants provided by SeCyT U.N.C. (Argentina). Research partially supported by LNLS - National Synchrotron Light Laboratory, Brazil.

---

[1] E V Bonzi, R A Barrea, *Experimental L X-ray fluorescence cross sections for elements with $45 \leq Z \leq 50$ at 7 keV by synchrotron radiation photoionization*, X-ray Spectrom. **34**, 253 (2005).

[2] E V Bonzi, N M Badiger, G B Grad, R A Barrea, R G Figueroa, *Measurement of L X-ray fluorescence cross sections for elements with $45 \leq Z \leq 50$ using synchrotron radiation at 8 keV*, Nucl. Instrum. Meth. B **269**, 2084 (2011).

[3] E V Bonzi, N M Badiger, G B Grad, R A Barrea, R G Figueroa, *L X-ray fluorescence cross sections experimentally determined for elements with $45 \leq Z \leq 50$ at 9 keV*, Appl. Radiat. Isotopes **70**, 632 (2012).

[4] S Puri, D Mehta, B Chand, N Singh, P N Trehan, *L shell fluorescence yields and costerkronig transition probabilities for the elements with $25 \leq Z \leq 96$*, X-ray Spectrom. **22**, 358 (1993).

[5] S Puri, B Chand, D Mehta, M L Garg, S Nirmal, P N Trehan, *K and L shell X-ray fluorescence cross sections*, Atom. Data Nucl. Data **61**, 289 (1995).

[6] M O Krause, *Atomic radiative and radiationless yields for K and L shells*, J. Phys. Chem. Ref. Data **8**, 307 (1979).

[7] M O Krause, C W Nestor, C J Sparks, E Ricci, *X-ray fluorescence cross sections for K and L-rays of the elements*, Oak Ridge National Laboratory, Report 5399 (1978).

[8] J H Scofield, *Theoretical photoionization cross sections from 1 to 1500 keV*, Lawrence Livermore National Laboratory, Report 51326 (1973).

[9] J H Scofield, *Relativistic Hartree Slater values for K and L X-ray emission rates*, Atom. Data Nucl. Data **14**, 121 (1974).

[10] C A Perez, M Radtke, H Tolentino, F C Vicentin, R T Neuenshwander, B Brag, H J Sanchez, M Rubio, M I S Bueno, I M Raimundo, J R Rohwedder, *Synchrotron radiation X-ray fluorescence at the LNLS: Beamline instrumentation and experiments*, X-ray Spectrom. **28**, 320 (1999).

[11] J M Jaklevic, R D Giauque, *Handbook of X-ray spectrometry: Methods and techniques*, Eds. R Van Grieken, A Markowicz, Marcel Dekker, New York (1993).

[12] M C Leypy, J Plagnard, P Stemmler, G Ban, L Beck, P Dhez, *Si(Li) detector efficiency and peak shape calibration in the low energy range using synchrotron radiation*, X-ray Spectrom. **26**, 195 (1997).

[13] D V Rao, R Cesareo, G E Gigante, *L X-ray fluorescence cross sections of heavy elements excited by 15.20, 16.02, 23.62 and 24.68 keV photons*, Nucl. Instrum. Meth. **83**, 31 (1993).

[14] J H Hubbell, S M Seltzer, *Tables of X-ray mass attenuation coefficients and mass-energy absorption coefficients from 1 KeV to 20 MeV for elements Z=1 to 92 and 48 additional substances of dosimetric interest*, NISTIR, Report 5632 (1995).

[15] Md R Khan, M Karimi, *$K\beta/K\alpha$ ratios in energy dispersive X-ray emission analysis*, X-ray Spectrom. **9**, 32 (1980).

3

Influence of surface tension on two fluids shearing instability

Rahul Banerjee,[1*] S. Kanjilal[1]

Using extended Layzer's potential flow model, we investigate the effects of surface tension on the growth of the bubble and spike in combined Rayleigh-Taylor and Kelvin-Helmholtz instability. The nonlinear asymptotic solutions are obtained analytically for the velocity and curvature of the bubble and spike tip. We find that the surface tension decreases the velocity but does not affect the curvature, provided surface tension is greater than a critical value. For a certain condition, we observe that surface tension stabilizes the motion. Any perturbation, whatever its magnitude, results stable with nonlinear oscillations. The nonlinear oscillations depend on surface tension and relative velocity shear of the two fluids.

I. Introduction

When two different density fluids are divided by an interface, the interface becomes unstable with exponential growth under the action of a constant acceleration acting in the direction perpendicular to the interface from the heavier to lighter fluid or under the action of relative velocity shear of two fluids. These two types of instabilities are known as Rayleigh-Taylor and Kelvin-Helmholtz instabilities, respectively. Temporal development of the nonlinear structure of the interface consequent to Rayleigh-Taylor or Kelvin-Helmholtz instability is currently a topic of interest both from theoretical and experimental points of view. The nonlinear structure is called a bubble if the lighter fluid penetrates across the unperturbed interface into the heavier fluid and it is called a spike if the opposite takes place. The instabilities arise in connection with a wide range of problems ranging from direct or indirect laser driven experiments in the ablation region at compression front during the process of inertial confinement fusion [1, 2] to mixing of plasmas in space plasma systems, such as boundary of planetary magnetosphere, solar wind and cluster of galaxies [3]. In high energy density physics(HEDP), formation of supernova remnant or formation of astrophysical jets [4–8] are also seen in these types of instabilities. In high energy density plasma experiments using Omega laser [9], Kelvin-Helmholtz instability growth has recently been observed .

There are several methods to describe the nonlinear structure of the interface of two constant density fluids under potential theory and the associated nonlinear dynamics has been studied by many authors [10–13]. Layzer [10] described the formation of the structure using an expansion near the tip of the bubble or the spike up to second order in the transverse coordinates in two dimensional motion and this approach was extended in Ref. [14] for Kelvin-Helmholtz instability. It is well known [15] that the surface tension reduces the linear Rayleigh-Taylor Growth rate. The lowering in the growth rate is seen to increase with increase in the wave number k up to a critical wave

*E-mail:rbanerjee.math@gmail.com

[1] St. Paul's Cathedral Mission College, 33/1, Raja Rammohan Roy, Sarani, 700 009 Kolkata, India.

number $k_c = \sqrt{\frac{(\rho_h-\rho_l)g}{T}}$, where T denotes surface tension, and ρ_h and ρ_l are the densities of the heavier and lighter fluids, respectively. The same effect has been described by Mikaelian [16] for Rayleigh–Taylor instability in finite thickness and Sung-Ik Sohn [17] described the effect using the Layzer nonlinear potential model. The nonlinear theory influence of surface tension was elaborately studied by Pullin [18] and Garnier *et al.* [19] using numerical methods.

The present paper addresses to the problem of the time development of the nonlinear interfacial structure caused by combined Rayleigh–Taylor and Kelvin–Helmholtz instability in presence of surface tension. It is shown that the growth rate of the instabilities is affected by the surface tension. The growth rate of the tip of the bubble or spike are significantly reduced due to the surface tension. We observed an oscillatory stabilization of the interface for large surface tension. This oscillation depends on the relative velocity shear also. Section II deals with the basic hydrodynamical equations together with the geometry involved. Here we assume that the fluids are inviscid and the motion is irrotational. The investigation of the nonlinear aspect of the structure of the two fluids interface is facilitated by Bernoull's equation together with the pressure balance equation at the interface. The long time asymptotic behavior of the bubble and spike tip for combined Rayleigh–Taylor and Kelvin–Helmholtz instabilities is derived in section III.A and III.B, respectively. We have also discussed the characteristics of the tip of the bubble and the spike derived analytically and numerically. Finally, we have concluded the results in section IV.

II. Basic mathematical model

We have considered two incompressible fluids separated by an interface located at $y = 0$ in a two-dimensional $x - y$ plane, where x axis lying normal to the unperturbed fluid interface. The fluid with density ρ_h is assumed to overlie the fluid with density ρ_l and gravity is taken along negative y-axis. In the following discussion, we shall denote the properties of the fluid above the interface by the subscript h and below the interface by the subscript l. After perturbation, the nonlinear interface is assumed to take up a parabolic shape, given by

$$y = \eta(x,t) = \eta_0(t) + \eta_2(t)(x - \eta_1(t))^2 \tag{1}$$

The perturbed interface forms a bubble or spike according to $\eta_0(t) > 0$, $\eta_2(t) < 0$ or $\eta_0(t) < 0$, $\eta_2(t) > 0$. Functions $\eta_0(t)$ and $\eta_1(t)$ are related to the position of the tip of the bubble from the unperturbed interface, i.e, at time t the position of the bubble tip is $(\eta_1(t), \eta_0(t))$ and $\eta_2(t)$ is related to the bubble curvature.

In our previous works [14, 20–23], we have considered $\eta_1(t) = 0$ due to the absence of velocity shear parallel to the unperturbed interface. However, in presence of streaming motion of the fluids, the tip of the bubble moves parallel to unperturbed interface with velocity $\dot{\eta}_1(t)$.

According to the extended Layzer model [10, 11, 14, 20], the velocity potentials describing the motion for the upper (heavier) and lower (lighter) fluids are assumed to be given by

$$\begin{aligned}\phi_h(x,y,t) &= a_1(t)\cos(k(x-\eta_1(t))e^{-k(y-\eta_0(t))}\\ &\quad + a_2(t)\sin(k(x-\eta_1(t))e^{-k(y-\eta_0(t))}\\ &\quad - xU_h\end{aligned} \tag{2}$$

$$\begin{aligned}\phi_l(x,y,t) &= b_0(t)y\\ &\quad + b_1(t)\cos(k(x-\eta_1(t))e^{k(y-\eta_0(t))}\\ &\quad + b_2(t)\sin(k(x-\eta_1(t))e^{k(y-\eta_0(t))}\\ &\quad - xU_l\end{aligned} \tag{3}$$

where U_h and U_l are streaming velocities of upper and lower fluids, respectively, and k is the perturbed wave number.

The evolution of the interface $y = \eta(x,t)$ can be determined by the kinematical and dynamical boundary conditions. The kinematical boundary conditions are

$$\frac{\partial\eta}{\partial t} - \frac{\partial\eta}{\partial x}\frac{\partial\phi_h}{\partial x} = -\frac{\partial\phi_h}{\partial y} \tag{4}$$

$$\frac{\partial\eta}{\partial x}\left(\frac{\partial\phi_h}{\partial x} - \frac{\partial\phi_l}{\partial x}\right) = \frac{\partial\phi_h}{\partial y} - \frac{\partial\phi_l}{\partial y} \tag{5}$$

and the dynamical boundary condition (first integral of the momentum equation) is of the form

$$-\rho_{h(l)}\frac{\partial\phi_{h(l)}}{\partial t}+\frac{1}{2}\rho_{h(l)}(\vec{\nabla}\phi_{h(l)})^2+\rho_{h(l)}gy = -p_{h(l)}+f_{h(l)}(t) \quad (6)$$

The pressure boundary condition at two fluid interface including surface tension [17, 22] is

$$p_h - p_l = \frac{T}{R} \quad (7)$$

where T is the surface tension and R is the radius of curvature.

Plugging the condition (7) at the interface $y = \eta(x,t)$ in Eq. (6), we obtain the following equation.

$$\rho_h[-\frac{\partial\phi_h}{\partial t}+\frac{1}{2}(\vec{\nabla}\phi_h)^2]-\rho_l[-\frac{\partial\phi_l}{\partial t}+\frac{1}{2}(\vec{\nabla}\phi_l)^2] +g(\rho_h-\rho_l)y = -\frac{T}{R}+f_h-f_l \quad (8)$$

We have restricted our study near the peak of the perturbed structure where $|k(x-\eta_1(t))| \ll 1$. Thus, we can neglect the terms of $O(|x-\eta_1|^i)$ $(i \geq 3)$ [14]. With this point of view, we have

$$\frac{1}{R} = 2\eta_2\left(1+4\eta_2^2(x-\eta_1)^2\right)^{-\frac{3}{2}} \approx 2\eta_2\left(1-6\eta_2^2(x-\eta_1)^2\right) \quad (9)$$

We substitute all the parameters η, ϕ_h and ϕ_l in the kinematic and dynamic boundary conditions represented by Eqs. (4), (5), (8) and (9), and equate coefficients of $(x-\eta_1)^i$,$(i = 0, 1, 2)$ and neglect terms $O(|x-\eta_1|^i)$ $(i \geq 3)$. This yields the following equations.

$$\frac{d\xi_1}{d\tau} = \xi_4 \quad (10)$$

$$\frac{d\xi_2}{d\tau} = V_h - \frac{\xi_5(2\xi_3+1)}{2\xi_3} \quad (11)$$

$$\frac{d\xi_3}{d\tau} = -\frac{1}{2}(6\xi_3+1)\xi_4 \quad (12)$$

$$\frac{kb_0}{\sqrt{kg}} = -\frac{12\xi_3\xi_4}{6\xi_3-1} \quad (13)$$

$$\frac{k^2b_1}{\sqrt{kg}} = \frac{6\xi_3+1}{6\xi_3-1}\xi_4 \quad (14)$$

$$\frac{k^2b_2}{\sqrt{kg}} = \frac{(2\xi_3+1)\xi_5-2\xi_3(V_h-V_l)}{2\xi_3-1} \quad (15)$$

$$\begin{aligned}\frac{d\xi_4}{d\tau} = {} & \frac{N_1(\xi_3,r)}{D_1(\xi_3,r)}\frac{\xi_4^2}{(6\xi_3-1)} \\ & +\frac{2(1-r)\xi_3(6\xi_3-1)}{D_1(\xi_3,r)}\left(1-12\xi_2^2\frac{k^2}{k_c^2}\right) \\ & +\frac{N_2(\xi_3,r)}{D_1(\xi_3,r)}\frac{(6\xi_2-1)\xi_5^2}{2\xi_3(2\xi_3-1)^2} \\ & +\frac{2(4\xi_3-1)(6\xi_3-1)}{D_1(\xi_3,r)(2\xi_3-1)^2} \\ & \times[(V_h-V_l)^2\xi_3-(V_h-V_l)(2\xi_3+1)\xi_5] \quad (16)\end{aligned}$$

and

$$\begin{aligned}\frac{d\xi_5}{d\tau} = {} & -\frac{(2\xi_3-1)r\xi_4\xi_5}{2\xi_3D_2(\xi_3,r)} \\ & +\frac{\xi_4(6\xi_3+1)}{2D_2(\xi_3,r)(6\xi_3-1)(2\xi_3-1)} \\ & \times\left[4(V_h-V_l)(4\xi_3-1)-\frac{\xi_5}{\xi_3}(28\xi_3^2-4\xi_3-1)\right] \quad (17)\end{aligned}$$

where $r = \frac{\rho_h}{\rho_l}$; $\xi_1 = k\eta_0$; $\xi_2 = k\eta_1$; $\xi_3 = \frac{\eta_2}{k}$; $\xi_4 = \frac{k^2a_1}{\sqrt{kg}}$; $\xi_5 = \frac{k^2a_2}{\sqrt{kg}}$; $\tau = t\sqrt{kg}$; $k_c^2 = \frac{(\rho_h-\rho_l)g}{T}$ and $V_{h(l)} = \frac{kU_{h(l)}}{\sqrt{kg}}$ are corresponding dimensionless quantities. The function $N_{1,2}(\xi_3,r)$ and $D_{1,2}(\xi_3,r)$ are given by

$$\begin{aligned}N_1(\xi_3,r) &= 36(1-r)\xi_3^2+12(4+r)\xi_3+(7-r); \\ D_1(\xi_3,r) &= 12(r-1)\xi_3^2+4(r-1)\xi_3-(r+1)\end{aligned} \quad (18)$$

and

$$\begin{aligned}N_2(\xi_3,r) &= 16(1-r)\xi_3^3+12(1+r)\xi_3^2-(1+r); \\ D_2(\xi_3,r) &= 2(1-r)\xi_3+(r+1)\end{aligned} \quad (19)$$

The temporal development of the combined effect of Rayleigh–Taylor and Kelvin–Helmholtz instability is given by Eqs. (10)–(12), (16) and (17).

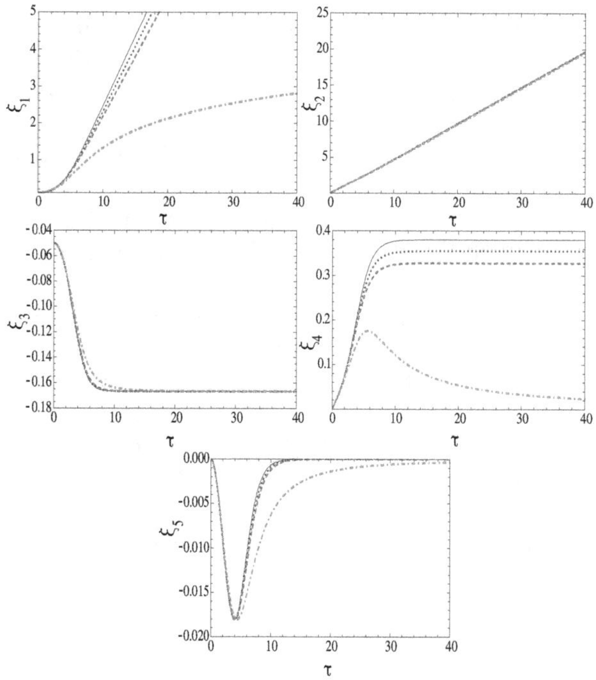

Figure 1: Bubble- variation of ξ_1, ξ_2, ξ_3, ξ_4 and ξ_5 with τ. Initial value ξ_1 =0.1, ξ_2 =0, ξ_3 =-0.05, $\xi_4 = 0$, and $\xi_5 = 0$ with $\rho_h = 3$, $\rho_l = 2$, V_h=0.5, $V_l = 0.1$, $\frac{k^2}{k_c^2}$=0 (line), 0.5 (dot), 1 (dash), 3.9 (dash-dot).

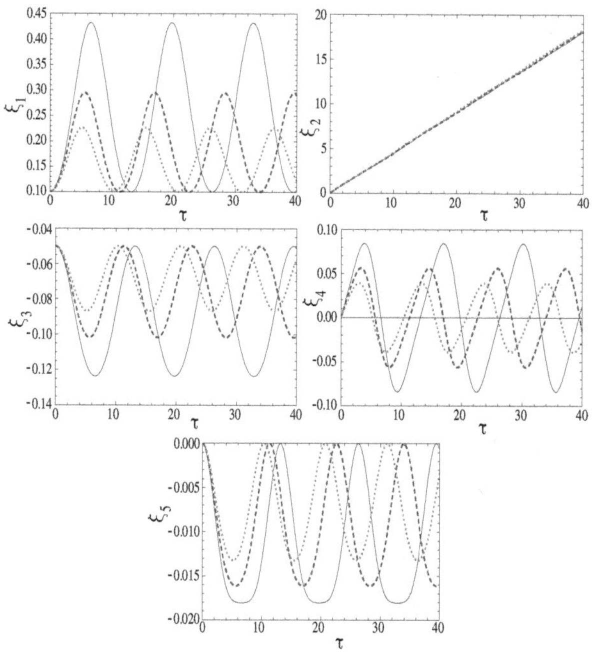

Figure 2: Bubble- variation of ξ_1, ξ_2, ξ_3, ξ_4 and ξ_5 with τ. Initial value ξ_1 =0.1, ξ_2 =0, ξ_3 =-0.05, $\xi_4 = 0$, and $\xi_5 = 0$ with $\rho_h = 3$, $\rho_l = 2$, V_h=0.5, $V_l = 0.1$, $\frac{k^2}{k_c^2}$=10 (line), 15 (dot), 20 (dash).

III. Numerical results and discussions

i. Effect of surface tension on bubble growth

In this section, we present the effect of surface tension on the nonlinear growth rate of the bubble tip for combined Rayleigh–Taylor and Kelvin–Helmholtz instability. To describe the dynamics of the bubble tip, it is essential to integrate Eqs. (10)–(12), (15) and (16) by numerical simulation. To obtain the initial conditions of the numerical integration, we assume that the initial interface is given by $y = \eta_0(t = 0)cos(kx)$. The expansion of the cosine function gives $(\xi_2)_{initial} = 0$ and $(\xi_3)_{initial} = -\frac{1}{2}(\xi_1)_{initial}$, where $(\xi_1)_{initial}$ is the arbitrary initial amplitude. Since the perturbation starts from rest, we may often choose $(\xi_4)_{initial} = (\xi_5)_{initial} = 0$. The non-dimensionalized time development plots of ξ_1, ξ_2, ξ_3,ξ_4 and ξ_5 are shown in Figs. 1, 2 and 3.

Before we describe the nature of the bubble tip, consider the asymptotic behavior of the tip. As $\tau \to \infty$, the asymptotic values of ξ_3, ξ_4 and ξ_5 for bubble are obtained by setting $\frac{d\xi_3}{d\tau} = 0$, $\frac{d\xi_4}{d\tau} = 0$ and $\frac{d\xi_5}{d\tau} = 0$. Note that, if $k^2 < 3\left(1 + \frac{15}{16}\frac{\rho_l}{\rho_h - \rho_l}(\Delta V)^2\right)k_c^2$, where $\Delta V = V_h - V_l$, the asymptotic values are

$$[(\xi_3)_{asymp}]_{bubble} = -\frac{1}{6} \tag{20}$$

$$[(\xi_4)_{asymp}]_{bubble} \tag{21}$$
$$= \sqrt{\frac{2A}{3(1+A)}\left(1 - \frac{k^2}{3k_c^2}\right) + \frac{5}{16}\frac{1-A}{1+A}(\Delta V)^2}$$

and

$$[(\xi_5)_{asymp}]_{bubble} = 0 \tag{22}$$

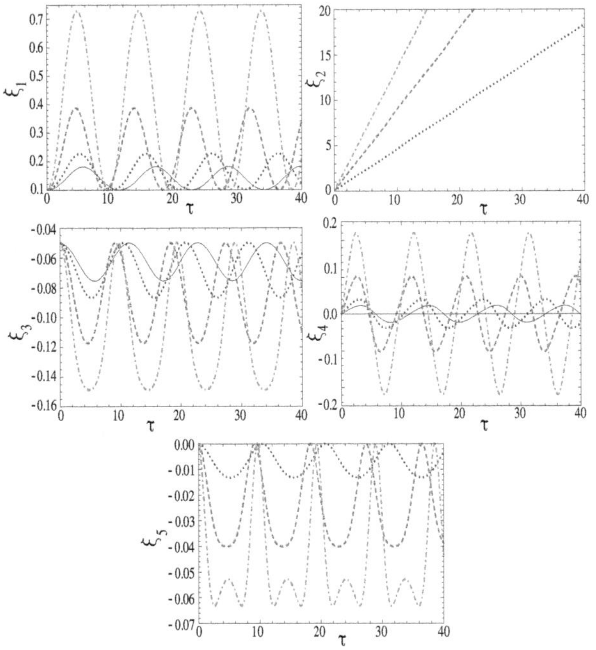

Figure 3: Bubble- variation of ξ_1, ξ_2, ξ_3, ξ_4 and ξ_5 with τ. Initial value ξ_1 =0.1, ξ_2 =0, ξ_3 =-0.05, $\xi_4 = 0$, and $\xi_5 = 0$ with $\rho_h = 3$, $\rho_l = 2$, $\frac{k^2}{k_c^2}$=20, V_h=0, $V_l = 0$ (line), V_h=0.5, $V_l = 0.1$(dot), V_h=1, $V_l = 0.1$ (dash), V_h=1.5, $V_l = 0.1$ (dash-dot).

where, $A = \frac{\rho_h - \rho_l}{\rho_h + \rho_l}$ is the Atwood number.

It is clear form Fig. 1 that surface tension suppresses the velocity and growth of the bubble tip significantly, provided surface tension is larger than a critical threshold, $T < T_c^{bubble}$, where

$$T_c^{bubble} = 3\left((\rho_h - \rho_l) + \frac{15}{16}\rho_l(\Delta V)^2\right)\frac{g}{k^2} \tag{23}$$

Here the critical value depends on the magnitude of relative velocity shear of two fluids and the growth and velocity of the tip reduced if $T < T_c^{bubble}$.

When there is no tangential velocity difference (i.e., $V_h = V_l$) between the two fluids initially, the fluids are purely prone to the Rayleigh–Taylor instability and the critical value becomes $\frac{3(\rho_h - \rho_l)g}{k^2}$. These results agree with the argument in Ref. [17]. In absence of surface tension, the asymptotic values coincide with the results obtained in our previous work [14].

Further, if $T > T_c^{bubble}$, oscillatory state emerges even for $r > 1$. Figures 2 and 3 describe the oscillatory state of the motion. The amplitude and the period of oscillation decrease monotonically for large surface tension (Fig. 2), while the amplitude of oscillation increases for large relative velocity shear (Fig. 3). In this respect, Figs. 2 and 3 show that there always exists a self generated oscillatory transverse velocity component ($-\xi_5$) due to perturbation and this depends upon surface tension as well as the relative velocity shear ΔV at the two fluids interface. For negative velocity shear (i.e, $\Delta V < 0$), the self generated oscillatory transverse velocity of the bubble peak acts opposite to the direction of V_h and the amplitude of oscillation increases for large surface tension.

If $T = T_c^{bubble}$, equilibrium is attained, i.e,

$$\dot{\xi}_3 = \dot{\xi}_4 = \dot{\xi}_5 = 0$$
$$\text{when } \xi_3 = -\frac{1}{6} \text{ and } \xi_4 = \xi_5 = 0 \tag{24}$$

and the equilibrium becomes unstable. This feature is shown with a dot-dash line in Fig. 1. Thus, the combined Rayleigh–Taylor and Kelvin–Helmholtz instability is stabilized when

$$k^2 > 3\left(1 + \frac{15}{16}\frac{\rho_l}{\rho_h - \rho_l}(\Delta V)^2\right)k_c^2,$$
$$\text{i.e., } T > T_c^{bubble} \tag{25}$$

while the instability however persists but with reduced growth rate for

$$k^2 \le 3\left(1 + \frac{15}{16}\frac{\rho_l}{\rho_h - \rho_l}(\Delta V)^2\right)k_c^2,$$
$$\text{i.e., } T \le T_c^{bubble} \tag{26}$$

According to the condition (25), for $\rho_h = 3$, $\rho_l = 2$, $V_h = 0.5$ and $V_l = 0.1$, the motion is stabilized when $\frac{k^2}{k_c^2} > 3.9$. These results are exhibited in Fig. 2, where $\frac{k^2}{k_c^2} > 3.9$. For $\frac{k^2}{k_c^2} = 3.9$, the growth rate of the instability is asymptotically diminished and becomes 0 (dash-dot line of Fig. 1). However, Fig. 1 shows the suppression of growth rate of the instability due to surface tension, when $\frac{k^2}{k_c^2} < 3.9$.

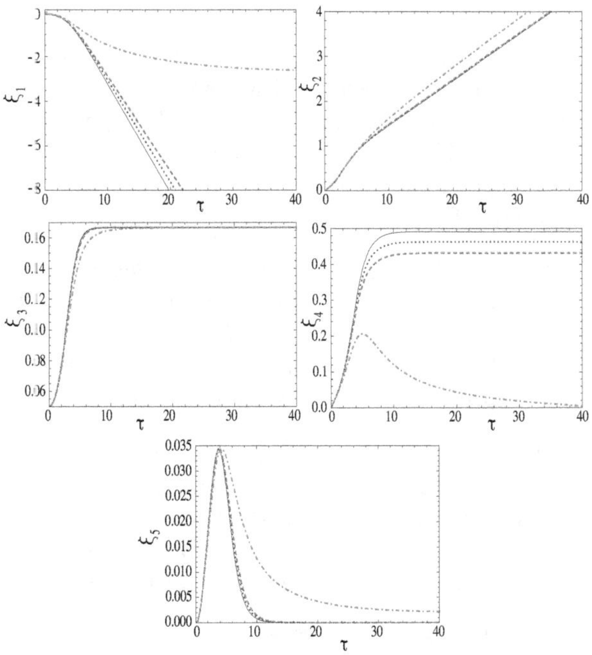

Figure 4: Spike- variation of ξ_1, ξ_2, ξ_3, ξ_4 and ξ_5 with τ. Initial value ξ_1 =-0.1, ξ_2 =0, ξ_3 =0.05, $\xi_4 = 0$, and $\xi_5 = 0$ with $\rho_h = 3$, $\rho_l = 2$, V_h=0.5, $V_l = 0.1$, $\frac{k^2}{k_c^2}$=0 (line), 0.5 (dot), 1 (dash), 4.35 (dash-dot).

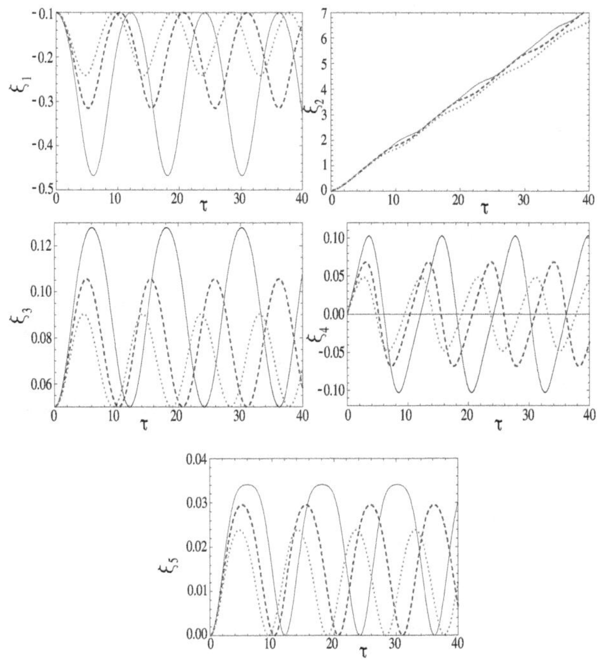

Figure 5: Spike- variation of ξ_1, ξ_2, ξ_3, ξ_4 and ξ_5 with τ. Initial value ξ_1 =-0.1, ξ_2 =0, ξ_3 =0.05, $\xi_4 = 0$, and $\xi_5 = 0$ with $\rho_h = 3$, $\rho_l = 2$, V_h=0.5, $V_l = 0.1$, $\frac{k^2}{k_c^2}$=10 (line), 15 (dot), 20 (dash).

ii. Effect of surface tension on spike growth

The temporal evolution of spike state is exhibited in Figs. 4, 5 and 6; the results follow from the numerical integration of Eqs. (10)–(12), (15) and (16) using the transformation $\xi_1 \to -\xi_1$, $\xi_3 \to -\xi_3$, $g \to -g$, $r \to \frac{1}{r}$ and $V_h \rightleftharpoons V_l$. The saturation curvature and velocity of the spike tip are given by

$$[(\xi_3)_{asymp}]_{spike} = \frac{1}{6} \tag{27}$$

$$[(\xi_4)_{asymp}]_{spike} = \sqrt{\frac{2A}{3(1-A)}\left(1-\frac{k^2}{3k_c^2}\right)+\frac{5}{16}\frac{1+A}{1-A}(\Delta V)^2} \tag{28}$$

and

$$[(\xi_5)_{asymp}]_{spike} = 0 \tag{29}$$

provided $k^2 < 3\left(1+\frac{15}{16}\frac{\rho_h}{\rho_h-\rho_l}(\Delta V)^2\right)k_c^2$.

Figure 4 describes that large surface tension suppresses the growth rate of the spike tip, as well as the bubble. The nonlinear oscillation of the spike tip is observed for $k^2 > 3\left(1+\frac{15}{16}\frac{\rho_h}{\rho_h-\rho_l}(\Delta V)^2\right)k_c^2$ and the equilibrium state arises when equality holds. The pattern of amplitude and period of oscillation are identical to that for the bubble (Figs. 5 and 6). Figure 5 shows the oscillatory behavior of the spike structure for different values of surface tension while the dependency of the relative velocity shear is demonstrated in Fig. 6.

IV. Conclusion

In this paper, we have studied a potential flow model to describe the nature of the nonlinear structure of a two-fluid interface under the combined action of Rayleigh–Taylor and Kelvin–Helmholtz instabilities due to surface tension. The analytic expressions for bubble and spike growth rates at

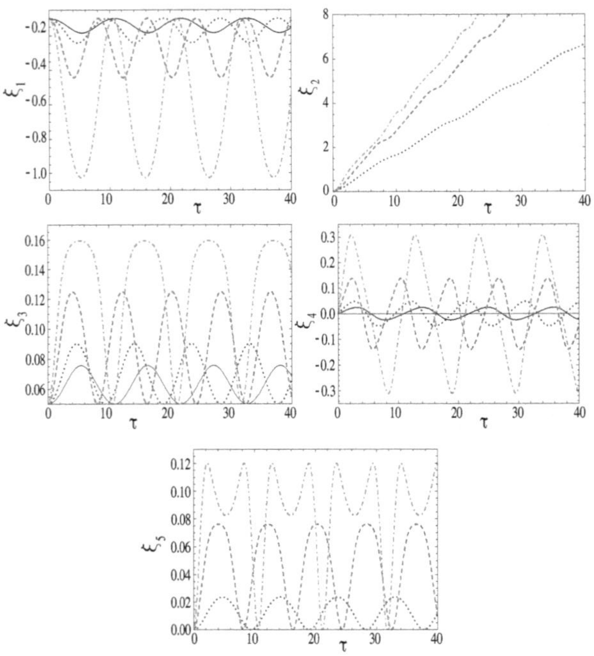

Figure 6: Spike- variation of ξ_1, ξ_2, ξ_3, ξ_4 and ξ_5 with τ. Initial value ξ_1 =-0.1, ξ_2 =0, ξ_3 =0.05, $\xi_4 = 0$, and $\xi_5 = 0$ with $\rho_h = 3$, $\rho_l = 2$, $\frac{k^2}{k_c^2}$=20, V_h=0, $V_l = 0$ (line), V_h=0.5, $V_l = 0.1$ (dot), V_h=1, $V_l = 0.1$ (dash), V_h=1.5, $V_l = 0.1$ (dash-dot).

asymptotic stage are obtained for arbitrary Atwood number and velocity shear. Surface tension becomes a stabilizing factor of the instability, provided it is larger than a critical value. In this case, oscillatory behavior of motion described by numerical integration of governing equations. The nature of oscillations depends on both surface tension and relative velocity shear of two fluids. On the other hand, below the critical value, surface tension dominates the growth and growth rate of the instability. This result is expected to improve the understanding of the stabilization factor for the astrophysical instability.

Acknowledgements - This work was supported by the University Grant Commission, Government of India under Ref. No. PSW-43/12-13 (ERO).

---

[1] D K Bradley, D G Barun, S G Glendinning, M J Edwards, J L Milovich, C M Sorce, G W Collins, S W Hann, R H Page, R J Wallace, *Very-high-growth-factor planar ablative Rayleigh–Taylor experiments*, Phys. Plasmas **14**, 056313 (2007).

[2] K S Budil, B A Remington, T A Peyser, K O Mikaelian, P L Miller, N C Woolsey, W M Wood-Vasey, A M Rubenchik, *Experimental comparison of classical versus ablative Rayleigh–Taylor instability*, Phys. Rev.Lett. **76**, 4536 (1996).

[3] H Hasegawa, M Fujimoto, T-D Phan, H Reme, A Balogh, M W Dunlop, C Hashimoto, R TanDokoro, *Transport of solar wind into Earth's magnetosphere through rolled-up Kelvin–Helmholtz vortices*, Nature **430**, 755 (2004).

[4] R P Drake, *Hydrodynamic instabilities in astrophysics and in laboratory high-energydensity systems*, Plasma Phys. Control. Fusion **47**, B419 (2005).

[5] J O Kane, H F Robey, B A Remington, R P Drake, J Knauer, D D Ryutov, H Louis, R Teyssier, O Hurricane, D Arnett, R Rosner, A Calder, *Interface imprinting by a rippled shock using an intense laser*, Phys. Rev. E **63**, 055401R (2001).

[6] D D Ryutov, B A Remington, *Scaling astrophysical phenomena to high-energy-density laboratory experiments*, Plasma Phys. Control. Fusion **44**, B407 (2002).

[7] L Spitzer, *Behavior of matter in space*, Astrophys. J. **120**, 1 (1954).

[8] B A Remington, R P Drake, H Takabe, D Arnett, *A review of astrophysics experiments on intense lasers*, Phys. Plasma **7**, 1641 (2000).

[9] E C Harding, J F Hansen, O A Hurricane, R P Drake, H F Robey, C C Kuranz, B A Remington, M J Bono, M J Grosskopf, R S Gillespie, *Observation of a Kelvin–Helmholtz instability in a high-energy-density plasma on the omega laser*, Phys. Rev.Lett. **103**, 045005 (2009).

[10] D Layzer, *On the Instability of Superposed Fluids in a Gravitational Field*, Astrophys. J. 122, 1 (1955).

[11] V N Goncharov, *Analytical model of nonlinear, single-mode, classical Rayleigh–Taylor instability at arbitrary Atwood numbers*, Phys. Rev. Lett. **88**, 134502 (2002).

[12] Sung-Ik Sohn, *Simple potential-flow model of Rayleigh–Taylor and Richtmyer–Meshkov instabilities for all density ratios*, Phys. Rev. E **67**, 026301 (2003).

[13] Q Zhang, *Analytical solutions of Layzer-type approach to unstable interfacial fluid mixing*, Phys. Rev. Lett. **81**, 3391 (1998).

[14] R Banerjee, L Mandal, M Khan, M R Gupta, *Effect of viscosity and shear flow on the nonlinear two fluid interfacial structures*, Phys. Plasmas **19**, 122105 (2012).

[15] S Chandrasekhar, *Hydrodynamic and hydromagnetic stability*, Dover, New York (1961).

[16] K O Mikaelian, *Rayleigh–Taylor instability in finite-thickness fluids with viscosity and surface tension*, Phys. Rev. E **54**, 3676 (1996).

[17] Sung-Ik Sohn, *Effects of surface tension and viscosity on the growth rates of Rayleigh–Taylor and Richtmyer–Meshkov instabilities*, Phys. Rev. E **80**, 055302(R) (2009).

[18] D I Pullin, *Numerical studies of surface tension effects in nonlinear KelvinHelmholtz and RayleighTaylor instability*, J. Fluid Mech. **119**, 507 (1982).

[19] J Garnier, C Cherfils-Clerouin, A P Holstein, *Statistical analysis of multimode weakly nonlinear Rayleigh–Taylor instability in the presence of surface tension*, Phys. Rev. E **68**, 036401 (2003).

[20] R Banerjee, L Mandal, S Roy, M Khan, M R Gupta, *Combined effect of viscosity and vorticity on single mode RayleighTaylor instability bubble growth*, Phys. Plasmas **18**, 022109 (2011).

[21] M R Gupta, R Banerjee, L K Mandal, R Bhar, H C Pant, M Khan, M K Srivastava, *Effect of viscosity and surface tension on the growth of RayleighTaylor instability and RichtmyerMeshkov instability induced two fluid interfacial nonlinear structure*, Indian J. Phys. **86**, 471 (2012).

[22] R Banerjee, L Mandal, M Khan, M R Gupta, *Bubble and spike growth rate of Rayleigh Taylor and Richtmeyer Meshkov instability in finite layers*, Indian J. Phys. **87**, 929 (2013).

[23] R Banerjee, L Mandal, M Khan, M R Gupta, *Spiky Development at the Interface in Rayleigh–Taylor Instability: Layzer Approximation with Second Harmonic*, J. Morden Phys. **4**, 174 (2013).

4

Nonlinearity arising from noncooperative transcription factor binding enhances negative feedback and promotes genetic oscillations

Iván M. Lengyel,[1] Daniele Soroldoni,[2,3] Andrew C. Oates,[2,3] Luis G. Morelli[1*]

We study the effects of multiple binding sites in the promoter of a genetic oscillator. We evaluate the regulatory function of a promoter with multiple binding sites in the absence of cooperative binding, and consider different hypotheses for how the number of bound repressors affects transcription rate. Effective Hill exponents of the resulting regulatory functions reveal an increase in the nonlinearity of the feedback with the number of binding sites. We identify optimal configurations that maximize the nonlinearity of the feedback. We use a generic model of a biochemical oscillator to show that this increased nonlinearity is reflected in enhanced oscillations, with larger amplitudes over wider oscillatory ranges. Although the study is motivated by genetic oscillations in the zebrafish segmentation clock, our findings may reveal a general principle for gene regulation.

I. Introduction

Cells can generate temporal patterns of activity by means of genetic oscillations [1, 2]. Genetic oscillations are biochemical oscillations in the levels of gene products [3–14]. They can be produced by negative feedback regulation of gene expression, in which a gene product inhibits its own production directly or indirectly [15]. Such autoinhibition is often performed by transcriptional repressors, proteins or protein complexes that bind the promoter of a gene and inhibit the transcription of new mRNA molecules [16,17], see Fig. 1. A theoretical description of biochemical oscillations requires a delayed negative feedback, together with sufficient nonlinearity and a balance of the timescales of the different processes involved [18]. Delays may occur naturally in transcriptional regulation, since the assembly of gene products involves intermediate steps [16,19,20]. Nonlinearity refers to the presence of nonlinear terms in the equations describing the dynamics. Such nonlinear terms may occur in the equations due to the presence of cooperative biochemical processes, where cooperativity is understood as a phenomenon in which several components act together to orchestrate some collective behavior [21]. Although some processes giving rise to nonlinear terms are known, in general it is still an open question how nonlinearity is built into genetic oscillators.

A compelling model system for genetic oscillations is the vertebrate segmentation clock [22–25]. This is a tissue-level pattern generator that controls the formation of vertebrate segments during embryonic development [25, 26]. The spatiotemporal patterns generated by the segmentation clock are

*Email: morelli@df.uba.ar

[1] Departamento de Física, FCEyN UBA and IFIBA, CONICET, Pabellón 1, Ciudad Universitaria, 1428 Buenos Aires, Argentina.

[2] MRC-National Institute for Medical Research, The Ridgeway, Mill Hill, NW7 1AA London, UK.

[3] Department of Cell and Developmental Biology, University College London, Gower Street, WC1E 6BT London, UK.

thought to be initiated by a genetic oscillator at the single cell level [7,9,10,27–29]. In this genetic oscillator, negative feedback is provided by genes of the Her family, which code proteins that form dimers and can act as transcriptional repressors [30–34]. The time taken from transcription, translation and splicing may introduce the necessary feedback delays in zebrafish and mouse [19,34–37]. One source of nonlinearity in the segmentation clock oscillator is the dimerization of gene products that bind the promoter of cyclic genes [32]. However, this may be insufficient to generate the observed oscillations in the levels of gene products. One way to increase nonlinearity would be cooperative binding of repressors to regulatory binding sites at the promoter [21, 38, 39]. Cooperative binding to multiple binding sites can make the negative feedback steeper [40]. A similar effect occurs with ultrasensitivity in phosphorylation cascades [41]. The presence of clusters of binding sites for transcription factors may be a common motif in gene regulation [42], and cooperativity in transcription factor binding has been reported for some systems [43].

In zebrafish, multiple binding sites for Her dimers have been identified in the promoter region of Her1, Her7, and other genes of the Notch pathway [32, 44, 45]. However, there is no evidence for cooperative binding of transcriptional regulators in the case of the zebrafish segmentation clock. Therefore, although cooperative binding is not ruled out, this lack of evidence raises the general question of what contribution could be expected from the multiple binding sites reported in the promoter of segmentation clock cyclic genes. Here we use theory to study how multiple binding sites affect nonlinearity and biochemical oscillations in a generic description of a genetic oscillator.

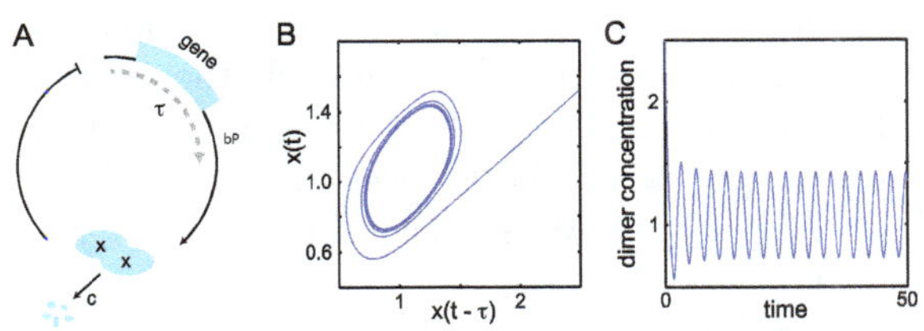

Figure 1: Delayed autoinhibition can produce genetic oscillations. (A) The gene (light blue box) is transcribed and translated into gene products x, with a delay τ. In this example, gene products form dimers that act as transcriptional repressors, inhibiting transcription of the gene (blunted arrow), and decay at a rate c. (B) and (C) Numerical solutions of Eq. (9) describing the oscillator in A. (B) Phase space: monomer concentration $x(t)$ vs. the delayed concentration $x(t-\tau)$ settled to a limit cycle. (C) Dimer concentration oscillates as a function of time. Parameters in B, C: $bP = 2$, $\tau = 1$, $c = 1$, $x_0 = 1$, $N = 12$, $M = 6$.

II. A promoter with multiple binding sites

We first evaluate the regulatory function of a promoter that contains N binding sites for a transcriptional repressor, see Fig. 2. We consider transcriptional repressors although the results in this section are more general and would apply to other types of transcription factors. We focus on a single transcriptional repressor for the sake of simplicity, and assume that all binding sites are identical. That is, binding and unbinding of repressors to the different sites occur at the same rates for all sites. Moreover, we assume that there is no cooperativity in repressor binding. This means that binding and unbinding rates are not affected by the presence of bound factors to any of the other sites.

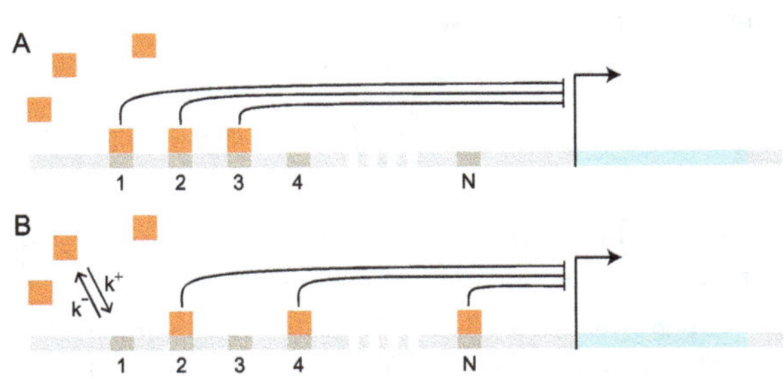

Figure 2: Schematic representation of a promoter with multiple binding sites, numbered from 1 to N. Transcriptional repressors (orange squares) bind and unbind from binding sites (numbered platforms) at the promoter of a gene (light blue stretch). (A) and (B) illustrate two equivalent configurations of the promoter with identical number of bound transcriptional repressors having the same inhibiting strength.

The state $\mathcal{P}_i$ of the promoter at any time can be characterized by the number i of bound factors, which goes from 0 to N. With a rate k^+, a free repressor binds to an empty site at the promoter, which steps from state $\mathcal{P}_i$ to state $\mathcal{P}_{i+1}$; with a rate k^-, a bound repressor falls off the promoter, which steps from state $\mathcal{P}_i$ to state $\mathcal{P}_{i-1}$:

$$\mathcal{P}_0 \underset{k^-}{\overset{k^+}{\rightleftharpoons}} \cdots \rightleftharpoons \mathcal{P}_{i-1} \underset{k^-}{\overset{k^+}{\rightleftharpoons}} \mathcal{P}_i \underset{k^-}{\overset{k^+}{\rightleftharpoons}} \mathcal{P}_{i+1} \rightleftharpoons \cdots \underset{k^-}{\overset{k^+}{\rightleftharpoons}} \mathcal{P}_N. \quad (1)$$

Denoting by P_i the promoter occupation probabilities which describe the fraction of time that the promoter spends at state $\mathcal{P}_i$ in the thermodynamic limit [46], the kinetics of binding and unbinding of transcriptional repressors to the binding sites is given by the set of ordinary differential equations

$$\begin{aligned}
\dot{P}_0 &= -k^+NxP_0 + k^-P_1 \\
&\vdots \\
\dot{P}_i &= -k^+(N-i)xP_i - k^-iP_i \quad (2)\\
&\quad +k^+(N-i+1)xP_{i-1} + k^-(i+1)P_{i+1} \\
&\vdots \\
\dot{P}_N &= -k^-NP_N + k^+xP_{N-1},
\end{aligned}$$

together with the conservation law

$$P = \sum_{i=0}^{N} P_i, \quad (3)$$

where x is the repressor concentration and P is proportional to the number of gene copies.

We assume here that binding and unbinding of repressors to the promoter occur much faster than other processes like transcription, translation, transport and decay of molecules. This means that the promoter occupation probabilities quickly reach equilibrium with a given concentration of transcriptional repressors [46,47]. This situation is described by $\dot{P}_i = 0$ for all i in Eq. (3), and the resulting algebraic equations can be solved by induction to obtain

$$P_i = \binom{N}{i}\left(\frac{x}{x_0}\right)^i P_0, \quad (4)$$

where $x_0 = k^-/k^+$ is the equilibrium constant for binding of factors. Using the constraint Eq. (3) we express P_0 in terms of P

$$P_0 = P\left(1 + \frac{x}{x_0}\right)^{-N}. \quad (5)$$

Equation (4) and Eq. (5) describe the equilibrium occupation of the promoter in terms of the concentration x of the transcriptional repressor.

III. Abrupt inhibition

In the previous section, we evaluated the kinetics of noncooperative binding to a promoter with multiple binding sites. How does the presence of bound transcriptional repressors affect the transcription rate of the gene downstream of the promoter? In general, the strength of inhibition will depend on the number of bound repressors, and the regulatory function $f(x)$ will have the form

$$f(x) = \sum_{i=0}^{N} a_i P_i, \quad (6)$$

where $a_0 = b$ is the basal transcription rate in the absence of bound repressors and a_i is the transcription rate in the presence of i bound repressors. Here we assume, for the sake of simplicity, that transcription proceeds at its basal rate b while there are M or less sites occupied by the repressors, and drops to zero when the number of occupied sites is larger than M, see Fig. 3. We shall consider an alternative scenario below.

In this situation, Eq. (6) becomes

$$f(x) = b\sum_{i=0}^{M} P_i. \quad (7)$$

Using Eq. (4) and Eq. (5), the resulting regulatory function for a promoter with N binding sites is

$$f_{N,M}(x) = bP\left(1 + \frac{x}{x_0}\right)^{-N} \times \sum_{i=0}^{M}\binom{N}{i}\left(\frac{x}{x_0}\right)^i. \quad (8)$$

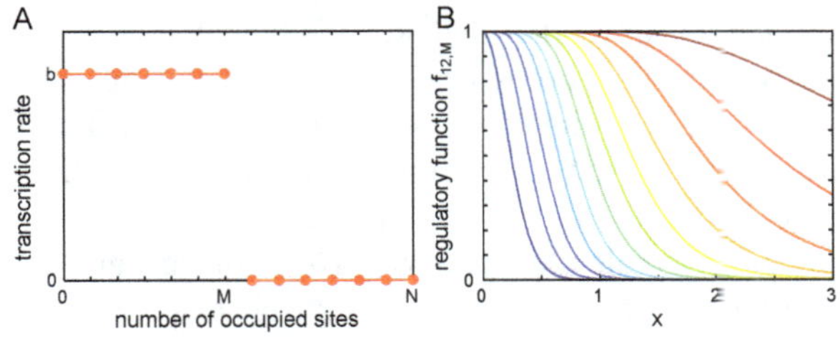

Figure 3: Abrupt inhibition. Full inhibition occurs when more than M sites are bound by transcriptional repressors. (A) Transcription rate as a function of the number of bound transcriptional repressors. (B) Normalized regulatory function, Eq. (8), as a function of repressor concentration x, with $N = 12$ and $M = 0, \ldots, 11$ (dark blue to dark red).

The zebrafish segmentation clock may be an interesting system to evaluate these results. In the *her1/her7* locus of zebrafish, an estimated number of about 12 binding sites has been reported [32,45]. The regulatory functions resulting from Eq. (8) for $N = 12$ are displayed in Fig. 3B. Although it is clear that inhibition of the promoter for a given level of repressors shifts to the right as M increases, it is less obvious how the value of M affects the steepness —that is the nonlinearity— of the negative feedback. We use Hill functions to parametrize the regulatory functions Eq. (8) in a more transparent form, see Appendix. Hill functions are parametrized by a Hill coefficient h characterizing the steepness of the curve, and an inhibition threshold K that describes the concentration of repressors that halves the production rate. We fit Hill functions to $f_{N,M}$ and obtain an effective Hill coefficient h and effective inhibition threshold K, for each value of N and M, Fig. 4A,B. Increasing the number of binding sites N while keeping M fixed can increase the Hill coefficient. For fixed N, increasing M changes the Hill coefficient in a nonmonotonic way: there is an optimal value of M that maximizes the Hill coefficient and therefore nonlinearity. The effective inhibition threshold K changes in a simpler form, increasing both with N and M. In conclusion, multiple binding sites can effectively increase the nonlinearity of the feedback via the regulatory function.

As discussed above, nonlinearity is an essen-

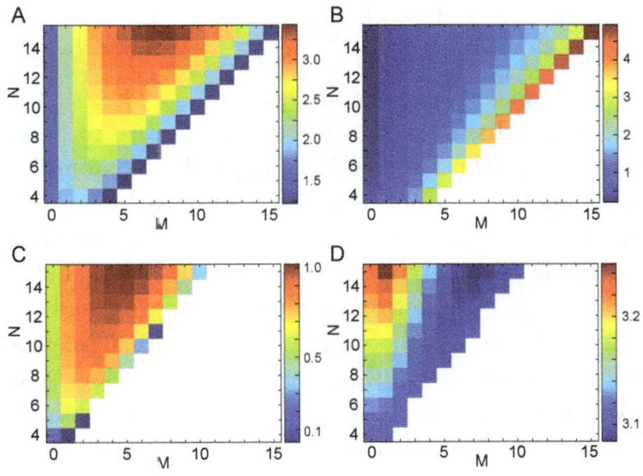

Figure 4: Multiple binding sites can increase nonlinearity and enhance oscillations. (A) and (B) Effective Hill parameters for the regulatory function $f_{N,M}$, Eq. (8), with $b\,P = 1$ and $x_0 = 1$. (A) Effective Hill coefficient h. (B) Effective inhibition threshold K. (C) and (D) Oscillations described by Eq. (9) with $b\,P = 2$, $\tau = 1$, $c = 1$ and $x_0 = 1$. (C) Amplitude of oscillations. (D) Period of oscillations. In C and D, the white region represents the nonoscillatory regime, in which the system settles to a fixed point. Color bar labels indicate values in each panel.

tial ingredient in a theory of biochemical oscillations [18]. We therefore ask how multiple binding sites in the promoter affect a biochemical oscillator. We use the regulatory function Eq. (8) in a generic model for genetic oscillations. We consider a gene that encodes a protein that forms dimers, and these dimers can bind to multiple binding sites at the promoter to inhibit transcription, Fig. 1A. We introduce an explicit delay τ to account for transcription, translation, splicing, and other processes involved in the assembly of the gene product and its dimerization. We assume that dimerization is a fast reaction, with a separation of timescales from other processes. Therefore, at any time the dimer concentration can be approximated by that of the monomers squared. The dynamics of the product concentration $x(t)$ is given by

$$\frac{dx}{dt} = bP \frac{\sum_{i=0}^{M} \binom{N}{i} \left(\frac{x(t-\tau)}{x_0}\right)^{2i}}{\left(1 + \left(\frac{x(t-\tau)}{x_0}\right)^2\right)^N} - cx(t), \quad (9)$$

where b is the basal production rate, c is the decay rate of products, P relates to the number of gene copies, τ is the total delay for product assembly, and x_0 is the product concentration that halves the basal transcription rate. The regulatory function Eq. (8) is parametrized by N and M. This genetic oscillator is a reduction of the models proposed by Lewis [19], and other authors [48, 49]. It describes the protein concentration, but does not include the mRNA; the duration of transcription and translation are both included in the delay τ. Furthermore, it does not describe effects present in theories that include more than one regulator [32, 50, 51], but here it is enough to illustrate the effects of multiple binding sites in a simpler context.

We integrate Eq. (9) numerically and evaluate the resulting dynamics by calculating the amplitude and period of oscillations. In all numerical simulations, we use the function `dde23` from MATLAB [52]. Scanning the values of N and M, we determine whether the system oscillates in steady state: when the difference in the maxima over the last ten cycles falls below 0.01, the simulation is stopped and we record the output.

The amplitude of oscillations grows with the number of binding sites N, Fig. 4C. The range where the system oscillates grows with the number of binding sites N. The amplitude is nonmonotonic in M: for a fixed number of binding sites N, there is an optimal value for M that maximizes the amplitude of oscillations, Fig.4C. The period of oscillations grows with N and decreases with M. These results show how the change in nonlinearity observed in the regulatory functions is reflected in the oscillations as the number of binding sites change.

IV. Gradual inhibition

There is strong evidence for the products of some cyclic genes of the zebrafish segmentation clock binding their own promoters and acting as transcriptional repressors [23,27,32,33,35,44]. However, we do not have detailed knowledge of how these transcriptional repressors affect transcription rates when bound to the promoters of cyclic genes. There is evidence from transcriptional analysis of the *Hes1* gene in mouse indicating that inhibition is gradual [30]. While the wildtype *Hes1* promoter containing all three N Box elements that are bound by Hes1 proteins showed a 30-fold inhibition of transcription in the presence of Hes1, mutations in one, two, and three of the N Box elements showed impaired inhibition with 14-, 7-, and 2-fold inhibition, respectively [30].

Motivated by these results, we consider here a scenario where additional bound repressors gradually reduce transcription rate until it drops to zero, Fig. 5A. For the sake of simplicity, we assume that transcription rate drops linearly as a function of bound repressors, from a basal rate b, to zero for $k+1$ occupied sites

$$a_i = \begin{cases} b\,(1 - i/(k+1)) & \text{if } i \le k+1 \\ 0 & \text{if } i > k+1\,, \end{cases} \quad (10)$$

see Fig. 5A. Using this gradual inhibition in Eq. (6) together with Eq. (4) and (5), we obtain regulatory functions $f_{N,k}(x)$

$$f_{N,k}(x) = bP\left(1 + \frac{x}{x_0}\right)^{-N} \times \sum_{i=0}^{k+1}\left(1 - \frac{i}{k+1}\right)\binom{N}{i}\left(\frac{x}{x_0}\right)^i, \quad (11)$$

see Fig. 5B. As in the previous case, it is clear that the effective inhibition threshold shifts to the right as k increases, but it is not so clear if the steepness of the regulatory function changes and, if so, how. Performing fits to Hill functions, we find that the nonmonotonic behavior of the effective Hill exponent h is observed again as k increases, Fig. 6A,B. Oscillations are similarly affected by noncooperative binding with gradual inhibition, Fig. 6C,D. These results show that the prediction of an optimal value for the number of bound repressors that fully inhibits the promoter is robust with respect to the details of how multiple bound repressors reduce the transcription rate.

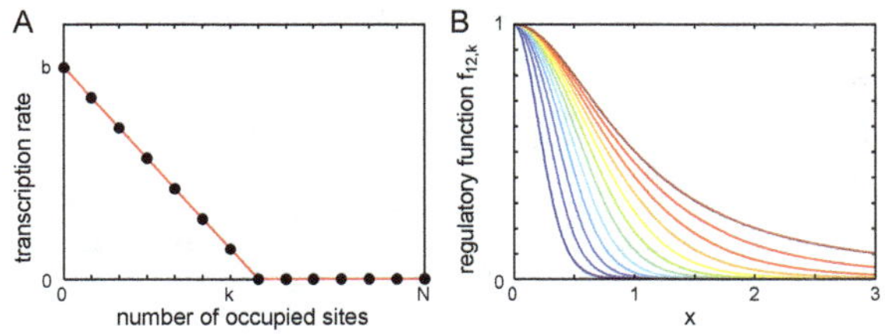

Figure 5: Gradual inhibition. Binding of multiple transcriptional repressors gradually inhibits transcription. (A) Transcription rate as a function of the number of bound transcriptional repressors. Transcription occurs at the basal rate b in the absence of bound repressors, and decreases linearly to zero for more than k bound repressors. (B) Normalized regulatory function Eq. (11) as a function of repressor concentration x, with $N = 12$ and $k = 0, \ldots, 11$ (dark blue to dark red).

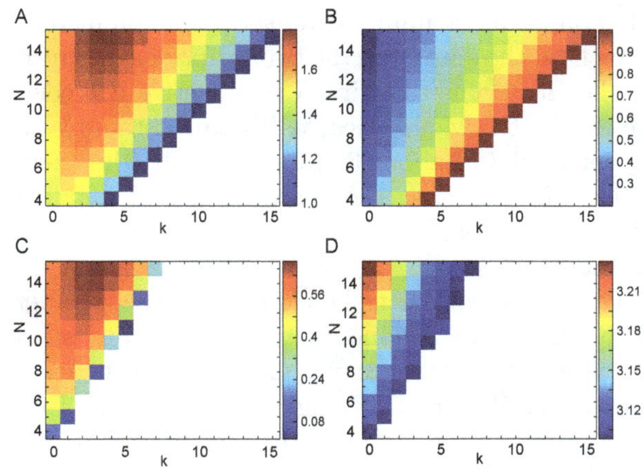

Figure 6: Multiple binding sites can increase nonlinearity and enhance oscillations in the presence of gradual inhibition. (A) and (B) Effective Hill parameters for the regulatory function $f_{N,k}$, Eq. (11), with $bP = 1$ and $x_0 = 1$. (A) Effective Hill coefficient h. (B) Effective inhibition threshold K. (C) and (D) Oscillations obtained using regulatory functions Eq. (11) with $bP = 2$, $\tau = 1$, $c = 1$ and $x_0 = 1$. (C) Amplitude of oscillations. (D) Period of oscillations. In C and D the white region represents the nonoscillatory regime, in which the system settles to a fixed point. Color bar labels indicate values in each panel.

V. Discussion

We studied the effects of multiple noncooperative binding sites in the promoter of a genetic oscillator. We evaluated the behavior of a promoter with multiple binding sites when binding of transcriptional repressors is noncooperative, Fig. 2. We considered two different hypotheses for how bound transcriptional repressors affect transcription rates, Figs. 3 and 5. In both cases, we calculated how the number of binding sites and the number of bound repressors required to produce full inhibition affect the nonlinearity of regulatory functions. We showed that there is an optimal value of the number of repressors needed to fully inhibit transcription that maximizes nonlinearity, Figs. 4AB and 6AB. This increased nonlinearity is reflected in the behavior of a genetic oscillator controlled by such regulatory functions, Figs. 4CD and 6CD.

Cooperative binding is a well known means to increase the nonlinearity of a biological dynamical system [21,39]. Here we show that the nonlinearity of a regulatory function can be increased by multiple binding sites, even if binding is noncooperative. This idea may have application in other biological control systems as well. Using the same formulation as we did here, one can also describe nonlinearity in the transcriptional activation of gene expression, thereby creating effective on-switches in developmental and physiological regulatory networks. A similar effect has been reported in a theoretical study of enzymes with multiple phosphorylation sites [53–55]. It was found that nonessential phosphorylation sites give rise to an increase in effective Hill coefficients, enhancing ultrasensitivity in signal transduction [55].

Previous work on the segmentation clock addressed the case of three binding sites [40], motivated by experimental observations of the *Hes7* mouse promoter [31]. The authors assumed that a single bound dimer inhibits transcription completely, corresponding to the particular case $M = 1$ of the theory developed here. They considered both noncooperative and cooperative binding, and showed that cooperativity increases the effective Hill coefficient as expected. In a follow-up [56], the authors addressed the case of *Hes1* regulation based on the report of four binding sites in the *Hes1* mouse promoter [30]. Again assuming that a single bound dimer inhibits transcription completely, they used the data from the transcriptional analysis of the *Hes1* gene to estimate an effective Hill coefficient for the Mouse Hes1 oscillator, obtaining an upper bound of about 3. More recently, the effect of two binding sites as compared to a single binding site was discussed together with differential decay of the monomers [57].

Our theory predicts how changing the number of binding sites N, and the number M of bound repressors that produce full inhibition affects a single cell oscillator. Although an experiment that changes the value of M may currently be challenging in a cellular system, the number of binding sites N is more amenable to experimental manipulation. For example, binding sites could be mutated [30], or deleted from the promoter, or they could be interfered with using genome editing strategies such as TALEN [58] or CRISPR [59] to alter or delete specific binding sites. To assess the effects of these perturbations in experiments may also pose some challenges. Dropping the number of binding sites from $N = 12$ to $N = 6$ introduces a period change of about 2.5%, while amplitude halves over the same range, Fig. 4C,D. Experiments will require at least such precision to reliably detect changes.

Our results suggest a possible evolutionary mechanism to increase nonlinearity in gene regulatory systems. In this mechanism, point mutations in the promoter that increase the number of binding sites for transcription factors may increase the steepness of regulatory functions. If the resulting steeper regulation performs some function better, such mutations would have a good chance to be conserved by natural selection. In the case of the segmentation clock, the amplitude of oscillations could increase with the number of binding sites, possibly reducing the signal to noise ratio. Furthermore, the range for oscillations would be wider, making the oscillatory regime less sensitive to slow extrinsic fluctuations of parameter values. Remarkably, it may happen that after an increase in N, an increase in M also raises nonlinearity, see Fig. 4A. This means that weaker repressors would result in better oscillations. This evolutionary mechanism would provide a simple way to gradually increase the nonlinearity of a feedback or other regulatory function.

The theory for the zebrafish regulatory function could be refined using the experimentally measured relative affinities of the binding sites at the *her1* and *her7* promoters [32]. Apart from the effects reported here, the number of binding sites may have additional roles. For example, it could serve as a buffer for fluctuations in gene expression [20,60–64], augmenting the precision of genetic oscillations. This will be the topic of future work.

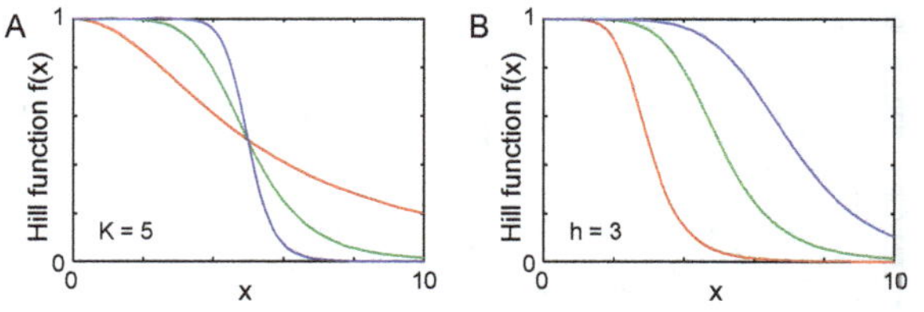

Figure 7: Hill functions are characterized by two parameters, the Hill coefficient h and the inhibition threshold K. (A) Hill functions with $K = 5$ and $h = 1$ (red), $h = 3$ (green) and $h = 7$ (blue). (B) Hill functions with $h = 3$ and $K = 3$ (red), $K = 5$ (green) and $K = 7$ (blue).

Acknowledgements - We thank Saúl Ares, Luciana Bruno, Ariel Chernomoretz and Hernan Grecco for helpful comments on an early version of this work, and James Ferrell for calling our attention to Ref. [53]. Thanks to E. Aikau for inspiration. LGM acknowledges support from MINCyT Argentina (PICT 2012 1952). A.C.O. and D.S. were supported by the Medical Research Council UK (MC_UP_1202/3) and the Wellcome Trust (WT098025MA).

Appendix: Effective Hill functions

Hill functions are often used to describe the nonlinearities present in gene regulatory networks [17]. Hill functions are sigmoidal step functions defined by

$$f_H(x) = \frac{1}{1 + (x/K)^{2h}}, \tag{12}$$

where the steepnes of the step is characterized by the exponent $2h$, and the inhibition threshold K is the concentration of repressor that halves the production rate, here scaled to unity, Fig. 7. Here

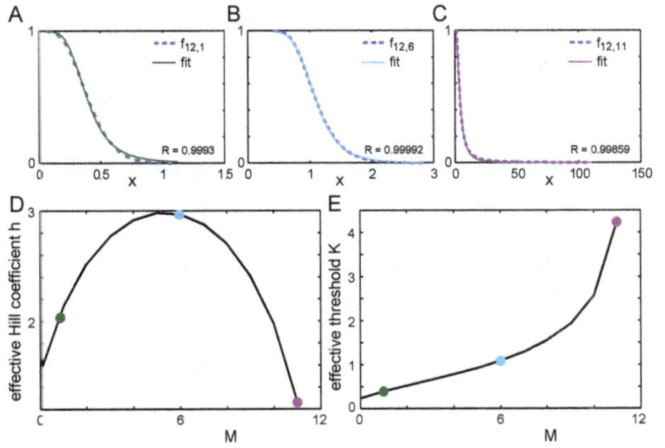

Figure 8: Effective Hill functions Eq. (12) can fit regulatory functions of the multiple binding site regulatory function, Eq. (8) for $N = 12$, with $bP = 1$ and $x_0 = 1$. (A-C) Examples of fits of Hill functions (solid lines) to regulatory functions (dashed lines) for (A) $M = 1$ (green), (B) $M = 6$ (light blue) and (C) $M = 11$ (purple). (D) Effective Hill coefficient h as a function of M. (E) Effective inhibition threshold K as a function of M. Dots in D and E correspond to fits from panels A, B and C.

we include an explicit factor 2 in the exponent to account for the dimerization of the transcriptional repressors. One advantage of Hill functions is that they are very simply parametrized. In the main text, we fit Hill functions to the more complex regulatory functions Eqs. (8) and (11). Some fits of Eq. (8) in the case $N = 12$ are displayed in Fig. 8 for illustration.

---

[1] A Goldbeter, *Biochemical Oscillations and Cellular Rhythms: The Molecular Bases of Periodic and Chaotic Behaviour*, Cambridge University Press, Cambridge (1997).

[2] L Glass, M C Mackey, *From Clocks to Chaos: The Rhythms of Life*, Princeton University Press, Princeton (1988).

[3] Y Liu, N F Tsinoremas, C H Johnson, N V Lebedeva, S S G M Ishiura, T Kondo, *Circadian orchestration of gene expression in cyanobacteria*, Gene. Dev. **9**, 1469 (1995).

[4] E Nagoshi, C Saini, C Bauer, T Laroche, F Naef, U Schibler, *Circadian gene expression in individual fibroblasts: Cell-autonomous and self-sustained oscillators pass time to daughter cells*, Cell **119**, 693 (2004).

[5] I Mihalcescu, W Hsing, S Leibler, *Resilient circadian oscillator revealed in individual cyanobacteria*, Nature **430**, 81 (2004).

[6] A Goldbeter, C Gérard, D Gonze, J C Leloup, G Dupont, *Systems biology of cellular rhythms*, FEBS Lett. **586**, 2955 (2012).

[7] I Palmeirim, D Henrique, D Ish-Horowicz, O Pourquié, *Avian hairy gene expression identifies a molecular clock linked to vertebrate segmentation and somitogenesis*, Cell **91**, 639 (1997).

[8] A Aulehla, W Wiegraebe, V Baubet, M B Wahl, C Deng, M Taketo, M Lewandoski, O Pourquié, *A β-catenin gradient links the clock and wavefront systems in mouse embryo segmentation*, Nat. Cell Biol. **10**, 186 (2008).

[9] Y Masamizu, T Ohtsuka, Y Takashima, H Nagahara, Y Takenaka, K Yoshikawa, H Okamura, R Kageyama, *Real-time imaging of the segmentation clock: Revelation of unstable oscillators in the individual presomitic mesoderm cells*, Proc. Natl. Acad. Sci. USA **103**, 1313 (2006).

[10] A J Krol, D Roellig, M L Dequéante, O Tassy, E Glynn, G Hattem, A Mushegian, A C Oates, O Pourquié, *Evolutionary plasticity of segmentation clock networks*, Development **138**, 2783 (2011).

[11] H Shimojo, T Ohtsuka, R Kageyama, *Oscillations in notch signaling regulate maintenance of neural progenitors*, Neuron **58**, 52 (2008).

[12] N Geva-Zatorsky, N Rosenfeld, S Itzkovitz, R Milo, A Sigal, E Dekel, T Yarnitzky, Y Liron, P Polak, G Lahav, U Alon, *Oscillations and variability in the p53 system*, Mol. Syst. Biol. **2**, 2006.0033 (2006).

[13] M B Elowitz, S Leibler, *A synthetic oscillatory network of transcriptional regulators*, Nature **403**, 335 (2000).

[14] J Stricker, S Cookson, M R Bennett, W H Mather, L S Tsimring, J Hasty, *A fast, robust and tunable synthetic gene oscillator*, Nature **456**, 516 (2008).

[15] J J Tyson, *Computational cell biology*, Chap. 9, Pag. 230, Springer, Berlin (2002).

[16] B Alberts, A Johnson, J Lewis, M Raff, K Roberts, P Walter, *Molecular biology of the cell*, 4th Ed., Garland Science, New York (2002).

[17] U Alon, *An introduction to systems biology: Design principles of biological circuits*, Chapman & Hall/CRC Press, Boca Raton, Florida (2006).

[18] B Novák, J J Tyson, *Design principles of biochemical oscillators*, Nat. Rev. Mol. Cell Biol. **9**, 981 (2008).

[19] J Lewis, *Autoinhibition with transcriptional delay: A simple mechanism for the zebrafish somitogenesis oscillator*, Curr. Biol. **13**, 1398 (2003).

[20] L G Morelli, F Jülicher, *Precision of genetic oscillators and clocks*, Phys. Rev. Lett. **98**, 228101 (2007).

[21] J E Ferrell, *Q&A: Cooperativity*, J. Biol. **8**, 53 (2009).

[22] O Pourquié, *Vertebrate segmentation: From cyclic gene networks to scoliosis*, Cell **145**, 650 (2011).

[23] A C Oates, L G Morelli, S Ares, *Patterning embryos with oscillations: Structure, function and dynamics of the vertebrate segmentation clock*, Development **139**, 625 (2012).

[24] Y Saga, *The synchrony and cyclicity of developmental events*, Cold Spring Harb. Perspect. Biol. **4**, a008201 (2012).

[25] D Roellig, L G Morelli, S Ares, F Jülicher, A C Oates, *Snapshot: The segmentation clock*, Cell **145**, 800 (2011).

[26] Y Saga, *The mechanism of somite formation in mice*, Curr. Opin. Genet. Dev. **22**, 331 (2012).

[27] A C Oates, R K Ho, *Hairy/E(spl)-related (Her) genes are central components of the segmentation oscillator and display redundancy with the Delta/Notch signaling pathway in the formation of anterior segmental boundaries in the zebrafish*, Development **129**, 2929 (2002).

[28] H Hirata, S Yoshiura, T Ohtsuka, Y Bessho, T Harada, K Yoshikawa, R Kageyama, *Oscillatory expression of the bHLH factor Hes1 regulated by a negative feedback loop*, Science **298**, 840 (2002).

[29] S A Holley, D Jülich, G J Rauch, R Geisler, C Nüsslein-Volhard, *her1 and the notch pathway function within the oscillator mechanism that regulates zebrafish somitogenesis*, Development **129**, 1175 (2002).

[30] K Takebayashi, Y Sasai, Y Sakai, T Watanabe, S Nakanishi, R Kageyama, *Structure, chromosomal locus, and promoter analysis of the gene encoding the mouse helix-loop-helix factor hes-1. negative autoregulation through the multiple n box elements*, J. Biol. Chem. **269**, 5150 (1994).

[31] Y Bessho, G Miyoshi, R Sakata, R Kageyama, *Hes7: A bhlh-type repressor gene regulated by notch and expressed in the presomitic mesoderm*, Genes Cells **6**, 175 (2001).

[32] C Schröter, S Ares, L G Morelli, A Isakova, K Hens, D Soroldoni, M Gajewski, F Jülicher, S J Maerkl, B Deplancke, A C Oates, *Topology and dynamics of the zebrafish segmentation clock core circuit*, PLoS Biol. **10**, e1001364 (2012).

[33] A Trofka, J Schwendinger-Schreck, T Brend, W Pontius, T Emonet, S A Holley, *The Her7 node modulates the network topology of the zebrafish segmentation clock via sequestration of the Hes6 hub*, Development **139**, 940 (2012).

[34] A Hanisch, M V Holder, S Choorapoikayil, M Gajewski, E M Özbudak, J Lewis, *The elongation rate of RNA polymerase ii in zebrafish and its significance in the somite segmentation clock*, Development **140**, 444 (2013).

[35] F Giudicelli, E M Özbudak, G J Wright, J Lewis, *Setting the tempo in development:*

An investigation of the zebrafish somite clock mechanism, PLoS Biol. **5**, 1309 (2007).

[36] E M Özbudak, J Lewis, *Notch signalling synchronizes the zebrafish segmentation clock but is not needed to create somite boundaries*, PLoS Genet. **4(2)**, e15 (2008).

[37] Y Harima, Y Takashima, Y Ueda, T Ohtsuka, R Kageyama, *Accelerating the tempo of the segmentation clock by reducing the number of introns in the hes7 gene*, Cell Rep. **3**, 1 (2013).

[38] J Keener, J Sneyd, *Mathematical physiology I: Cellular physiology*, 2nd Ed. Springer, Berlin (2008).

[39] H Qian, *Cooperativity in cellular biochemical processes: Noise-enhanced sensitivity, fluctuating enzyme, bistability with nonlinear feedback, and other mechanisms for sigmoidal responses*, Ann. Rev. Biophys. **41**, 179 (2012).

[40] S Zeiser, H V Liebscher, H Tiedemann, I Rubio-Aliaga, G K H Przemeck, M H de Angelis, G Winkler, *Number of active transcription factor binding sites is essential for the hes7 oscillator*, Theor. Biol. Med. Model. **3**, 11 (2006).

[41] J Gunawardena, *Multisite protein phosphorylation makes a good threshold but can be a poor switch*, P. Natl. Acad. Sci. USA **102**, 14617 (2005).

[42] V Gotea, A Visel, J M Westlund, M A Nobrega, L A Pennacchio, I Ovcharenko, *Homotypic clusters of transcription factor binding sites are a key component of human promoters and enhancers*, Genome Res. **20**, 565 (2010).

[43] D S Burz, R Rivera-Pomar, H Jäckle, S D Hanes, *Cooperative dna-binding by bicoid provides a mechanism for threshold-dependent gene activation in the drosophila embryo*, EMBO J. **17**, 5998 (1998).

[44] T Brend, S A Holley, *Expression of the oscillating gene her1 is directly regulated by hairy/enhancer of split, t-box, and suppressor of hairless proteins in the zebrafish segmentation clock*, Dev. Dynam. **238**, 2745 (2009).

[45] D Soroldoni, personal communication (2014).

[46] L Bintu, N E Buchler, H G Garcia, U Gerland, T Hwa, J Kondev, R Phillips, *Transcriptional regulation by the numbers: Models*, Curr. Opin. Genet. Dev. **15**, 116 (2005).

[47] H G Garcia, A Sanchez, T Kuhlman, J Kondev, R Phillips, *Transcription by the numbers redux: experiments and calculations that surprise*, Trends Cell Biol. **20**, 723 (2010).

[48] N A M Monk, *Oscillatory expression of Hes1, p53, and NF-κB driven by transcriptional time delays*, Curr. Biol. **13**, 1409 (2003).

[49] M H Jensen, K Sneppen, G Tiana, *Sustained oscillations and time delays in gene expression of protein Hes1*, FEBS Lett. **541**, 176 (2003).

[50] O Cinquin, *Repressor dimerization in the zebrafish somitogenesis clock*, PLoS Comp. Biol. **3**, e32 (2007).

[51] A Ay, S Knierer, A Sperlea, J Holland, E M Özbudak, *Short-lived her proteins drive robust synchronized oscillations in the zebrafish segmentation clock*, Development **140**, 3244 (2013).

[52] L F Shampine, S Thompson, *Solving ddes in Matlab*, Appl. Numer. Math. **37**, 441 (2001).

[53] L Wang, Q Nie, G Enciso, *Nonessential sites improve phosphorylation switch*, Biophys. J. **99**, L41 (2010).

[54] S Ryerson, G Enciso, *Ultrasensitivity in independent multisite systems*, J. Math. Biol. **69**, 977 (2014).

[55] G Enciso, *Nonautonomous and random dynamical systems in life sciences* Lecture Notes in Mathematics (Mathematical Biosciences Subseries) No 2102, Springer Verlag, Berlin (2013).

[56] S Zeiser, J Müller, V Liebscher, *Modeling the hes1 oscillator*, J. Comput. Biol. **14**, 984 (2007).

[57] M Campanelli, T Gedeon, *Somitogenesis clock-wave initiation requires differential decay and multiple binding sites for clock protein*, PLoS Comp. Biol. **6**, e1000728 (2010).

[58] V M Bedell, Y Wang, J M Campbell, T L Poshusta, C G Starker, R G K II, W Tan, S G Penheiter, A C Ma, A Y H Leung, S C Fahrenkrug, D F Carlson, D F Voytas, K J Clark, J J Essner, S C Ekker, *In vivo genome editing using a high-efficiency talen system*, Nature **491**, 114 (2012).

[59] E Pennisi, *The crispr craze*, Science **341**, 833 (2013).

[60] N Barkai, S Leibler, *Biological rhythms: Circadian clocks limited by noise*, Nature **403**, 267 (2000).

[61] M B Elowitz, A J Levine, E D Siggia, P S Swain, *Stochastic gene expression in a single cell*, Science **297**, 1183 (2002).

[62] J Raser, E O'Shea, *Noise in gene expression: Origins, consequences, and control*, Science **309**, 2010 (2005).

[63] B Munsky, G Neuert, A V Oudenaarden, *Using gene expression noise to understand gene regulation*, Science **336**, 183 (2012).

[64] L S Tsimring, *Noise in biology*, Rep. Prog. Phys. **77**, 026601 (2014).

5

Granular discharge rate for submerged hoppers

T. J. Wilson,[1,2] C. R. Pfeifer,[1,3] N. Meysingier,[1,4] D. J. Durian[1*]

The discharge of spherical grains from a hole in the bottom of a right circular cylinder is measured with the entire system underwater. We find that the discharge rate depends on filling height, in contrast to the well-known case of dry non-cohesive grains. It is further surprising that the rate increases up to about twenty five percent, as the hopper empties and the granular pressure head decreases. For deep filling, where the discharge rate is constant, we measure the behavior as a function of both grain and hole diameters. The discharge rate scale is set by the product of hole area and the terminal falling speed of isolated grains. But there is a small-hole cutoff of about two and half grain diameters, which is larger than the analogous cutoff in the Beverloo equation for dry grains.

I. Introduction

The flow of granular materials is of widespread practical [1, 2] and fundamental [3, 4] interest. An important example is the gravity-driven flow of grains from an hourglass or flat-bottomed hopper. For ordinary fluids, the discharge rate is proportional to the pressure head, which is set by filling height, and decreases continuously to zero with vanishing hole size. The discharge of grains is strikingly different. In particular, for dry non-cohesive grains, the mass per unit time discharged from a hole in the bottom of a hopper is accurately described by the empirical Beverloo equation [5]:

$$W = C\rho_b g^{1/2}(D - kd)^{5/2}. \quad (1)$$

Here ρ_b is the mass density of the bulk granular medium, $g = 980$ cm/s^2, D is the hole diameter, d is the grain diameter, and C and k are dimensionless fitting parameters [5]. The filling height plays no role. The container diameter, and hence the grain-grain contact pressure at the bottom of the container, as given by the Janssen argument [6], also plays no role. Intuitively, "transient arches" intermittently form and break over the hole, shielding the exiting grains from any sort of pressure head. Then the rough scale for discharge is the free-fall speed $\sqrt{gD}$ times hole area, i.e., $W \sim \rho_b g^{1/2} D^{5/2}$. Beverloo et al. plotted their data as $W^{2/5}$ versus D and found a straight line, but such that W vanishes at a nonzero small-hole cutoff diameter of kd [7]. This has been rationalized by an "empty annulus" around the perimeter of the hole, through which grain centers cannot pass. Nevertheless, fundamental justification of Eq. (1) remains a topic of on-going interest [8, 9].

The Beverloo Eq. (1) is supported by a large number of experiments, as reviewed by Nedderman and Savage et al. [7]. But discrepancies of up to forty percent have been reported when the hole size is increased more widely than usual [10].

*Email: djdurian@physics.upenn.edu

[1] Department of Physics & Astronomy, University of Pennsylvania, Philadelphia, PA 19104-6396, USA.

[2] Department of Physics, Illinois Wesleyan University, Bloomington, IL 61702-2900, USA.

[3] Department of Physics & Astronomy, Carleton College, Northfield, MN 55057, USA.

[4] Strath Haven High School, Wallingford, PA 19086, USA.

Recently, we have found excellent agreement with the Beverloo equation for up to three [11] and four [12] decades in discharge rate for spherical grains. Typical ranges for the numerical coefficients are $0.5 < C < 0.7$ and $1.2 < k < 3$; near the low end for spherical grains. Discharge rates can be increased, paradoxically, by placing an obstacle over the hole [17–19]. And for small enough holes, stable arches can form and cause a clog [20]. Related behavior has been reported for the upward discharge of bubbles in an underwater silo [13], particles on a conveyor belt [14], and for disks floating on a fluid that flows through an orifice [15, 16]. An important challenge is to relate all such phenomena to the velocity and density fields, which can be measured in quasi-2D [9,21–25] and index-matched [26] systems.

The physical intuition behind the transient arch and empty annulus concepts is appealing, and helps explain the contrast between granular and Newtonian fluids. But it has proven difficult to translate into a first-principles theory of hopper discharge and clogging. This could be because the jamming and unjamming of grains in the converging flow near the boundary is not only collective, but is also even more difficult to model than jamming in uniform systems [27]. As an experimental approach to alter transient arching and grain free-fall, we previously explored the effects of tilting the hopper and, hence, the plane of the hole away from horizontal [11, 12]. Now in this paper, we perturb transient arching in a totally different way by submerging the entire system underwater. Here the grain dynamics become overdamped rather than inertial, and the characteristic grain exit speed becomes set by the fluid and the grain size, rather than by free-fall and the hole size. We find not only a change in the scale of Eq. (1), but also in the small-hole cutoff. This is our main focus. In addition, we find an unexpected dependence of discharge rate on filling height, opposite in sign to that for ordinary liquids, which is to be the subject for further experiments [28].

II. Materials and methods

The granular media consist of monodisperse spherical beads, primarily of glass (Potter Industries), but also of lead (McMaster-Carr). The average and standard deviation of the grain diameters d, the grain material density ρ_g, and the resulting terminal falling speed in water v_t, are collected in Table 1. The latter are computed using an accurate empirical formula [29] for the dimensionless drag coefficient c_d versus Reynolds number $\mathrm{Re} = \rho v_t d/\eta$, where ρ and η are respectively the density and viscosity of water. This is done by equating gravity and drag forces, $\Delta mg = c_d \rho {v_t}^2 A/2$ where $\Delta m = (4/3)(\rho_g - \rho)\pi(d/2)^3$ and $A = \pi(d/2)^2$, and solving numerically for v_t. The packings are further characterized by the draining angle of repose, θ_r, as measured in air and under water. The results in Table 1 show that θ_r is larger for glass beads when dry, but larger for lead beads when wet. Also, θ_r is noticeably larger for the $d = 0.11$ mm glass beads. Finally, note that the polydispersity is about 20% for the $d = 0.11$ mm glass beads and the lead beads, but is $1-6\%$ for the other three glass bead samples.

Two different hoppers types are used. The first is constructed from 93 mm diameter transparent cylindrical plastic (polyethylene terephthalate) jars with screw-on plastic lids of equal diameter. For these, an outlet hole of desired diameter D is drilled into a plastic lid. Nearly two dozen different hole diameters are used, each drilled into a different lid. The hole diameters are measured by calipers with an uncertainty of ± 0.02 cm. An inlet for water is opened on the other end of the jar in order to prevent a back-flow of water up into the hopper to replace the lost volume of discharged grains. The other hopper type consists of plastic graduated cylinders, diameter 50 mm or 38 mm, with a single hole drilled into the bottom.

All discharge measurements are conducted with the hopper fixed to a sturdy aluminum stand, all completely underwater in a large aquarium. Prior to use, the glass beads are submerged and repeatedly poured back and forth between two containers in the same aquarium in order to allow all air to escape. The grains are then poured slowly into the hopper, without exposure to air, and allowed to settle with the outlet hole blocked. As such, the packings have a solids volume fraction ϕ near random-close packing (Table 1). For large holes, flow commences immediately after the hole is unblocked. For small holes, gentle tapping is required to start the discharge. In either case, the discharge rate is measured only after the flow has proceeded long enough that a conical depression fully develops at the top of the packing. The height H of the pack-

Grains	d (mm)	ρ_g (g/mL)	v_t (cm/s)	Re	ϕ	θ_r dry (deg)	θ_r wet (deg)
glass	0.11 ± 0.02	2.499 ± 0.007	0.94	1	0.57 ± 0.01	29.8 ± 0.2	26.5 ± 0.3
glass	0.31 ± 0.02	2.504 ± 0.003	4.0	12	0.59 ± 0.01	24.8 ± 0.3	22.7 ± 0.3
glass	0.96 ± 0.05	2.519 ± 0.007	15	150	0.62 ± 0.01	24.6 ± 0.3	21.5 ± 0.3
glass	3.00 ± 0.04	2.590 ± 0.010	36	1100	0.57 ± 0.01	24.2 ± 0.3	19.8 ± 0.8
lead	1.0 ± 0.2	10.9 ± 0.2	49	520	0.62 ± 0.01	17 ± 1	22 ± 3

Table 1: Granular materials properties: d is the average grain diameter, ρ_g is the density of individual grains, v_t is the terminal falling speed of individual grains in water, Re is the corresponding Reynolds number, ϕ is the volume fraction of grains in the packing, θ_r is the draining angle of response as measured for both dry and submerged packings. Measurement uncertainties are also given, except for the grain diameter where the standard deviation of the size distribution is reported.

ing is then measured as the level of the grains at the container wall. The height of the packing over the hole is then estimated as $h = H - (D/2)\tan\theta_r$.

The mass per unit time discharge rate, W, is found by starting/stopping a timer while inserting/removing a cup from the discharge stream. The collection cup is then removed from the aquarium and weighed, carefully topped off with water. The mass of the same cup, filled with water but no grains, is also determined. The difference between the two mass measurements is $\Delta M = V_g(\rho_g - \rho)$ where V_g is the volume of grains. This gives the mass of the collected grains as $M_g = V_g\rho_g = \Delta M\rho_g/(\rho_g - \rho)$. The discharge rate, W, is then computed as M_g divided by the duration of the collection time. The collection times range from a few seconds to several hours, and M_g ranges from about 10 g to 250 g for the glass beads. Collection of grains is begun only after a steady-state depression is formed at the top surface of packing; thus the possibility of initial transients, which could depend on initial packing fraction [30], is not investigated.

III. Versus Packing Height

A prudent preliminary task is to measure whether or not the discharge rate W for submerged grains is independent of the packing height h, as in the case of dry grains. So in Fig. 1 we show W versus h data for $d = 1$ mm beads, of glass and lead, for holes of various diameter. The rates all appear to approach a constant for very large packing heights. Therefore, the driving pressure on the grains at the outlet must be shielded from the weight of the packing due to transient arch formation as for dry grains. For smaller packing heights, however, the discharge rate data in Fig. 1 are not constant. Rather, the rates all *increase* as the hopper empties. This is surprising, because the sign of the effect is opposite to that for ordinary fluids where the rate decreases with the diminishing pressure head as the container empties. Similar behavior has been noticed for submerged hoppers [31], and also for dry grains in a cylindrical tube with a conical orifice citeLePennecPF98. Preliminary data, with a new apparatus, show that the discharge increase upon emptying is much smaller for dry grains, on the order of one percent [28]. We are unaware of any theoretical explanation. In fact, both the $\mu(I)$ flow law and discrete contact dynamics simulations for two dimensional dry grains predict a decrease in discharge rate as the hopper empties [33, 34]. The former approach has been extended to suspensions [35], but not yet applied to hopper flows.

The variation of discharge rate with packing height is roughly exponential. Fits to the form $W(h) = W_o\{1 + 0.25\exp[-h/(10\text{ cm})]\}$ are shown by solid curves in Fig. 1, where the asymptotic discharge rate W_o is the only fitting parameter. This function has no theoretical basis, as yet, but is a simple form that allows us to both estimate W_o and to compare the displayed data sets. The fits are quite good, and thus illustrate (i) that the size of the surge is about 25% at most, and (ii) that the characteristic height is about 10 cm. These two numbers represent the average of individual fitting results for all hole sizes for the 93 mm hoppers, when all three parameters are allowed to float. For the 93, 50 and 38 mm diameter hoppers, shown in

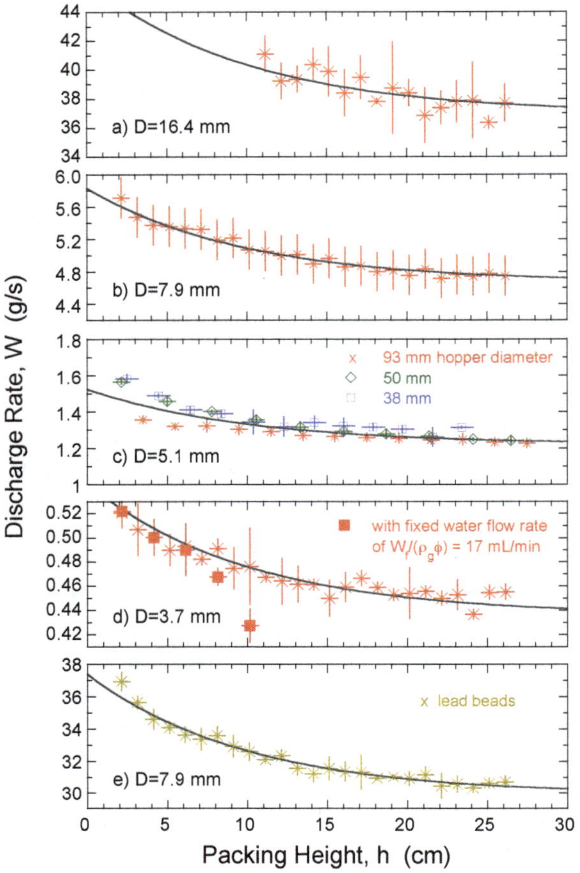

Figure 1: Discharge rate vs packing height for (a-d) $d = 1$ mm diameter glass beads and (e) $d = 1$ mm diameter lead beads, and different hole diameters D as labeled. Note that time proceeds right to left. The hopper diameter is 93 mm, except as noted in (c). In (d), data are also included where the container is sealed and a fixed flow of water is pumped into the top at the same volumetric rate at which grains exit the hole at bottom. The horizontal error bars indicate the range of packing heights over which the discharge rate is measured. The vertical error bars indicate the standard deviation of 3-5 repetitions. The solid curves are fits to $W(h) = W_o\{1 + 0.25\exp[-h/(10\text{ cm})]\}$, where W_o is the only adjustable parameter, to show that the height-dependence is similar for different discharge geometries.

Fig. 1(c), the individually-fitted decay lengths are 14.6 ± 3, 8.7 ± 0.4 and 4.9 ± 0.8 cm, respectively. Thus, the decay length appears to be on the order of ten times the hopper diameter. A factor of order one might have been expected if Janssen-type wall effects were primarily responsible. Nonetheless, one hypothesis for the surge would be that the grain-grain contact pressure over the outlet decreases as the hopper empties, and a resulting subtle decrease in packing fraction allows for greater fluidity and hence greater discharge rate. However, this seems ruled out by Ref. [36], where data show a decrease of pressure but no change in discharge rate.

Whatever the cause of the packing height dependence of discharge rate, the fluid clearly plays a role. So it is logical to consider interstitial flow between the grains. Certainly, such flow is generated as the grains move apart and out of the hole. A second hypothesis would be that some liquid is drawn down through the whole packing, even more easily and rapidly as the hopper empties —which would cause the grain discharge rate to increase. For large packing heights, such fluid flow would vanish and the grain discharge rate would become constant. As a test, we sealed off the top of a 93 mm diameter hopper and connected it to a gear pump for enforcing a constant interstitial liquid flow rate. The liquid pump rate was set to $W_o/(\rho_g\phi)$, so that grains and fluid can flow inside the hopper together, in unison, with the same macroscopic velocity field. This would be the condition for very tall packings, but is now enforced for all packing heights. Grain discharge rate data are plotted in Fig. 1(d), on top of prior data for an open-topped hopper with no liquid pumping. Except for one perhaps-spurious point, the two data sets are indistinguishable in showing the same rate increase as the hopper empties. Therefore, interstitial fluid flow down through the packing is not responsible for the surprising variation of discharge rate with packing height. The fluid and the walls probably play a role, but the mechanism is not known. Further experiments are now in progress to address this issue [28].

IV. Versus Hole Diameter

As our main task, we now turn to the variation of discharge rate with the hole diameter D, in the limit of large packing heights where the rate has a constant value W_o. For this, we collect data for 93 mm diameter hoppers with packing heights over the range 25 cm $< H <$ 28 cm; this procedure agrees well with results from exponential fits illustrated in Fig. 1. The simplest scale for discharge rate is $W_o \propto \rho_g v_t D^2$, where ρ_g is the density of the grain material, v_t is the terminal falling speed for an individual grain in water. Note that v_t depends on fluid and grain properties, and scales with grain diameter as d^2 at low Re but as $d^{1/2}$ at high Re (see Section II.). This contrasts with dry grains, where the discharge speed depends on grain size only for very small holes. Imposing a Beverloo-like cutoff by the substitution $D \to (D - kd)$, we arrive at the expectation

$$W_o = C\rho_g v_t (D - kd)^2, \quad (2)$$

$$= C(\rho_g v_t d^2)[(D/d) - k]^2. \quad (3)$$

Similar Beverloo-type forms with cutoff have been found for the upward discharge of bubbles in a quasi-2D silo [13] and for particles on a conveyor belt [14]. Consideration of v_t based on a single grain warrants caution, since there must be a more complex flow of the interstitial fluid in response to granular shear and dilation near the aperture.

For comparison of data with Eq. (3), Fig. 2 shows a dimensionless plot of discharge rate measurements as $[W_o/(\rho_g v_t d^2)]^{1/2}$ versus D/d. Indeed, for glass beads of diameter 0.3, 1, and 3 mm, we find excellent collapse of the data onto a single line that vanishes at non-zero D/d. This analysis is analogous to the famous Beverloo plot of $W^{2/5}$ versus D/d for dry grains [7]. Here the main plot is double-logarithmic to illustrate a range of roughly two decades in dimensionless hole diameter and five decades in dimensionless discharge rate; the inset has linear axes with a smaller range so that the functional form versus hole diameter is evident. Data for the lead and the 0.1 mm glass beads show similar linear behavior, but do not collapse onto the results for the larger glass beads. Perhaps this is because their polydisersities are significantly larger, as seen in Table 1.

Two lines are shown in Fig. 2 that correspond to Eq. (3) and match the two groupings of data. The proportionality constants, C, are close to 2/3 and 1/2. It is reassuring that these values are of order one. But it is more interesting that the lines both vanish at the same ratio of hole to grain diameter. Based on individual linear fits to the separate data sets, this small-hole cutoff is $k = 2.4 \pm 0.1$. By contrast, for dry spherical grains, the Beverloo cutoff for Eq. (1) is $k = 1.5$; and for air bubbles in water it is $k = 0.66$. Therefore, the cutoff involves dynamics and cannot be explained by purely geometrical concepts like the "empty annulus". Alternatively, our data could also be taken as further evidence for the point of view that discharge rate should be described by a functional form in which $k = 1$ is enforced [9, 10].

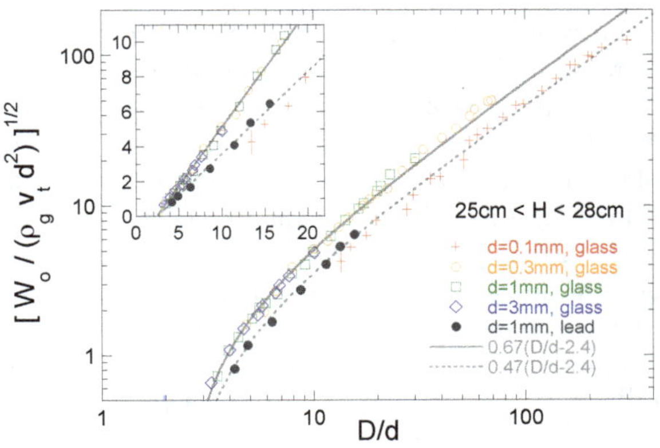

Figure 2: Mass discharge rate W_o vs hole diameter D, in the large packing-height limit, for several grain types as labelled. The axes are made dimensionless by grain diameter d, density ρ_g of individual grains, and terminal falling speed v_t of individual grains in water. The packing height range is 25 cm $< H <$ 28 cm, and the hopper diameter is 93 mm. Vertical error bars, given by the standard deviation of 3-5 repetitions, are smaller than symbol size except for the $d = 0.1$ mm glass beads as shown. The solid curves represent lines that vanish at $D/d = 2.4$, as specified in the legend. Fits to ${W_o}^{1/2} \propto (D/d - k)$ give the average cutoff as $k = 2.4 \pm 0.1$ grain diameters.

V. Conclusion

In summary, we have observed a striking dependence of the discharge rate on packing height, such that the rate surges as the hopper empties. The mechanism is not yet understood, but may involve wall effects and the flow of liquid in between grains as they come apart near the exit. The latter is reminiscent of an increase of flux due to dilation of grains under a fixed obstacle placed above the hole [17–19]. But here the relevant length scale appears to be the hopper diameter, rather than the hole diameter, and the effect vanishes for very tall packing heights where the discharge rate is constant. In this regime, for two spherical grain types and a wide range of grain diameters, we find excellent agreement between discharge rate data and a modified Beverloo-like equation over five decades in dimensionless discharge rate. This empirical result could be of practical use, in the same vein as the Beverloo equation. While the basic scale is set by dimensional analysis, we find the existence of a small-hole cutoff such that the discharge rate extrapolates to zero as the hole diameter is decreased to 2.4 times the grain diameter. For dry grains, the cutoff is significantly smaller; this has been known for over fifty years but remains unexplained based on the underlying microscopic physics. The contrast with our findings suggests a missing ingredient: grain dynamics, rather than geometry alone, play a crucial role.

Acknowledgements - We thank Ted Brzinski, Charles Thomas, and Adam Roth for helpful discussions and assistance. This work was supported by the NSF through grants DMR-1305199 and MRSEC/DMR-112090 (REU program - TJW and CRP).

---

[1] R M Nedderman, *Statics and kinematics of granular materials*, Cambridge University Press, NY (1992).

[2] F J Muzzio, T Shinbrot, B J Glasser, *Powder technology in the pharmaceutical industry: The need to catch up fast*, Powder Technol. **124**, 1 (2002).

[3] H M Jaeger, S R Nagel, R P Behringer, *Granular solids, liquids, and gases*, Rev. Mod. Phys. **68**, 1259 (1996).

[4] J Duran, *Sands, powders, and grains: An introduction to the physics of granular materials*, Springer, NY (2000).

[5] W A Beverloo, H A Leniger, J van de Velde, *The flow of granular solids through orifices*, Chem. Eng. Sci. **15**, 260 (1961).

[6] L Vanel, E. Clément, *Pressure screening and fluctuations at the bottom of a granular column*, Eur. Phys. J. B **11**, 525 (1999).

[7] R M Nedderman, U Tuzun, S B Savage, G T Houlsby, *The flow of granular materials-1: Discharge rates from hoppers*, Chem. Eng. Sci. **37**, 1597 (1982).

[8] J E Hilton, P W Cleary, *Granular flow during hopper discharge*, Phys. Rev. E **84**, 011307 (2011).

[9] A Janda, I Zuriguel, D Maza, *Flow rate of particles through apertures obtained from self-similar density and velocity profiles*, Phys. Rev. Lett. **108**, 248001 (2012).

[10] C Manoc, A Janda, R Arévalo, J M Pastor, I Zuriguel, A Garcimartín, D Maza, *The flow rate of granular materials through an orifice*, Granul. Matter **9**, 407 (2007).

[11] H G Sheldon, D J Durian, *Granular discharge and clogging for tilted hoppers*, Granul. Matter **12**, 579 (2010).

[12] C C Thomas, D J Durian, *Geometry dependence of the clogging transition in tilted hoppers*, Phys. Rev. E **87**, 052201 (2013).

[13] Y Bertho, C Becco, N Vandewalle, *Dense bubble flow in a silo: An unusual flow of a dispersed medium*, Phys. Rev. E **73**, 056309 (2006).

[14] M A Aguirre, J G Grande, A Calvo, L A Pugnaloni, J-C Géminard, *Pressure independence of granular flow through an aperture*, Phys. Rev. Lett. **104**, 238002 (2010).

[15] A Guariguata, M A Pascall, M W Glimer, A K Sum, E D Sload, C A Koh, D T Wu, *Jamming of particles in a two-dimensional fluid-driven flow*, Phys. Rev. E **86**, 061311 (2012).

[16] P G Lafond, M W Glimer, C A Koh, E D Sloan, D T Wu, A K Sum, *Orifice jamming of fluid-driven granular flow*, Phys. Rev. E **87**, 042204 (2013).

[17] I Zuriguel, A Janda, A Garcimartín, C Lozano, R Arévalo, D Maza, *Silo clogging reduction by the presence of an obstacle*, Phys. Rev. Lett. **107**, 278001 (2011).

[18] C Lozano, A Janda, A Garcimartín, D Garza, I Zuriguel, *Flow and clogging in a silo with an obstacle above the orifice*, Phys. Rev. E **86**, 031306 (2012).

[19] F Alonso-Marroquin, S I Azeezullah, S A Galindo-Torres, L M Olsen-Kettle, *Bottlenecks in granular flow: When does an obstacle increase the flow rate in an hourglass?* Phys. Rev. E **85**, 020301 (2012).

[20] K To, P Y Lai, H K Pak, *Jamming of granular flow in a two-dimensional hopper*, Phys. Rev. Lett. **86**, 71 (2001).

[21] J Choi, A Kudrolli, R R Rosales, M Z Bazant, *Diffusion and mixing in gravity-driven dense granular flows*, Phys. Rev. Lett. **92** 174301 (2004).

[22] J Choi, A Kudrolli, M Z Bazant, *Velocity profile of granular flows inside silos and hoppers*, J. Phys.: Cond. Matter **17**, S2533 (2005).

[23] J F Wambaugh, R R Hartley, R P Behringer, *Force networks and elasticity in granular silos*, Eur. Phys. J. E **32**, 135 (2010).

[24] D D Chen, K W Desmond, E R Weeks, *Topological rearrangements and stress fluctuations in quasi-two-dimensional hopper flow of emulsions*, Soft Matter **8** 10486 (2012).

[25] C C Kuo, M Dennin, *Buckling-induced jamming in channel flow of particle rafts*, Phys. Rev. E **87**, 030201 (2013).

[26] A V Orpe, A Kudrolli, *Velocity correlations in dense granular flows observed with internal imaging*, Phys. Rev. Lett. **98**, 238001 (2007).

[27] A J Liu, S R Nagel, *The jamming transition and the marginally jammed solid*, Ann. Rev. Cond. Matt. Phys. **1**, 347 (2010).

[28] J Koivisto, D J Durian, unpublished (2014).

[29] F A Morrison, *Data correlation for drag coefficient for sphere*, Department of Chemical Engineering, Michigan Technological University, Houghton, MI, www.chem.mtu.edu/~fmorriso/DataCorrelationForSphereDrag2010.pdf (accessed June 26, 2012). This work gives a convenient empirical function that accurately matches experimental data between the limits of $c_d = 24/\mathrm{Re}$ at small Re and $c_d \approx 1/2$ at large Re.

[30] L Rondon, O Pouliquen, P Aussillous, *Granular collapse in a fluid: Role of the initial volume fraction*, Phys. Fluids **23**, 073301 (2011).

[31] A Kudrolli, *Private communication regarding experiments done with A. Orpe*, private communication (2012).

[32] T Le Pennec, K J Maloy, E G Flekkoy, J C Messager, M Ammi, *Silo hiccups: Dynamic effects of dilatancy in granular flow*, Phys. Fluids **10**, 3072 (1998).

[33] L Staron, P Y Lagrée, S Popinet, *The granular silo as a continuum plastic flow: The hourglass vs the clepsydra*, Phys. Fluids **24** 103301 (2012).

[34] L Staron, P Y Lagrée, S Popinet, *Continuum simulation of the discharge of the granular silo. A validation test for the mu(I) visco-plastic flow law*, Eur. Phys. J. E **37**, 5 (2014).

[35] F Boyer, E Guazzelli, O Pouliquen, *Unifying suspension and granular rheology*, Phys. Rev. Lett. **107**, 188301 (2011).

[36] C Perge, M A Aguirre, P Gago, L A Pugnaloni, D Le Tourneau, J-C Géminard, *Evolution of pressure profiles during the discharge of a silo*, Phys. Rev. E **85**, 021303 (2012).

6

Flow rate of polygonal grains through a bottleneck: Interplay between shape and size

Ezequiel Goldberg,[1] C. Manuel Carlevaro,[1–3] Luis A. Pugnaloni[3,4*]

We report two-dimensional simulations of circular and polygonal grains passing through an aperture at the bottom of a silo. The mass flow rate for regular polygons is lower than for disks, as observed by other authors. We show that both the exit velocity of the grains and the packing fraction are lower for polygons, which leads to the reduced flow rate. We point out the importance of the criteria used to define when two objects of different shape are considered to be of the same size. Depending on this criteria, the mass flow rate may vary significantly for some polygons. Moreover, the particle flow rate is non-trivially related to a combination of mass flow rate, particle shape and particle size. For some polygons, the particle flow rate may be lower or higher than that of the corresponding disks depending on the size comparison criteria.

I. Introduction

The flowing properties of granular materials are key in a myriad of industrial processes. In this respect, the discharge of a silo or a hopper constitutes the most ubiquitous operation. Considerable attention has been paid to silo discharge in the literature. Since the initial works in the '60s [1, 2], a number of studies have reproduced the essential aspects highlighted in the pioneering works. These are related to the little or no effect played by the height of the granular column (see, however, a recent experimental report on submerged silos [3]), and the 5/2 power dependency of the flow rate with the size of the aperture (see, however, a study over a wide range of apertures [4]). The irrelevance of the column height for the flow rate has been recently shown not to be connected with the Janssen's effect [5–7]. Less attention has been paid to the material properties and particle shape. In general, material properties seem to have a marginal effect on flow rate [8–10]. However, the flow pattern may change from funnel flow to mass flow to mixed flow as friction is varied either for the particle–particle interaction or the particle–wall interaction [9, 10].

Particle shape is also a factor that has been considered by a number of authors. However, unlike material properties, shape is not defined by a finite number of parameters. For a particular basic shape, for instance rectangles, one can assess the effect of the few parameters that describe that shape (e.g., long side and short side). However, this cannot tell if a different basic shape (e.g., ellipses) will yield different results. As a consequence, the effect

*E-mail: luis.pugnaloni@frlp.utn.edu.ar

1 Universidad Tecnológica Nacional - FRBA, UDB Física, Mozart 2300, C1407IVT Buenos Aires, Argentina.

2 Instituto de Física de Líquidos y Sistemas Biológicos (CONICET La Plata, UNLP), Calle 59 Nro 789, 1900 La Plata, Argentina.

3 Consejo Nacional de Investigaciones Científicas y Técnicas, Argentina.

4 Dpto. Ingeniería Mecánica, Facultad Regional La Plata, Universidad Tecnológica Nacional, Av. 60 Esq. 124, 1900 La Plata, Argentina.

of shape has to be surveyed necessarily in a limited manner and the selection of what basic shapes to consider has to be led by the application of interest or by practical issues such as the limitations in fabrication of some shapes or their computer modeling.

Studies on circular grains and on supercircles (rounded squares) in 2D show a significantly lower mass flow rate for the supercircles [11]. These authors compare particles of different shape that have the same long diameter. We will see that this selection of matching size may bias the conclusions. A number of ulterior works have shown the same overall trend that mass flow rate is reduced in comparison with rounded particles if they present sharp edges [12–14]. A contrasting result was found by some works in which elongated rounded particles flow faster than circles (2D) or spheres (3D); however, they considered frictionless grains in one occasion [15] and, in the attempt to validate with experiments, they found difficulties in producing particles of same mass and different shape [16]. One particularly interesting exception to the reduced flowability of sharp particles is the case of tetrahedrons that seem to flow faster than spheres [14]. Unfortunately, this study compared spheres and tetrahedrons of different mass/volume, and the conclusion cannot be warranted to be general.

There is a significant point generally overlooked in many of these studies. Since the relative size between the particle and the aperture is the main factor affecting the flow rate, comparison with different shapes has to be done by fixing the particle size. For example, some workers have claimed that hexahedrons and ellipsoids flow faster than spheres based on experiments with different seeds, without paying much attention to the fact that the seeds with non-spherical shape used were also smaller in size [17]. The problem was recognized in other works when comparing tetrahedrons with spheres [14]. There is an arbitrary decision to make as to what defines if two particles of different shape are of the same size. The canonical solution to this is to define that particles must have the same mass to be considered of the same size. Notice, however, that some studies use a characteristic length of the particles (e.g., maximum feret diameter) [11,13,14]. Since the shape is being changed, keeping a characteristic length constant will lead to different masses. There is no correct selection in this respect. The parameter that defines the effective size has to be chosen according to the particular application, for example. We will show that this choice has a significant impact on the mass flow rate observed in particles of different shapes.

An additional aspect to the flow of grains is the actual number of particles outpouring per unit time. Although for a given shape this is proportional to the mass flow rate, in comparing results for different shapes the particle flow rate has a different proportionality constant for each shape. Moreover, the criteria used to define if particles of same size are compared will strongly affect the particle flow rate. We will present results that show that to deliver a fixed number of particles per unit time, the use of different particle shapes requires the increase of the aperture in some cases and the reduction in others.

II. Simulation

We simulate a 2D silo by means of a discrete element method (DEM) using the Box2D library [18]. This package uses a constraint solver. At each time step of the dynamics, a series of iterations (typically 20) are used to resolve penetrations between bodies through a Lagrange multiplier scheme [19]. After detecting overlaps, the inelastic collision at each contact is solved and new linear and angular velocities are assigned. The equations of motion are integrated through a symplectic Euler algorithm. The time step δt used to integrate the equations of motion is 0.025 d/g; with d the diameter of our circular grains and g the acceleration of gravity. Solid friction is also handled by means of a Lagrange multiplier scheme that implements the Coulomb criterion. Box2D is highly efficient when handling complex bodies such as polygons.

The silo consists of a box $20d$ wide and $200d$ high. Gravity acts in the negative vertical direction. 2000 grains are typically used in each discharge, and these are not reinjected after they leave the silo. After grains initially placed at random without overlaps settle, an aperture opened at the center of the flat base of the silo allows the grains to discharge. For small apertures, clogging may be observed. In such cases, the arrested discharge is discarded from the analysis. Only those discharges that did not present clogging are considered. Some

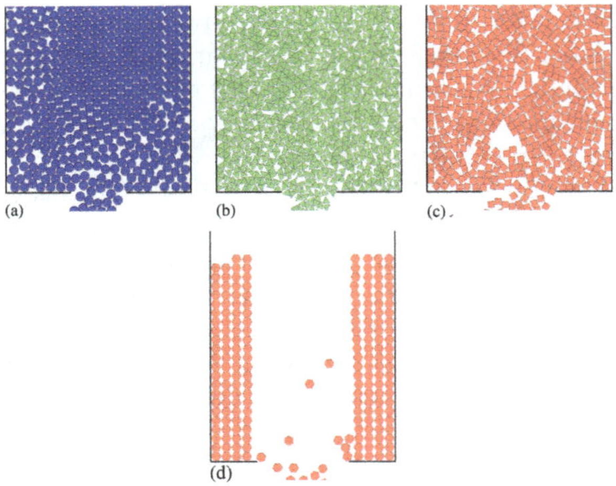

Figure 1: Snapshots of the simulations for: (a) disks, (b) triangles and (c) squares. Only part of each system is shown. (d) Sample snapshot of a "rat-hole" observed in one of the discharges for hexagons.

material may be left over at the stagnant zones on each side of the orifice.

Grains of different shapes corresponding to regular polygons (triangles, squares, pentagons, and hexagons) are used in the simulations apart from the circular grains. All grains have the same material density. Each simulation contains only one type of grain and these are monosized. Figure 1 shows some snapshots of systems. The restitution coefficient is set in all cases to $\epsilon = 0.01$ and the friction coefficient μ is varied between 0.1 and 2.0. The significantly low restitution coefficient allows for a rapid dissipation of the kinetic energy during the initial filling of the silo. The material density ρ of the grains is set to 0.01 $\mathrm{Kg/m^2}$

After the aperture is opened, we allow any transient to fade before we calculate the flow rate and other properties. For any simulation, 15 independent realizations with different initial particle positions are carried out to estimate error bars. It is important to mention that, for squares and hexagons, some discharges with large apertures lead to a "rat-hole" effect where particles at each side of the opening order in a stack [see for example Fig. 1(d)]. Although the phenomenon is interesting in itself, these instances have been detected and removed from any analysis.

One may be concerned by the ordering typically observed in monosized samples of grains, specially in disks [see the upper region of the system in Fig. 1(left)]. Notice, however, that the structure next to the aperture is rather disordered during the flow. Hence, even if a very ordered packing is observed initially in the silo (corresponding to a high packing fraction), the flow may induce a lowering of the order and density connected with Reynolds dilatancy. One strategy to avoid ordering is using polydisperse systems. However, this would complicate our analysis regarding what is the effective size of the particles. In this respect, a recent study has shown that an effective size cannot be defined for the flow rate of bidisperse mixtures [20].

III. Results and discussion

i. Mass flow rate

We first consider the mass flow rate Q of grains as a function of the aperture size D for particles of same mass. Figure 2 shows Q for different grain shapes. Since simulations are in 2D, the grains have been set with the same material density and area to yield the same mass m. Circular grains present values of Q between 30% and 90% above those of polygonal grains. Squares and hexagons show larger fluctuation between independent realizations in comparison with other polygons and disks. In particular squares, which show the lowest flow rate, tend to clog rather often and data on Q for orifices $\leq 8d$ could not be collected. Overall, the flow rate increases in the following order: squares, pentagons, triangles, hexagons and disks.

We have fitted Q using the 2D Beverloo's rule

$$Q = C\rho\sqrt{g}(D - kd)^{3/2}. \tag{1}$$

Where ρ is the material density of the grains, and C and k are fitting parameters. Notice that the bulk density ρ_b is generally used instead of the material density ρ which does not take into account the voidage of the system. In this case, we use ρ and let C carry this information to avoid taking an arbitrary decision on the region where ρ_b has to be measured. Table I shows the values of C and k obtained for each particle shape and Fig. 2 includes the fitting curves. Disks, as well as polygons, comply with the 3/2 exponent rule, at least within this narrow range of D explored. The values of C obtained confirm the general trend that the mass flow

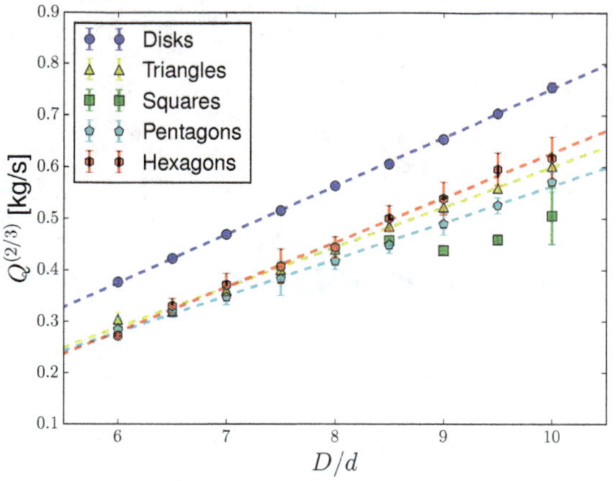

Figure 2: Mass flow rate Q as a function of the aperture size D for different particle shapes with $\mu = 0.5$. All particle shapes have the same area (mass). The continuous lines are fits to the 3/2-power of $D - kd$ (see fitting parameters in Table I).

rate increases with shape in the order described in the previous paragraph. The k-value indicates the intercept with zero flow rate and constitutes an extrapolated estimate of the minimum size of the orifice necessary to observe flow. In practice, this estimate obtained with data at relatively large values of D is not very reliable. It has been shown that a more careful analysis of the flow at smaller D where intermittent flow can still be analyzed suggests that a different equation to that of Beverloo is more appropriate by considering that only at $k = 1$ would Q become strictly zero [4]. Nevertheless, the value of k gives an indication of how soon the flow rate drops as D is decreased. For disks, we find $k \approx 2.0$, which is consistent with experiments where a similar range of D was used [5] (if smaller D are considered k may drop down to ≈ 1 [4, 8]). Table I indicates that disks and pentagons are predicted to stop flowing at $D \approx 2d$. However, triangles and hexagons would only flow for somewhat larger D. Again, these are only rough indicators, since measurements at lower values of D may lead to different estimates, particularly if states of intermittent flow are included in the analysis.

These results show that shape matters in granular flow as shown by others [11–16]. However, the variability of Q between different faceted grains is

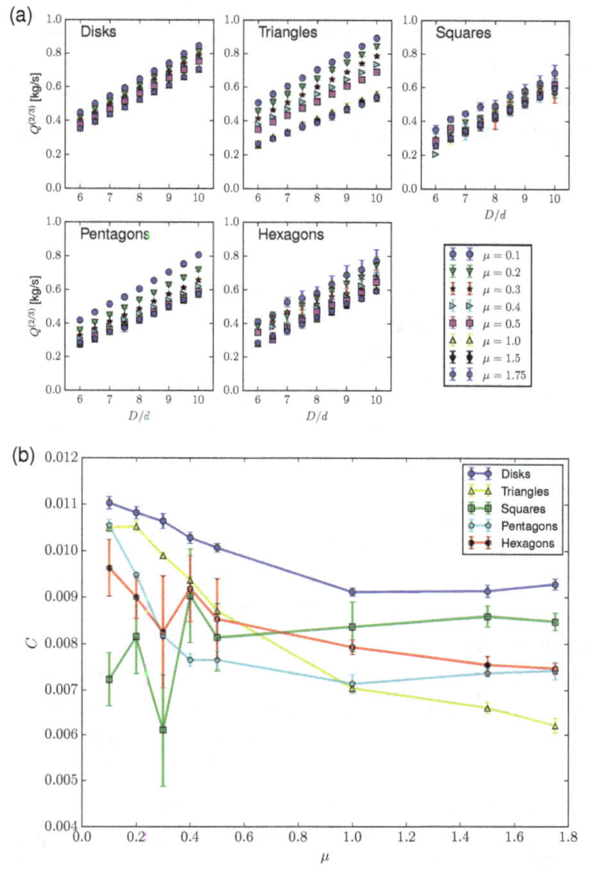

Figure 3: (a) Mass flow rate as a function of D for different friction coefficients for each particle shape: disks, triangles, squares, pentagons and hexagons. All particles have the same area (mass) irrespective of their shape. (b) Value of the proportionality constant C in Eq. (1) after fitting the data in part (a). The error bars correspond to one standard deviation estimated from the covariance matrix of the fit parameters.

rather limited in comparison with the significantly higher mass flow rate of disks. In general, setting apart triangles, the more vertices a polygon has, the higher Q is. This is consistent with the idea that polygons with many facets should tend to have the behavior of circular grains.

One may speculate that the use of faceted particles changes particularly the frictional character of the interactions. While disks may roll on their contacts, polygons will mostly slide side-to-side on each other. A polygon may be forced to "roll"

Table 1: Fitting parameters for the Beverloo equation in 2D as defined in Eq. (1). For squares, the poor statistics due to clogging events prevent us from extracting reliable fits.

Shape	C	k
Disks	0.92	2.01
Triangles	0.70	2.30
Squares	–	–
Pentagons	0.61	2.09
Hexagons	0.82	2.77

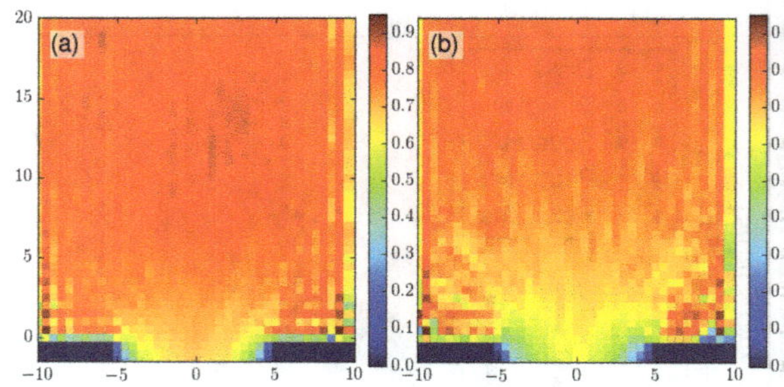

Figure 4: Mean packing fraction distribution for disks (a) and triangles (b) averaged over a discharge. The aperture is $D = 10d$ and the friction coefficient $\mu = 0.5$. Disks and triangles have the same area (mass).

(rather than slide) on top of flat inclined surface if the friction angle θ (i.e., $\theta = \tan\mu$) exceeds the maximum angle of stability θ_{stab} (defined as the angle of inclination of a flat supporting base at which the vertical line that passes through the center of mass of the polygon meets one of its vertexes). Hence, it is expected that changing the friction coefficient may lead to different effects on the polygons dynamics and hence on the flow rate. The corresponding values of μ beyond which a polygon would roll before sliding are $\mu > \sqrt{3} \approx 1.732$ for triangles, $\mu > 1$ for squares, and $\mu > \sqrt{5 + 2\sqrt{5}} \approx 0.727$ for pentagons.

Figure 3 shows Q for different μ for each grain shape. As we can see, friction does not induce any significant change in flow rate if $\mu > 1.0$. However, a considerable increase in Q is observed if we lower μ below 0.5. The one exception is found in squares that are little affected by any change in μ. The increase in Q with decreasing μ is consistent with previous simulations of disks [8]. Interestingly, there is not clear evidence of a qualitative change when μ is higher than the maximum stability angle for each polygon. We presume that the interaction with other touching grains is as important as the interaction with any eventual grain that serves as primary support for a given polygon. Since the stability angle defined is only based on the sole interaction with a base, the expectation that a qualitative feature may be observed at these μ values may be unjustified due to the complex multiparticle interactions.

For large orifices, the flow rate has a marginal dependence on d. Therefore, the actual choice of comparing results from grains of different shapes having the same mass is less relevant. Hence, the proportionality constant C in Eq. (1) when fitting to the data in Fig. 3(a) characterizes the flow of grains of a given shape independently of their size. In Fig. 3(b), we show C as a function of μ for each particle shape. This plot makes it clear that the flow rate decreases as μ increases and plateaus out for $\mu > 1$. Triangles have a much pronounced decay in C as μ increases, with values above those of other polygons for low μ but somewhat below for high friction. This shows that for large orifices, the flow rate for triangles is the highest among the polygons at low friction, but the lowest at high friction. The order observed for the flowability in Table I is therefore dependent on friction coefficient.

ii. Velocity, density and energy profiles

The reduced flow rate of polygonal grains may be caused by a drop in the packing fraction ϕ and/or a fall in the actual velocity of the grains. The packing fraction of polygonal grains in a rectangular box subjected to tapping has been previously considered [21]. It was shown that regular polygons able to tessellate the plane (triangles, squares and hexagons) in general arrange in higher packing fractions as it may be expected. Pentagons, on the other hand, yield lower packing fractions in comparison with monosized disks. Despite this, the packing fraction during flow does not need to be connected with the values found in static systems. Flow implies shear, which induces Reynolds dilatancy, which in turn reduces ϕ if the initial compaction is above the dilatancy onset.

We have measured ϕ profiles across the silo during a discharge for the case of disks and triangles of the same mass (see Fig. 4). The calculations are done by using the package Shapely for polygon intersections [22]. The time average of ϕ during the discharge measured in cells of $1d \times 1d$ are considered. As we can see, ϕ is lower for triangles than it is for disks in general. This is in contrast with the higher packing fractions found for triangles in tapped static systems [21]. It is clear that the reduction in ϕ observed in the area of the outlet stretches farther into the silo for triangles. At the aperture, ϕ is smaller overall for triangles and the effective "empty annulus" (i.e., the depleted region next to the edges of the aperture) looks thicker for triangles too. This explains the high values of k found in the fits of the Beverloo rule for triangles (see Table I), which is consistent with experimental results on sharp grains like some sand samples [1]. The reduced mass flow rate observed for polygons may be then caused by the reduction in ϕ. However, the reduction with respect to circular particles is less than 15%, which is insufficient to explain the at least 30% change in Q. Hence, one should expect that also the mean velocity of the exiting grains should be higher for disks.

The connection between packing fraction and flow regime in sheared granular samples of disks and pentagonal grains has been recently considered [23]. In this study, the flow regime is characterized by the inertial number I [24]. The authors show that as I approaches zero, the system reaches a quasistatic limit and the packing fraction does not depend on shear rate, but it does depend on particle shape. A similar analysis can be, in principle, done for the flow of the discharging grains in our simulations, focusing on the region just above the outlet. However, the flow regime, even for the smaller apertures, is far from the quasistatic limit in our simulations. We estimated $I > 0.1$ in all cases, which is one order of magnitude above the highest values considered in Ref. [23].

Figure 5 shows the profiles for the vertical and horizontal velocity of disks and triangles. The mean horizontal velocity is different from zero only close the aperture in a V-shaped region consistent with convergent streamlines towards the outlet. Triangles show a more stretched region of non-zero horizontal velocity in agreement with the stretched region of low ϕ [see Fig. 4(b)]. Disks present two hotspots above each edge of the outlet with high horizontal mobility in clear contrast with the triangles that present a smoother profile. From the mean vertical velocity one can observe clearly a more significant stagnant zone for triangles on each side of the aperture. This is consistent with results of several authors (see for example, [11, 14]) and with the intuition that sharp edged grains will display a higher effective angle of repose. This taller stagnant zone seems to be responsible for inducing shear on a larger area for triangles, leading to the reduced ϕ. Also, the vertical velocity at the aperture is in average higher for disks; roughly 9%. The combined effect of reduced ϕ and reduced vertical velocity adds to a $\approx 25\%$ change in Q when comparing disks and triangles at this opening width.

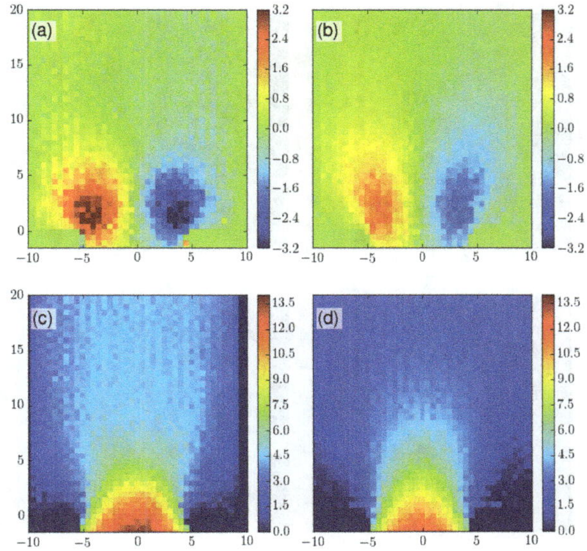

Figure 5: Mean particle velocity: (a) v_x for disks, (b) v_x for triangles, (c) v_y for disks and (d) v_y for triangles. The aperture is $D = 10d$ and the friction coefficient $\mu = 0.5$. Disks and triangles have the same area (mass). The color scale indicates the velocity in ms^{-1}.

In Fig. 6, we show the translational k_t and rotational k_r kinetic energy profile for disks and triangles. k_t is dominated by the vertical component of the velocity and therefore the features observed in

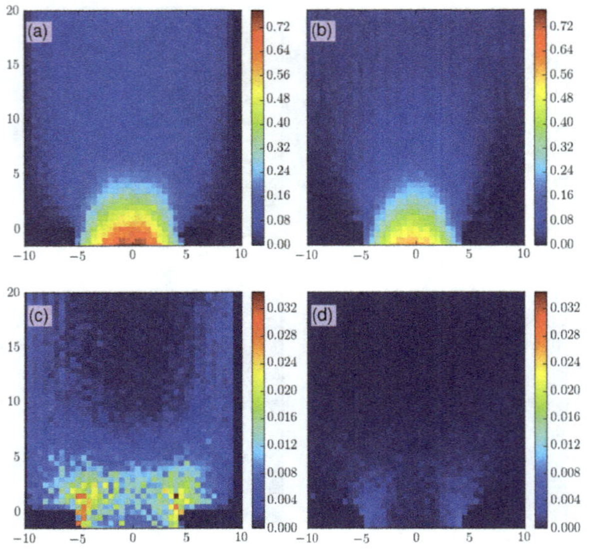

Figure 6: Mean kinetic energy of the particles: (a) translational for disks, (b) translational for triangles, (c) rotational for disks and (d) rotational for triangles. The aperture is $D = 10d$ and the friction coefficient $\mu = 0.5$. Disks and triangles have the same area (mass). The color scale indicates the energy in J.

Fig. 5(c) and Fig. 5(d) are reproduced. An interesting feature observed in k_r is that disks present a wide region over which grains rotate. This region extends all along the orifice width (and beyond, over the flat bottom base) at a height about $2d$ above the aperture. Triangles, in marked contrast, only acquire rotational energies less than half of those of disks and over a limited region next to each edge of the aperture. This might be due to the frustration of any initial spin a particle may acquire since collisions of its sharp edges with neighboring grains will tend to stop any rotational motion.

At first sight, the reduced rotational energy of triangles seems counter-intuitive. If triangles have a reduced translational energy, one should expect that part of the energy input provided by the work done by the force of gravity might go into the rotational degrees of freedom. Notice, however, that the energy is also dissipated by collisions and friction. It is clear then that triangles provide a much efficient dissipation of the energy for all degrees of freedom.

iii. The effective size of a grain

It is important to notice that comparing the behavior of grains of different shapes conveys an arbitrary decision on the parameter that decides when two grains are considered of the same size. In the previous section, we have chosen to define as particles of the same size those that have the same mass, irrespective of the shape. However, in some applications, one may be interested in comparing the flow of different polygons that have the same radius, side or perimeter.

Figure 7 shows the mass flow rate for triangular particles of different sizes chosen to match the radius, perimeter or area of the reference disks. We have also included data for triangles whose side is the same as the diameter of the disks. As we can see, since the actual radius of the triangle depends on the parameter used to define if the polygon is "of the same size as the disk", the mass flow rate varies up to 25%.

The importance of the arbitrary definition of the effective size of the grains when comparing different shapes was previously recognized [14]. However, the extent of the effect that this may cause in the mass flow rate was not studied. The 25% maximum difference observed in Q for triangles shows that this cannot be neglected, although other shapes may be less affected.

Notice that it is usual to compare particles of different shape that have the same mass. However, there exist several works where authors use a characteristic length instead of the mass [11,13,14]. In some applications, due to design, economic cost, practical constraints, etc., fabricated objects of different shape may or may not have the same mass. Estimations of the relative mass flow rate of such objects delivered by hoppers in comparison with circular counterparts would need to take into account the relative size.

iv. Particle flow rate

It is worth mentioning that in some cases, the number of grains flowed per unit time may be the important parameter rather than the mass flow rate. Although this is proportional to the mass flow rate, the proportionality constant depends on the size of the grains and the mean ϕ. Hence, depending on whether polygons are compared by their radius,

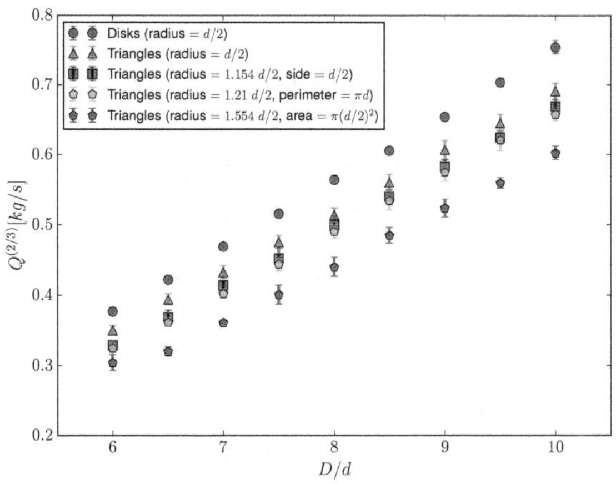

Figure 7: Mass flow rate as a function of the aperture size D for disks and triangles with $\mu = 0.5$. Different triangles that match the diameter, perimeter and area of the disks as well as having the side equal to the diameter of a disk are considered.

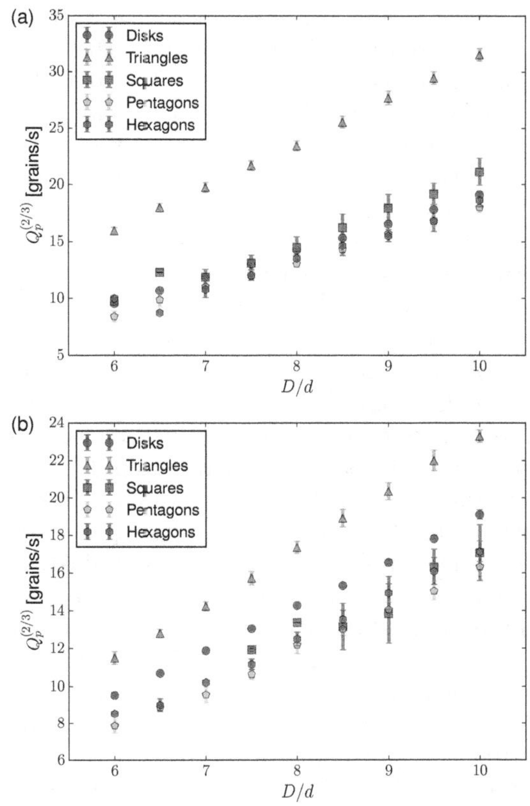

Figure 8: (a) Particle flow rate as a function of the aperture size D for different particle shapes with $\mu = 0.5$ and the same particle radius. (b) Particle flow rate as a function of the aperture size D for different particle shapes with $\mu = 0.5$ and the same particle perimeter.

side or perimeter, the actual particle flow rate, Q_p varies. Naturally, if polygonal grains are compared by their area in 2D, the particle flow rate yields the same results as in Fig. 2, simply scaled by the same factor (the particle density) for all shapes. However, this is not the case if grains have, for example, the same diameter.

In Fig. 8, we present Q_p for different polygons if the radius (a) or the perimeter (b) is used as the reference to compare sizes. As we can see, the relative position of the curves changes significantly depending on the parameter used to compare the particle sizes since the relative masses of the grains are different. If particles have the same radius [Fig. 8(a)], pentagons and hexagons yield a slightly lower particle flow rate than disks, while squares deliver a larger number of particles per unit time. Triangles present up to 100% higher particle flow rates than disks. This is in dramatic contrast with the mass flow rate observed in Fig. 7. If we use the perimeter to compare grain sizes, most polygons have a lower particle flow rate than disks while triangles still display a larger flow in number of grains per seconds.

The results of Fig. 8 show that the particle flow rate is particularly sensitive to the effective size of the grains. Squares, for example, may be delivered at a higher or lower rate than disks, in terms of number of particles, depending on whether they have the same radius or the same perimeter as disks. However, the mass flow rate of squares is lower than that of disks in both cases.

IV. Conclusions

We have considered the discharge of circular and polygonal grains from a flat bottomed silo. We have seen, in agreement with other workers, that for grains of the same mass, the mass flow rate is markedly lower for polygons in comparison with disks. We have found that this is caused by a combination of reduced mean packing fraction for the flowing polygons and also a reduced mean vertical velocity. The rotational kinetic energy of the polygons is also significantly smaller than that of disks.

When comparing results for different shapes, it is necessary to arbitrarily select a parameter to define when two grains can be said to be of equal size. We found that this selection, which may be constrained by specific applications, may lead to significant changes in the actual mass flow rates.

We have discussed that the ratio of the particle flow rate to the mass flow rate is a non-trivial function of the grain shape since it does not only depend on the mass of the grains but also on the actual packing fraction. As a consequence, it is difficult to predict the particle flow rate from the sole knowledge of the mass flow rate. We have shown that the particle flow rate for a given shape, which can be the measure of interest in many industrial applications, can result higher or lower than for a different particle shape depending on the parameter used to define their effective size. Hence, in a specific application, in order to deliver the same number of grains per unit time for a different particle shape, the aperture must be either reduced or widened depending on the parameter used to compare their sizes (e.g., diameter, mass, perimeter, etc).

Acknowledgements - This work has been supported by ANPCyT (Argentina) through grant PICT 2012-2155 and UTN (Argentina) through grants PID MAUTNLP0002184 and PID IFI1871.

---

[1] W Beverloo, H Leniger, J Van de Velde, *The flow of granular solids through orifices*, Chem. Eng. Sci. **15**, 260 (1961).

[2] R L Brown, J C Richards, *Exploratory study of the flow of granules through apertures*, Trans. Inst. Chem. Eng. **37**, 108 (1959).

[3] T J Wilson, C R Pfeifer, N Mesyngier, D J Durian, *Granular discharge rate for submerged hoppers*, Papers in Physics **6**, 060009 (2014).

[4] C Mankoc, A Janda, R Arévalo, J Pastor, I Zuriguel, A Garcimartín, D Maza, *The flow rate of granular materials through an orifice*, Granular Matter **9**, 407 (2007).

[5] M A Aguirre, J G Grande, A Calvo, L A Pugnaloni, J-C Géminard, *Pressure independence of granular flow through an aperture*, Phys. Rev. Lett. **104**, 238002 (2010).

[6] M A Aguirre, J G Grande, A Calvo, L A Pugnaloni, J-C Géminard, *Granular flow through an aperture: Pressure and flow rate are independent*, Phys. Rev. E **83**, 061305 (2011).

[7] C Perge, M A Aguirre, P A Gago, L A Pugnaloni, D Le Tourneau, J-C Géminard, *Evolution of pressure profiles during the discharge of a silo*, Phys. Rev. E. **85**, 021303 (2012).

[8] L Kondic, *Simulations of two dimensional hopper flow*, Granular Matter **16**, 235 (2014).

[9] R M Nedderman, *Statics and kinematics of granular materials*, Cambridge University Press, Cambridge (2005).

[10] R L Brown, J C Richards, *Principles of powder mechanics: essays on the packing and flow of powders and bulk solids*, Pergamon Press Oxford, London (1970).

[11] P W Cleary, M L Sawley, *DEM modelling of industrial granular flows: 3D case studies and the effect of particle shape on hopper discharge*, Appl. Math. Model. **26**, 89 (2002).

[12] F Y Fraigea, P A Langston, G Z Chen, *Distinct element modelling of cubic particle packing and flow*, Powder Technol. **186**, 224 (2008).

[13] D Höhner, S Wirtz, V Scherer, *A numerical study on the influence of particle shape on hopper discharge within the polyhedral and multi-sphere discrete element method*, Powder Technol. **226**, 16 (2012).

[14] D Höhner, S Wirtz, V Scherer, *Experimental and numerical investigation on the influence of particle shape and shape approximation on hopper discharge using the discrete element method*, Powder Technol. **235**, 614 (2013).

[15] P A Langston, M A Al-Awamleh, F Y Fraige, B N Asmar, *Distinct element modelling of non-spherical frictionless particle flow*, Chem. Eng. Sci. **59**, 425 (2004).

[16] J Li, P A Langston, C Webba, T Dyakowskia, *Flow of sphero-disc particles in rectangular hoppers: A DEM and experimental comparison in 3D*, Chem. Eng. Sci. **59**, 5917 (2004).

[17] B Jin, H Tao, W Zhong, *Flow. behaviors of non-spherical granules in rectangular hopper*, Chinese J. Chem. Eng. **18**, 931 (2010).

[18] Box2D Physics Engine, www.box2d.org

[19] E Catto, *Iterative dynamics with temporal coherence*, available at http://box2d.googlecode.com/files/GDC2005 ErinCatto.zip (retrived on October 2010)

[20] M. Benyamine, *Discharge flow of a bidisperse granular media from a silo*, Phys. Rev. E **90**, 032201 (2014).

[21] C M Carlevaro, L A Pugnaloni, *Steady state of tapped granular polygons*, J. Stat. Mech. P01007 (2011).

[22] Shapely package, https://pypi.python.org/pypi/Shapely

[23] E Azáema, F Radjï, J-N. Roux, *Internal friction and absence of dilatancy of packings of frictionless polygons*, Phys. Rev. E **91**, 010202(R) (2015).

[24] F da Cruz, S Emam, M Prochnow, J N Roux, F Chevoir, *Rheophysics of dense granular materials: Discrete simulation of plane shear flows*, Phys. Rev. E **72**, 021309 (2005).

7

Ergodic-nonergodic transition in tapped granular systems: The role of persistent contacts

Paula A. Gago,[1–3] Diego Maza,[4] Luis A. Pugnaloni[1,2*]

Static granular packs have been studied in the last three decades in the frame of a modified equilibrium statistical mechanics that assumes ergodicity as a basic postulate. The canonical example on which this framework is tested consists in the series of static configurations visited by a granular column subjected to taps. By analyzing the response of a realistic model of grains, we demonstrate that volume and stress variables visit different regions of the phase space at low tap intensities in different realizations of the experiment. We show that the tap intensity beyond which sampling by tapping becomes ergodic coincides with the forcing necessary to break all particle–particle contacts during each tap. These results imply that the well-known "reversible" branch of tapped granular columns is only valid at relatively high tap intensities.

I. Introduction

Granular matter is ubiquitous in nature. However, due to the complexity of the real particle–particle interactions, the standard approaches of continuum mechanics and thermodynamics are still limited in providing meaningful descriptions of the states in which these systems can be. Edwards and Oakeshot introduced a tentative approach inspired by the ideas of equilibrium statistical mechanics to formally describe the global properties of a static granular pack. Since the introduction of this theory —where the entropy of the systems is governed by the spatial disorder of the grains [1]—, a number of studies have used it to frame the interpretation of the results of specific experiments. The most relevant case is the so-called "Chicago experiment", where a column of grains was repeatedly tapped following an annealing-type protocol [2, 3]. The main outcome of this experiment is that a stationary state can be reached, where the mean volume fraction, ϕ, is a well defined function of the tap amplitude, Γ. Others have also obtained seemingly reproducible states without the need of annealing [4]. However, it has been shown recently, by simulation of frictionless grains, that these stationary states are not necessarily ergodic [5]. At low Γ, different members of an ensemble of steady-states prepared with a well defined protocol may sample a different region of the phase space, as the fluctuations of ϕ indicate.

In this paper, we demonstrate that not only the volume but also the force moment tensor, Σ, are sampled in a non-ergodic fashion and that ergod-

*E-mail: luis.pugnaloni@frlp.utn.edu.ar

[1] Dpto. Ingeniería Mecánica, Facultad Regional La Plata, Universidad Tecnológica Nacional, Av. 60 Esq. 124, 1900 La Plata, Argentina.

[2] Consejo Nacional de Investigaciones Científicas y Técnicas, Argentina.

[3] Current address: Department of Earth Science and Engineering, Imperial College London, South Kensington Campus, London SW7 2AZ, UK.

[4] Departamento de Física y Matemática Aplicada, Facultad de Ciencias, Universidad de Navarra, Navarra, Spain.

icity seems to be recovered if all particle–particle contacts are lost during each tap. This sets a clear limit to the range of driving forces able to generate a sequence of configurations for which the Edwards framework can be applied.

II. Numerical protocol

We simulated using the LAMMPS package [6] a quasi-two-dimensional cell containing $N = 1000$ spherical particles of diameter d. The cell is 1.1 d thick and 27.8 d wide (the granular column is about 35 layers deep) to have a one to one representation of a previously introduced experimental device [7,8]. We use a model for soft frictional spheres described in Refs. [9,10]. The normal component, F_n, of the contact interaction is given by an elastic repulsive force proportional to the overlap of the interacting spheres and a dissipative term proportional to the normal component of the relative velocity. The tangential term, F_t, implements an elastic shear force and a damping force. The shear force takes into account the cumulated tangential displacement between the particles while they remain in contact. Whenever $F_t > \mu F_n$ (μ is the friction coefficient), this lower dynamic friction force is used. In this work we use the same interaction parameters as in Ref. [11,12]. The wall–particle interaction is defined with the same parameters as the particle–particle force. Tapping is simulated by imposing an external vertical motion to the cell. This pulse is a single sinusoidal cycle $A\sin(\omega t)$. We fix $\omega = 2\pi \times 33$ Hz and use the tap amplitude A as control parameter. The tap intensity is characterized by $\Gamma = A\omega^2/g$. The mechanical equilibrium after each tap is deemed achieved if the kinetic energy of each particle has fallen (in average) below $10^{-6}mgd$. Where m is the mass of one particle and g the acceleration of gravity.

We study 20 independent realizations of a decreasing ramp of the tap amplitude. We initially fill the cell by placing the spheres at random positions before letting them deposit under the action of gravity. In each realization, we decrease Γ in small steps, from $\Gamma = 20.0$ down to $\Gamma = 0.8$, and apply 200 taps for each Γ. Note that for $\Gamma < 1.0$, the column of grains does not detach from the base during a tap. The 200 taps at each value of Γ are enough to reach a steady-state. We do not observe any drift of the mean values of ϕ or Σ after the initial 100 taps, which we will discard later in our analysis. Finally, we also study a cyclic annealing protocol: starting from the final configuration at $\Gamma = 0.8$ for each of the former realizations; the tap amplitude is cyclically increased and decreased every 200 taps in the range $0.8 < \Gamma < 5.5$ in order to compare the steady-states reached with an alternative method.

III. Data analysis

To measure the packing fraction we use the 2D Voronoi tessellation (implemented in [13]) of the x–z plane projection of the particle positions, disregarding the third coordinate on the thin direction of the cell. Then, we associate to each particle a "local volume" fraction by dividing the particle area by the corresponding Voronoi area. In order to avoid boundary effects, we disregard particles closer than $2d$ to the lateral walls. Following the recommendations in Ref. [14], we analyze horizontal slices of the granular column 15 d thick measured at approximately the same depth with respect to the free surface in order to retrieve unbiased results for the force moment tensor due to the uneven free surface. Averaging over the N particles contained in the slice of interest, we obtain the volume fraction of each static configuration. To obtain the steady-state ϕ corresponding to a given Γ, we averaged this quantity over the last 100 configurations obtained for each tap intensity. We also obtain the force moment tensor $\sigma_i^{\alpha\beta} = \sum_c r_c^\alpha f_c^\beta$ of each particle in the slice of interest. Here, the sum runs over all the contacts c of the particle, $\vec{r}_c$ is the vector from the center of the grain to the contact c and $\vec{f}_c$ is the corresponding contact force. We apply the same averaging protocol used for ϕ to obtain the force moment tensor for the configuration and Σ for the steady-state of a given realization and Γ.

IV. Results

In Fig. 1(a) the ensemble average of the steady-state ϕ (i.e., averaged over the 20 independent realizations) is displayed as a function of Γ. The error bars correspond to the standard deviation over the 20 realizations. As observed by a number of authors, the curve seems to be very well defined with

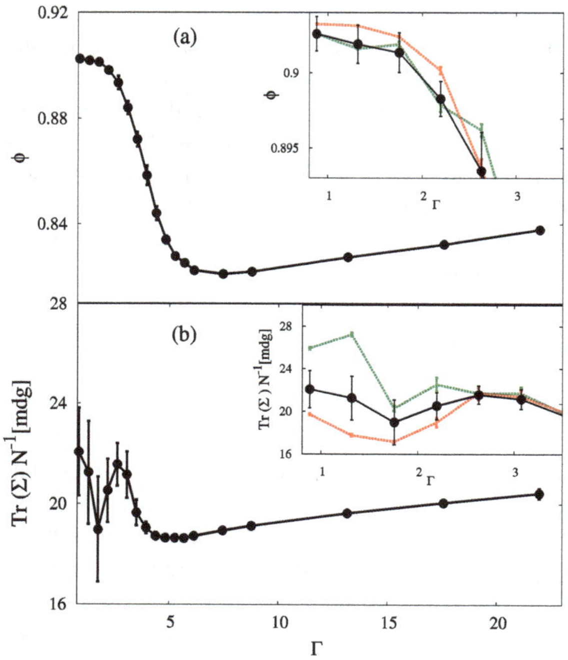

Figure 1: Ensemble average of the steady-state packing fraction ϕ (a) and trace of the force moment tensor $\mathrm{Tr}(\Sigma)$ (b) as a function of Γ. The error bars correspond to the standard deviation over the 20 averaged realizations. The insets show the same results and two of the 20 independent realizations (dashed lines) in the low-tap-intensity region. The error bars on the single realization data correspond to the estimated standard error of the mean.

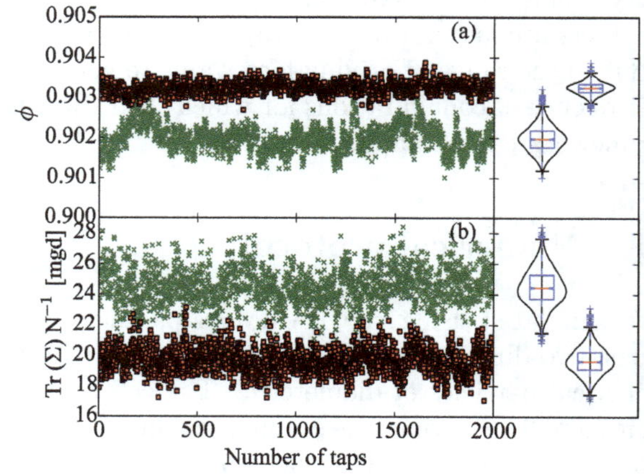

Figure 2: ϕ (a) and $\mathrm{Tr}(\Sigma)$ (b) as a function of the tap number for two of the 20 independent realizations at $\Gamma = 0.8$. The notched boxes and "violin" diagrams shown suggest that both realizations can be hardly considered as representing the same steady-state.

independent realizations falling within a very narrow range of ϕ values for any given Γ. In the past, this led to the conclusion that this was a truly reversible process, where lowering or raising Γ would lead to the same steady-state ϕ. In the inset of Fig. 1(a) two of the independent realizations are shown for the low-tap-intensity region. From this picture, it is clear that steady-states corresponding to a given Γ can differ from one realization to another. Notice that in the inset the error bars for the two isolated realizations correspond to the standard error of the mean (SEM), which gives an estimate of the uncertainty of the mean value reported rather than the size of the ϕ fluctuations. For these two realizations, although the mean ϕ seems to agree within the estimated error for intensities $\Gamma > 1.5$, it is clear that they are different for low Γ. This is consistent with the findings of Paillusson and Frenkel [5] for frictionless spheres under event-driven simulations. However, in our simulations we are able to extract the stress state of the system as well as the history of the contacts. These reveal valuable information, as we discuss below.

In Fig. 1(b), we show the trace $\mathrm{Tr}(\Sigma)$ of Σ averaged over all 20 realizations as a function of Γ. As before, the error bars indicate the standard deviation over the 20 realizations. As we can see, the variability of the mean stress is significantly large at low Γ. This is not due to the large fluctuations during a given series of taps but to the variations observed from one realization of the protocol to the other. Indeed, the inset in Fig. 1(b) shows that, for low Γ, the mean values of $\mathrm{Tr}(\Sigma)$ for two of the 20 realizations have a relatively small SEM (i.e., the fluctuations in each realization are small). However, realizations differ from each other. The difference between results corresponding to different realizations become much more evident here than in the case of the ϕ–Γ plot. We suggest that the stress tensor may be more sensitive and then more suitable to sense if ergodicity is fulfilled in experimental data. Overall, as Γ is decreased, different realizations explore non-overlapping ranges of volume/stress. Therefore, temporal averages (on

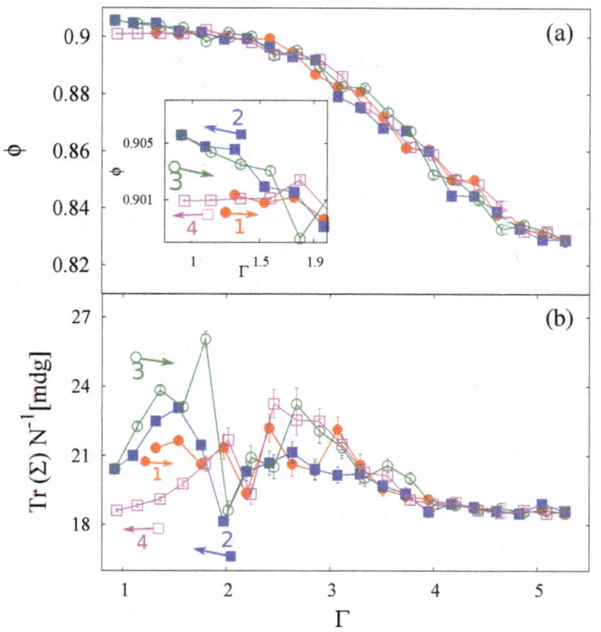

Figure 3: Steady-state ϕ (a) and $\mathrm{Tr}(\Sigma)$ (b) as a function of Γ corresponding to a full annealing protocol. Starting from the filled circles (red) increasing ramp, and following by filled squares (blue), open circles (green) and open squares (magenta). The error bars correspond to the estimated SEM.

a single time series) do not match with ensemble averages (over the realizations).

Since the number of taps we have explored for each Γ may be small to assume that the steady-state has been properly sampled, we carried out 2000 additional taps at $\Gamma = 0.8$ for each realization. Since some of the signals are not normally distributed, we confirmed the stationarity of these states by using a non-parametric test at a level of significance of 5% [15, 16]. Two of the normally distributed realizations are displayed in Fig. 2 as a function of the tap number. Comparing the corresponding notched boxes and "violin" diagrams of both signals, it is clear that the states do not match. Hence, the *reversible branch* found in Ref. [3] is not so at low Γ in our case since truly stationary states with distinguishable ϕ and $\mathrm{Tr}(\Sigma)$ may be obtained on different realizations of the annealing protocol. Of course, this is harder to detect in ϕ since the dispersion between realizations is much smaller compared with the range of ϕ values obtained at different Γ.

The former results also confirm the hypothesis that non-ergodicity is present in a typical tapping protocol beyond the special case reported in Ref. [5], where the steady-states obtained did not follow any annealing-type protocol. Hence, we observe this non-ergodic behavior even after annealing the system from high tapping strengths. To stress this point and assess if the speed of the annealing may prevent the system from reaching a unique steady-state on each realization, we apply a slower cyclic annealing protocol (similar to the one introduced in [2]) to each of the 20 final states at low Γ in order to reproduce the "reversible branch". In Fig. 3 we display a sequence of two successive up and down ramps applied to one of the 20 initial realizations using Γ-steps about one half of those used in Fig. 1. Although in the scale used for ϕ the steady-state packing fraction seems reversible, a close inspection shows that the states have distinguishable ϕ at low Γ [see Fig. 3(a) and the corresponding inset]. This is much more evident when the stress is analyzed [see Fig. 3(b)].

In order to set a criterion to decide if the steady-states are not ergodic for a given Γ, we show in Fig. 4 the p-values for the Kruskal–Wallis [17] one-way analysis of variance performed on the 20 realizations at each Γ. This simple non-parametric test allows for the rejection of the null hypothesis that all 20 data series are drawn from a unique distribution (which does not need to be normal), hence that they correspond to a unique steady-state. If the p-value is significant (in our case $p > 0.01$), then we cannot rule out the possibility that the 20 series come from the same steady-state. As we can see from Fig. 4(a), the test run on the data for ϕ indicates that for $\Gamma < 5.0$, the null hypothesis must be rejected and therefore there exist at least two out of the 20 steady-states that are not the same. However, for higher Γ, the test is significant and then the 20 realizations may correspond to the same steady-state. Interestingly, when the test is run on Σ [see Fig. 4(b)], the steady-state seems to be unique for all 20 realizations if $\Gamma > 3.75$. Although differences between realizations are simpler to detect on visual inspection of $\mathrm{Tr}(\Sigma)$, it is actually ϕ that sets a higher threshold for the Γ values needed to ensure an ergodic steady-state (i.e., $\Gamma > 5.0$).

The previous results indicate that the steady-states sampled at low tap intensities do not only

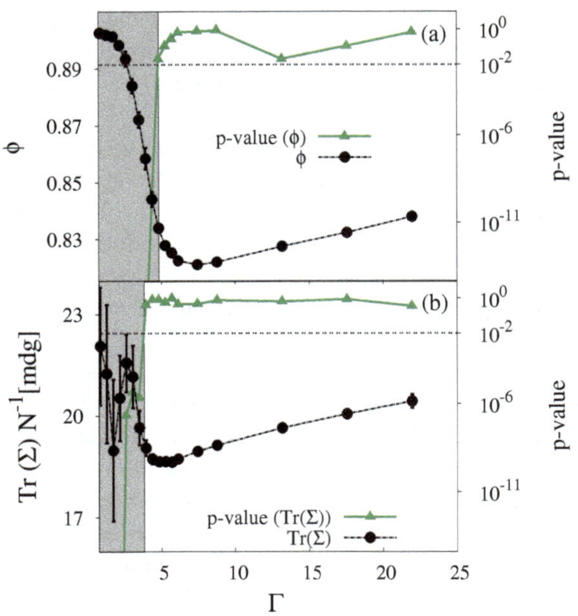

Figure 4: P-value (up-triangles, right axis) as a function of Γ for the Kruskal–Wallis [17] one-way analysis of variance for ϕ (a) and $\mathrm{Tr}(\Sigma)$ (b). The horizontal dotted line corresponds to the significance level used (1%). The black circles correspond to the ϕ and $\mathrm{Tr}(\Sigma)$ data from Fig. 1.

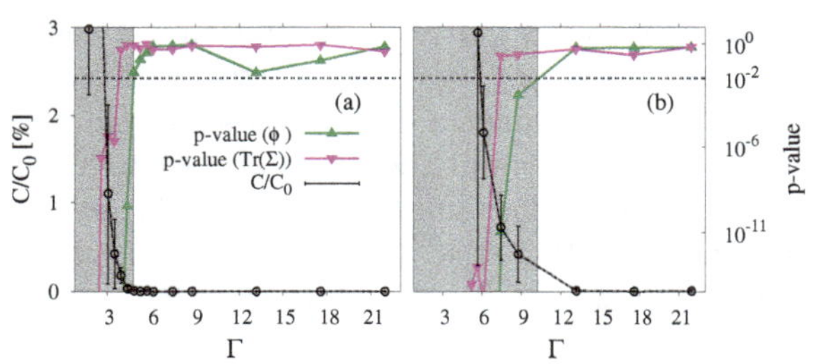

Figure 5: (a) Percentage $C/C_0 \times 100$ of persistent contacts (black circles) as a function of Γ, averaged over 5 taps on 6 independent realizations. The error bars correspond to the standard deviation over the 6 realizations. Up-triangles (green) correspond to the p-values in Fig. 4 for ϕ and down-triangles (magenta) to $\mathrm{Tr}(\Sigma)$. (b) Same as (a) for frictionless grains.

depend on Γ but on the particular history of each realization. Notice that this goes beyond the history dependent out-of-equilibrium trajectories already reported in tapped systems [18] since here we are focusing on the steady-states. One may hypothesize that the constraints imposed by the contacts is one of the reasons for the non-ergodic behavior at low tap. If a contact persists from one tap to the next, the contact force after coming back to rest will depend on the history of all contacts of that particular grain. In order to test this idea, we analyze the evolution of all contacts during each tap and identify those that persist (i.e., contacts that did not break at any time during the pulse of energy). Figure 5 shows the average ratio C/C_0 of persistent contacts, C, to the total number of contacts, C_0, as a function of Γ. For this calculation, each contact was tracked during the final 5 taps for each Γ on 6 of the independent realizations and only grains that fall within the layer of interest, as discussed above, where included in the analysis. The percentage of persistent contacts is very small but non-zero up to $\Gamma \approx 5.0$. As it is expected, when Γ is increased sufficiently all the contacts are broken and new ones are made during each tap (resulting in $C/C_0 = 0$). This transition coincides with the value of Γ where the realizations seem to sample the same steady-state (see the p-values included in Fig. 5). Therefore, when small taps are applied, the aging of some of the contacts seem to lead the system to sample different regions of the phase space during independent realizations. However, if all contacts are made anew at each tap, the sampling becomes compatible with the idea of ergodicity introduced in Fig. 4. In order to generalize this result, we also simulated frictionless grains. Interestingly, the same conclusion drawn for frictional grains is true for frictionless ones: different realizations seem to sample the same steady-state only if all contacts are made anew upon each tap [see Fig. 5 (b)].

V. Conclusions

Our analysis of the steady-states of tapped granular systems indicate that these states are history-dependent for tap intensities below a certain threshold. This is in contradiction with the general assumption that macroscopic time averages —such as the volume fraction— can be recovered when the amplitude of the perturbation applied to the system is tuned back and forth. The differences between independent realizations become particularly noticeable in the stress distribution. These findings show that the postulates of the equilibrium statistical thermodynamics may not be always fulfilled to describe the steady state of static granular systems (see also Ref. [19] for a discussion on the Boltzmann distribution failure for an analytically solvable model). Focusing on tap intensities that warrant that all contacts are made anew after each tap may allow exploring the available phase space in agreement with the ergodic hypothesis. However, gentle perturbations deserve an approach that includes memory effects to suitably describe the states. In that sense, non-equilibrium thermodynamic approaches may be a suitable alternative [20]. Further research on such alternative formalisms, the effect of other types of forcing mechanisms (e.g., shear), and possible extensions to other complex systems (e.g., active matter) become necessary.

Acknowledgements - This work has been partially supported by projects PICT-2012-2155 ANPCyT (Argentina), FIS2011-26675 MINECO (Spain) and PIUNA (Universidad de Navarra).

---

[1] S F Edwards, R B Oakeshot, *Theory of powders*, Physica **157D**, 1091 (1989).

[2] E R Nowak, J B Knight, E Ben-Naim, H M Jaeger, S R Nagel, *Density fluctuations in vibrated granular materials*, Phys. Rev. E **57**, 1971 (1998).

[3] E R Nowak, J B Knight, M L Povinelli, H M Jeager, S R Nagel, *Reversibility and irreversibility in the packing of vibrated granular material*, Powder Technol. **94**, 79 (1997).

[4] Ph Ribière, P Richard, P Philippe, D Bideau, R Delannay, *On the existence of stationary states during granular compaction*, Eur. Phys. J. E **22**, 249 (2007).

[5] F Paillusson, D Frenkel, *Probing ergodicity in granular matter*, Phys. Rev. Lett. **109**, 208001 (2012).

[6] S Plimpton, P Crozier, A Thompson, *LAMMPS-large-scale atomic/molecular massively parallel simulator*, Sandia National Laboratories (2007).

[7] L A Pugnaloni, I Snchez, P A Gago, J Damas, I Zuriguel, D Maza, *Towards a relevant set of state variables to describe static granular packings*, Phys. Rev. E **82**, 050301 (2010).

[8] S Ardanza-Trevijano, I Zuriguel, R Arévalo, D Maza, *Topological analysis of tapped granular media using persistent homology*, Phys. Rev. E **89**, 052212 (2014).

[9] N V Brilliantov, F Spahn, J-M Hertzsch, T Pöschel, *Model for collisions in granular gases*, Phys. Rev. E. **53**, 5382 (1996).

[10] L E Silbert, D Ertaş, G S Grest, T C Halsey, D Levine, S Plimpton, *Granular flow down an inclined plane: Bagnold scaling and rheology*, Phys. Rev. E **64**, 051302 (2001).

[11] L A Pugnaloni, J Damas, I Zuriguel, D Maza, *Master curves for the stress tensor invariants in stationary states of static granular beds. Implications for the thermodynamic phase space*, Papers in Physics **3**, 030004 (2011).

[12] L A Pugnaloni, M Mizrahi, C M Carlevaro, F Vericat, *Nonmonotonic reversible branch in four model granular beds subjected to vertical vibration*, Phys. Rev. E. **78**, 051305 (2008).

[13] C Rycroft, *VORO++: A three-dimensional Voronoi cell library in C++*, Chaos **19**, 041111 (2009).

[14] P A Gago, L A Pugnaloni, D Maza, *Relevance of system size to the steady-state properties of tapped granular systems*, Phys. Rev. E **91**, 032207 (2015).

[15] W Constantine, D Percival, *Fractal: Fractal Time Series Modeling and Analysis. R package version 2.0-0*, http://CRAN.R-project.org/package=fractal (2014).

[16] R Core Team, *R: A language and environment for statistical computing*, R Foundation for Statistical Computing, Vienna, Austria (2013).

[17] W H Kruskal, W A Wallis, *Use of ranks in one-criterion variance analysis*, J. Am. Stat. Assoc. **47**, 583 (1952).

[18] Ch Josserand, A V Tkachenko, D M Mueth, H M Jaeger, *Memory effects in granular material*, Phys. Rev. Lett. **85**, 3632 (2000).

[19] R M Irastorza, C M Carlevaro, L A Pugnaloni, *Exact predictions from the Edwards ensemble versus realistic simulations of tapped narrow two-dimensional granular columns*, J. Stat. Mech. P12012 (2013).

[20] G Lebon, D Jou, JC Casas-Vazquez, *Understanding Non-equilibrium Thermodinamics*, Springer-Verlag, Berlin Heidelberg (2008).

8

Bayesian regression of piecewise homogeneous Poisson processes

Diego J. R. Sevilla[1*]

In this paper, a Bayesian method for piecewise regression is adapted to handle counting processes data distributed as Poisson. A numerical code in *Mathematica* is developed and tested analyzing simulated data. The resulting method is valuable for detecting breaking points in the count rate of time series for Poisson processes.

I. Introduction

Bayesian statistics have revolutionized data analysis [1]. Techniques like the Generalized Lomb-Scargle Periodogram [2] allow us to obtain oscillation frequencies of time series with unprecedented accuracy. The Gregory and Loredo method [3] goes further allowing us to find and characterize periodic signals of any period and shape.

To detect non-periodical variations, the *Exact Bayesian Regression of Piecewise Constant Functions* by Marcus Hutter (hereafter Hutter's method) [4] is valuable. It permits to estimate the most probable partition of a data set in segments of constant signals, determining the number of segments and their borders, and in-segments means and variances. Hutter's method works with two continuous distributions: Normal, and Cauchy-Lorentz. The latter —the canonical example of a pathological distribution with undefined moments—, is also suitable to analyze data with other symmetric probability distributions, especially with heavy tails.

In the case of counting processes, especially for low rates, when data consist in non-negative small integers, methods specially designed to discrete probability distributions are necessary. Some regression methods, specially for non-homogeneous Poisson processes [5], were developed.

In this paper, Hutter's method is adapted for analyzing data distributed as Poisson. The results are summarized in a code in *Mathematica* [6]. It can be used to analyze data of several physical processes which follow the Poisson distribution (e.g., detection of photons in X-ray Astronomy, particles in nuclear disintegration, etc.), if sudden changes in detection rates are suspected.

II. Method

Hutter's method is summarized in Table 1 of Ref. [4] in a pseudo C code which is divided in two blocks. The first one calculates moments A_{ij}^k with $k = 0, 1, 2$ of the PDF of the statistical models for segments of data $D_{ij} := \{n_{i+1}, \ldots, n_j\}$. The second one performs the regression from moments A_{ij}^k. The code developed in this work is divided in three blocks.

As the members of the Poisson distributions family are identified by one parameter -the mean rate r of the Poisson process-, the PDF of the models

*E-mail: dsevilla@fceia.unr.edu.ar

[1] Departamento de Física y Química, Escuela de Formación Básica. Facultad de Ciencias Exactas, Ingeniería y Agrimensura. Universidad Nacional de Rosario, Av. Pellegrini 250, S2000BTP Rosario, Argentina.

for a segment D_{ij} is [1]

$$P(r|D_{ij}, I) = \frac{P(r|I)\,P(D_{ij}|r)}{P(D_{ij})}, \quad (1)$$

where $P(r|I)$ is the prior of parameter r, $P(D_{ij}|r)$ is the likelihood of segment D_{ij} for a given r, $P(D_{ij})$ is the global likelihood of the family, and I represents a prior information.

Usually, the prior information consists of global quantities calculated from $D := D_{0N}$, *i.e.*, from all the data set. For Poisson processes, only one quantity is necessary: the mean rate $\hat{r}$. Considering the conjugate prior of the Poisson distribution [7], the prior results

$$P(r|\hat{r}) = \frac{r^{\hat{r}-1}\,e^{-r}}{\Gamma(\hat{r})}. \quad (2)$$

For a Poisson process with rate r, the likelihood of a segment D_{ij} is

$$P(D_{ij}|r) = \prod_{t=i+1}^{j} \frac{r^{n_t}\,e^{-r}}{n_t!}. \quad (3)$$

So, the moments of the posterior can be expressed in an analytical form

$$A^k_{ij} = \frac{\Gamma(k + \hat{r} + \sum_{t=i+1}^{j} n_t)\prod_{t=i+1}^{j}\frac{1}{n_t!}}{\Gamma(\hat{r})\,(j-i+1)^{k+\hat{r}+\sum_{t=i+1}^{j} n_t}}. \quad (4)$$

Code block 1 calculates A^k_{ij}. It needs as input the time series to be analyzed (list `data`). The output are functions `A0[i,j]`, `A1[i,j]` and `A2[i,j]` and integer `n`, which is the length of `data`.

Code block 1: *Mathematica* code to calculate A^k_{ij}.

```
n=Length[data];
r=Mean[data];
Do[
  Do[
    d=j-i; m=r+Sum[data[[t]],{t,i+1,j}];
    A0[i,j]=(m-1)!/(Gamma[r]*(d+1)^m)*
      Product[1/data[[t]]!,{t,i+1,j}];
    A1[i,j]=m*A0[i,j]/(d+1);
    A2[i,j]=(m+1)*A1[i,j]/(d+1);
  ,{j,i+1,n}];
,{i,0,n}];
```

As the second block of Hutter's code only needs the moments A^k_{ij} as inputs, it could work properly with no changes. It computes the evidence, the probability for k segments and its MAP estimation $\hat{k}$, the probability of boundaries locations and the MAP locations of the $\hat{k}$ boundaries, the first and second in-segment moments, and an interesting regression curve that smooths the final result.

Nevertheless, for our specific problem, once the segments boundaries are obtained, we can estimate their means and variances straightforwardly, so we only use a part of Hutter's second block, which is shown in code block 2. The logical of the algorithm is explained in Ref. [4]. Code block 2 needs as inputs `A0[i,j]`, `A1[i,j]`, `A2[i,j]` and `n`, all calculated in code block 1, and integer `kmax`, which is the maximum number of segments to be considered. The outputs are the evidence (`e`), the probability for k segments (`c[k]`), its MAP (`khat`), the probability of boundaries locations (`B[i]`), and their MAP (`that[p]`).

Code block 2: *Mathematica* code to calculate breaking points.

```
Do[
  L[0,i]=KroneckerDelta[i,0];
  R[0,i]=KroneckerDelta[i,n];
,{i,0,n}];
Do[
  Do[
    L[k+1,i]=Sum[L[k,h]*A0[h,i],{h,k,i-1}];
    R[k+1,i]=Sum[R[k,h]*A0[i,h],{h,i+1,n-k}];
  ,{i,0,n}]
,{k,0,kmax-1}];
e=1/kmax*Sum[L[k,n]/Binomial[n-1,k-1],{k,1,kmax}];
Do[
  c[k]=L[k,n]/(Binomial[n-1,k-1]*kmax*e)
,{k,1,kmax}];
khat=1;
Do[If[c[khat]<c[k],khat=k];,{k,0,kmax}];
Do[B[i]=
  Sum[L[p,i]*R[khat-p,i]/L[khat,n],{p,0,khat}];
,{i,0,n}];
Do[
  that[p]=0;
  dummy=L[p,0]*R[khat-p,0];
  Do[
    If[dummy<L[p,h]*R[khat-p,h],
      that[p]=h;
      dummy=L[p,h]*R[khat-p,h];
    ];
  ,{h,1,n}];
,{p,0,khat}];
```

Finally, we calculate the in-segments means and estimate their statistical errors. If one segment of m elements has n counts, its mean rate is n/m, and its variance is approximately n/m^2.

To determine the accuracy of the fit, it is also useful to estimate the uncertainties of the boundaries locations. A reasonable estimation of the uncertainties can be obtained from the second moments of the probability distributions of boundaries locations in the neighborhoods of the breaking points. If the probabilities of the boundaries locations are given by function $B(k)$, the uncertainties approximately result

$$\epsilon_{\hat{k}} = \sqrt{\frac{\sum_{k=\hat{k}-a+1}^{\hat{k}+a} B(k)\,(k-\hat{k})^2}{\sum_{k=\hat{k}-a+1}^{\hat{k}-a} B(k)}}, \tag{5}$$

where a is large enough to consider all the width of the peak corresponding to boundary $\hat{k}$, but small enough to avoid including peaks of other boundaries. In practice, if $\hat{\jmath}$, $\hat{k}$ and $\hat{l}$ are the positions of consecutive boundaries, a good value for a is the minimum of $(\hat{k}-\hat{\jmath})/2$ and $(\hat{l}-\hat{k})/2$.

Code block 3 uses these approaches to calculate the constant piecewise regression of data, and to estimate its statistical errors. The inputs are `c[k]`, `khat`, `B[i]` and `that`. The outputs are list `reg`, which represents the best fit, and lists `re1` and `re2`, which represent the minimum and maximum estimations considering the statistical errors of `reg`.

Code block 3: *Mathematica* code to calculate regression.

```
TBP=DeleteDuplicates[Table[that[k],{k,0,khat}]];
NBP=Length[TBP]-1;
Do[BP[k]=TBP[[k+1]];,{k,0,NBP}];
ethat[0]=0;ethat[NBP]=0;
Do[
  l1=IntegerPart[(BP[k-1]+BP[k])/2]+1;
  l2=IntegerPart[(BP[k+1]+BP[k])/2];
  ethat[k]=
    Round[Sqrt[Sum[(j-BP[k])^2*B[j],{j,l1,l2}]/
    Sum[B[j],{j,l1,l2}]]];
,{k,1,NBP-1}];
Do[
  mm[k]=BP[k]-BP[k-1];
  nn[k]=Sum[data[[i]],{i,BP[k-1]+1,BP[k]}];
,{k,1,NBP}];
reg=Flatten[Table[Table[nn[k]/mm[k]
  ,{BP[k]-BP[k-1]}],{k,1,NBP}]];
BP1[0]=BP2[0]=BP[0];BP1[NBP]=BP2[NBP]=BP[NBP];
Do[
  s=If[reg[[BP[k]+1]]>reg[[BP[k]]],1,-1];
  BP1[k]=BP[k]+s*ethat[k];
  BP2[k]=BP[k]-s*ethat[k];
,{k,1,NBP-1}];
re1=Flatten[Table[Table[(nn[k]-Sqrt[nn[k]])/mm[k]
  ,{BP1[k]-BP1[k-1]}],{k,1,NBP}]];
re2=Flatten[Table[Table[(nn[k]+Sqrt[nn[k]])/mm[k]
  ,{BP2[k]-BP2[k-1]}],{k,1,NBP}]];
```

III. Applications and Discussion

Figure 1 (top) shows, in blue dots, data simulated using *Mathematica*. Data consist of 150 Poisson distributed elements, the first 50 with rate 1.5, the second 50 with rate 0.5, and the last 50 with rate 1.0. Applying the first 2 blocks of code on data,

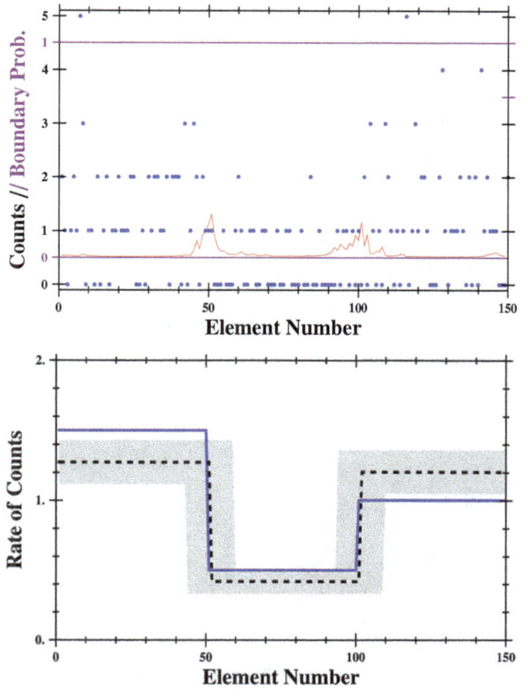

Figure 1: Top: Simulated data (blue dots) and boundaries location probability (red line). Bottom: Regression curve and its error estimation (black dashed line and gray zone), and the rate curve used in simulation (blue line).

we can see that the probability of having 2 breaking points is very high. Figure 1 (top) also shows, in red line, the probability for the boundaries locations. Applying code block 3, we obtain the regression [Fig. 1 (bottom), black dashed curve] and its error estimation [Fig. 1 (bottom), gray zone]. The continuous blue line in Fig. 1 (bottom) indicates the rates used in simulation.

The regression in the example above fits very well with the rate curve used in simulation. But sometimes regressions result qualitatively different to the rate curve, showing more or less breaking points, even for data simulated in the same conditions. This effect is due to chance. To show this issue, 2000 simulations with the same conditions were performed. In 992 of them, two breaking points were found. In the others, there were found zero (44), one (208), three (419), four (155), five (73), six (41), and seven or more (68) breaking points. For cases in which two breaking points were found, statistics of the most likely boundaries loca-

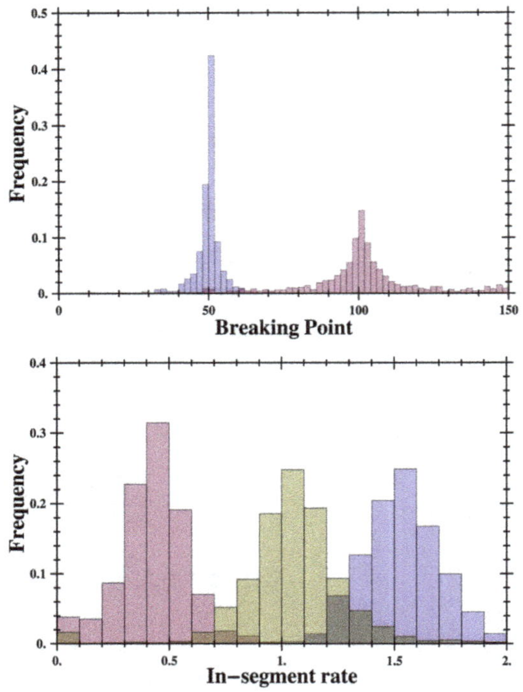

Figure 2: Top: Histogram of the boundaries locations. Bottom: Histogram of the in-segment mean rates. Both figures were calculated for a set of 2000 simulations similar to that shown in Fig. 1.

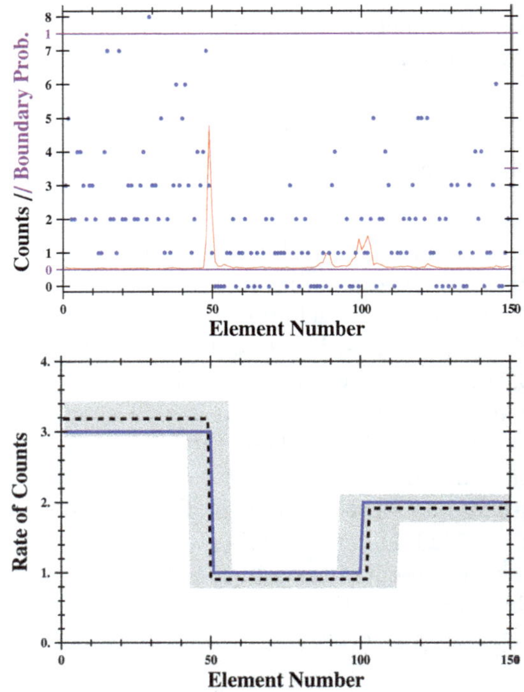

Figure 3: Top: Simulated data (blue dots) and boundaries location probability (red line). Bottom: Regression curve and its error estimation (black dashed line and gray zone), and the rate curve used in simulation (blue line).

tions and in-segment mean rates were calculated. Figure 2 shows histograms of those statistics.

Figure 2 (top) shows histograms for boundaries locations. It is clear that the bigger the step, the smaller the uncertainty on its location. Figure 2 (bottom) shows histograms of in-segment rates. It is clear that the greater the rate, the smaller its relative statistical error.

Figure 3 shows data and boundaries location probabilities for a simulation similar to the previous ones, but now with rates 3.0, 1.0 and 2.0. Comparing Fig. 1 (top) and Fig. 3 (top), we can see that in the latter one the boundaries locations are found more accurately.

Again, 2000 simulations with the same conditions were performed. In 1159 of them, two breaking points were found, while in the others, there were found zero (1), one (33), three (499), four (167), five (75), six (32), and seven or more (34) breaking points. Figure 4 shows histograms of the statistics of the most likely boundaries locations and in-segment rates for the simulations with two breaking points found. Comparing with Fig. 2, we can see that the histograms are now narrower. These results confirm what was stated above.

It is important to note that the probability for the *real curve* to be *completely* inside the region defined by the error estimations of the regression is significantly less than one. It is easy to see why: if the errors were independent and equal to the standard error, the probability of satisfying n error conditions simultaneously would be 0.68^n. But even the actual probability could be lower, since it is clear that the errors must be dependent. Nevertheless, the error estimations presented here are useful to get an idea of the accuracy of the regression.

Finally, the capability to detect a breaking point with this code was tested for different count rates. To do this, simulated data sets of a single step in the count rate were used. Data sets consist in 100 Poisson distributed elements, the first 50 for a rate r_1 and the last 50 for a rate r_2. 1000 simulations were

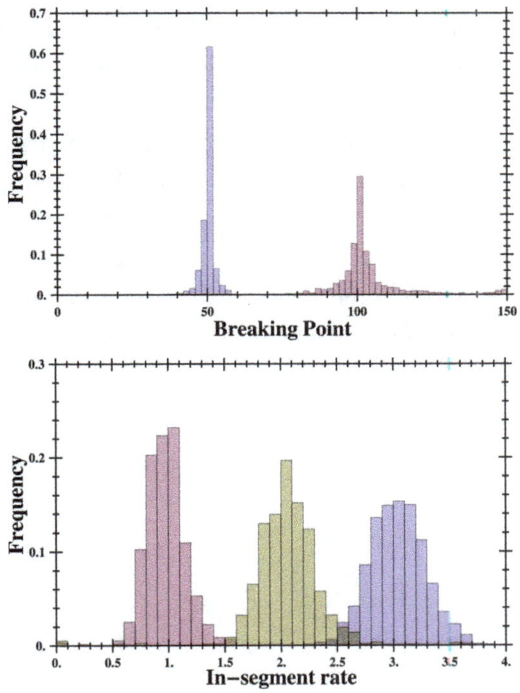

Figure 4: Top: Histogram of the boundaries locations. Bottom: Histogram of the in-segment rates. Both figures were calculated for a set of 2000 simulations similar to that shown in Fig. 3.

Table 1: Statistics of successful detections for a single step.

$r_1\backslash r_2$	3.0	2.0	1.6	1.2	0.8
0.4	0.86	0.74	0.70	0.63	0.46
0.8	0.77	0.70	0.60	0.36	
1.2	0.69	0.61	0.30		
1.6	0.65	0.28			
2.0	0.60				

done for each pair (r_1, r_2). Statistics of successful detections are presented in Table 1. A successful detection is considered when only one breaking point between elements 40 and 60 is detected.

Table 1 shows that the smaller is the mean rate difference and the smaller are the mean rates, the more difficult is the detection of the step. This result is expected because in a Poisson distribution the variance is equal to the mean.

IV. Conclusions

In this work, a code for Bayesian regression of piecewise constant functions was adapted to handle data from Poisson processes. For this purpose, equations for calculating the moments of the posteriors of segments of data were found through Bayes theorem, considering the conjugate prior of the Poisson distribution as prior. These results, as well as part of Hutter's method, were used to calculate the most probable number of segments and their boundaries. Procedures for calculating in-segments mean rates and the uncertainties of mean rates and boundaries locations are also provided. The resulting method is summarized in a code in *Mathematica*.

The code was applied to simulated data. Firstly, two examples with tree segments were analyzed. The code performed well in both cases considering the dispersion of data, and the results improved in the case of higher mean rates and mean rates differences. This occurs because of the statistical dispersion of Poisson distributed data, which is greater than the mean rate if the mean rate is lower than one.

Finally, simulations of data of a single step were analyzed for different rates, and statistics of the regressions with only one breaking point are presented in a table. This table shows the effect of the rates and rate differences in the regression accuracy, and, together with the errors estimations provided by the code, can serve as an indicator of the reliability of the method.

Supplementary material including the source code for the algorithms can be found at the journal website [8].

Acknowledgements - This work was partially supported by the National University of Rosario.

[1] P C Gregory, *Bayesian logical data analysis for the physical sciences*, Cambridge University Press, Cambridge, UK (2004).

[2] G L Bretthorst, *Lecture notes in statistics*, Springer, Berlin (1988).

[3] P C Gregory, T J Loredo, *A new method for the detection of a periodic signal of unknown shape and period*, Astrophys. J. **398**, 146 (1992).

[4] M Hutter, *Exact Bayesian regression of piecewise constant functions*, Bayesian Analysis **2**, 635 (2007).

[5] J F Lawless, *Regression methods for Poisson process data*, J. Am. Stat. Assoc. **82**, 399 (1987).

[6] Wolfram Research Inc., Mathematica version 9.0, Wolfram Research, Inc., Champaign, Illinois (2012).

[7] A Gelman, J B Carlin, H S Stern, D B Rubin, *Bayesian data analysis*, Taylor & Francis, UK (2014).

[8] *Mathematica* codes and examples by the author can be found at `http://www.papersinphysics.org`.

9

Topological quantum phase transition in strongly correlated Kondo insulators in 1D

Franco T. Lisandrini,[1,2] Alejandro M. Lobos,[1] Ariel O. Dobry,[1,2] Claudio J. Gazza[1,2*]

We investigate, by means of a field-theory analysis combined with the density-matrix renormalization group (DMRG) method, a theoretical model for a strongly correlated quantum system in one dimension realizing a topologically-ordered Haldane phase ground state. The model consists of a spin-1/2 Heisenberg chain coupled to a tight-binding chain via two competing Kondo exchange couplings of different type: a "s-wave" Kondo coupling (J_K^s), and a less common "p-wave" (J_K^p) Kondo coupling. While the first coupling is the standard Kondo interaction studied in many condensed-matter systems, the latter has been recently introduced by Alexandrov and Coleman [Phys. Rev. B **90**, 115147 (2014)] as a possible mechanism leading to a topological Kondo-insulating ground state in one dimension. As a result of this competition, a topological quantum phase transition (TQPT) occurs in the system for a critical value of the ratio J_K^s/J_K^p, separating a (Haldane-type) topological phase from a topologically trivial ground state where the system can be essentially described as a product of local singlets. We study and characterize the TQPT by means of the magnetization profile, the entanglement entropy and the full entanglement spectrum of the ground state. Our results might be relevant to understand how topologically-ordered phases of fermions emerge in strongly interacting quantum systems.

I. Introduction

The study of topological quantum phases of matter has become an area of great interest in present-day condensed matter physics. A topological phase is a quantum phase of matter which cannot be characterized by a local order parameter, and thus falls beyond the Landau paradigm. In particular, topological insulators (i.e., materials which are insulating in the bulk but support topologically protected gapless states at the edges) were first proposed theoretically for two- and three-dimensional systems with time-reversal symmetry [1–3], and soon after found in experiments on HgTe quantum wells [4] and in $Bi_{1-x}Sb_x$ [5], and Bi_2Se_3 [6], generating a lot of excitement and subsequent research. The electronic structure of a topological insulator cannot be smoothly connected to that of a trivial insulator, a fact that is mathematically expressed in the existence of a nonzero topological invariant, an integer number quantifying the non-local topological order in the ground state. A complete classification based on the dimensionality and underlying symmetries has been achieved in the form of a "periodic table" of topological insulators [7–9]. Nevertheless, this classification refers only to the gapped phases of noninteracting fermions, and leaves open the problem of characterizing and classifying strongly in-

*E-mail: gazza@ifir-conicet.gov.ar

[1] Instituto de Física Rosario (CONICET), Bv 27 de Febrero 210 bis, S2000EZP Rosario, Santa Fe, Argentina.

[2] Facultad de Ciencias Exactas, Ingeniería y Agrimensura, Universidad Nacional de Rosario, Argentina.

teracting topological insulators. This is a very important open question in modern condensed-matter physics.

Recently, Dzero *et al.* [10–12] proposed a new kind of topological insulator: the topological Kondo insulator (TKI), which combines features of both non-interacting topological insulators and the well-known Kondo insulators, a special class of heavy-fermion system with an insulating gap strongly renormalized by interactions. Within a mean-field picture, TKIs can be understood as a strongly renormalized f-electron band lying close to the Fermi level, and hybridizing with the conduction-electron $d-$bands [13–15]. At half-filling, an insulating state appears due to the opening of a low-temperature hybridization gap at the Fermi energy induced by interactions. Due to the opposite parities of the states being hybridized, a topologically non-trivial ground state emerges, characterized by an insulating gap in the bulk and conducting Dirac states at the surface [10]. At present, TKI materials, among which samarium hexaboride (SmB_6) is the best known example, are under intense investigation both theoretically and experimentally [16–19] .

From a theoretical point of view, TKIs are interesting systems arising from the interplay between strong interactions and topology. Although the large-N meanfield approach was successful in describing qualitatively heavy-fermion systems and TKIs in particular, it would be desirable to understand better how TKIs emerge. In order to shed more light into this question, in a recent work Alexandrov and Coleman proposed a one-dimensional model for a topological Kondo insulator [20], the "p-wave" Kondo-Heisenberg model (p-KHM). Such a model consists of a chain of spin-1/2 magnetic impurities interacting with a half-filled one-dimensional electron gas through a Kondo exchange (see Fig. 1). Using a standard mean-field description [13–15], which expresses the original interacting problem as an effectively non-interacting one, the authors mentioned above found a topologically non-trivial insulating ground state (i.e., topological class D [7–9]) which hosts magnetic states at the open ends of the chain. However, this system was studied recently using the Abelian bosonization approach combined with a perturbative renormalization group analysis, revealing an unexpected connection to the Haldane phase at low temperatures [21]. The Haldane phase is a paradigmatic example of a strongly interacting topological system [22–26]. The results in Ref. [21] indicate that 1D TKI systems might be much more complex and richer than expected with the naïve mean-field approach, as they are uncapable of describing the full complexity of the Haldane phase, and suggest that they must be reconsidered from the more general perspective of interacting symmetry-protected topological (SPT) phases [25–28]. More recently, two numerical studies using exact DMRG methods have confirmed that 1D TKIs belong to the universality class of Haldane insulators [29, 30]. These studies have extended the regime of validity of the results in Ref. [21].

In this work, we theoretically investigate the robustness of the Haldane phase in one-dimensional topological Kondo insulators, and study the effect of local interactions that destabilize the topological phase and drive the system to a non-topolgical phase. Our goal is to characterize the system at, and near to, the topological quantum phase transition (TQPT) from the perspective of symmetry-protected topological phases, using the concepts of entanglement entropy and entanglement spectrum to detect the topologically-ordered ground states. This is a novel perspective in the context of TKIs, which might shed new light on the emergence of topological order on strongly correlated phases of fermions, and makes our work interesting from the pedagogical and conceptual points of view.

II. Theoretical model

We describe the system depicted in Fig. 1 with the Hamiltonian $H = H_1 + H_2 + H_K^{(s)} + H_K^{(p)}$, where the conduction band is represented by a L-site tight-binding chain

$$H_1 = -t \sum_{j=1,\sigma}^{L-1} \left(c_{j,\sigma}^{\dagger} c_{j+1,\sigma} + \text{H.c.} \right), \qquad (1)$$

with $c_{j,\sigma}^{\dagger}$, the creation operator of an electron with spin σ at site j. The Hamiltonian

$$H_2 = J_H \sum_{j=1}^{L-1} \mathbf{S}_j \cdot \mathbf{S}_{j+1} \quad (J_H > 0), \qquad (2)$$

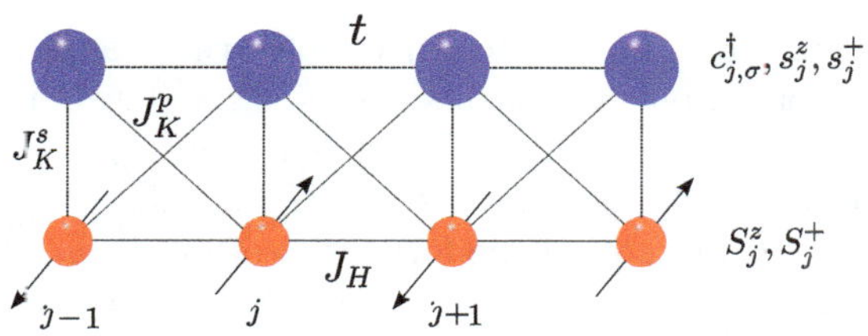

Figure 1: Sketch of the Kondo-Heisenberg model under consideration. The lower leg represents a spin-1/2 Heisenberg chain with $J_H > 0$. The upper leg represents a half-filled one-dimensional tight binding chain interacting with the lower leg through two different Kondo exchange couplings, a "s-wave" J^s_K and a "p-wave" J^p_K. We also show the fermionic and spin operators defined on each eight-dimensional "supersite" j (see text).

corresponds to a spin-1/2 antiferromagnetic Heisenberg chain. The terms $H^{(s)}_K$ and $H^{(p)}_K$ describe two different types of Kondo exchange couplings between H_1 and H_2, namely

$$H^{(s)}_K = J^s_K \sum_{j=1}^{L} \mathbf{S}_j \cdot \mathbf{s}_j, \tag{3}$$

$$H^{(p)}_K = J^p_K \sum_{j=1}^{L} \mathbf{S}_j \cdot \boldsymbol{\pi}_j, \tag{4}$$

with $J^a_K > 0$ ($a = s, p$). Eq. (3) describes the usual antiferromagnetic Kondo exchange coupling of a spin $\mathbf{S}_j$ in the Heisenberg chain to the local spin density in the fermionic chain at site j, defined as:

$$\mathbf{s}_j \equiv \sum_{\alpha,\beta} c^\dagger_{j,\alpha} \left(\frac{\boldsymbol{\sigma}_{\alpha\beta}}{2}\right) c_{j,\beta}, \tag{5}$$

where $\boldsymbol{\sigma}_{\alpha\beta}$ is the vector of Pauli matrices. On the other hand, Eq. (4) describes a "p-wave" Kondo interaction, which is unusual in that it couples the spin $\mathbf{S}_j$ to the p-wave spin density in the fermionic chain at site j, defined as: [20]

$$\boldsymbol{\pi}_j \equiv \sum_{\alpha,\beta} \left(\frac{c^\dagger_{j+1,\alpha} - c^\dagger_{j-1,\alpha}}{\sqrt{2}}\right) \left(\frac{\boldsymbol{\sigma}_{\alpha\beta}}{2}\right) \left(\frac{c_{j+1,\beta} - c_{j-1,\beta}}{\sqrt{2}}\right), \tag{6}$$

where the notation $c_{0,\sigma} = c_{L+1,\sigma} = 0$ is implied.

The case $J^s_K > 0$ and $J^p_K = 0$ corresponds to the so-called "Kondo-Heisenberg model" in 1D, which has been extensively studied in the past in connection to the stripe phase of high-T_c superconductors [31–37]. These previous works indicate that, at half-filling, that model does not support any topological phases. On the other hand, the case $J^s_K = 0$ and $J^p_K > 0$ is the "p-wave" Kondo-Heisenberg model proposed recently in Ref. [20], and subsequently studied in Refs. [21, 29, 30], where it was stablished that the ground state corresponds to the Haldane phase. In both cases, at half-filling the system develops a Mott-insulating gap in the fermionic chain due to "umklapp" processes (i.e., backscattering processes with $2k_F$ momentum transfer) originated in the Kondo interactions J^s_K or J^p_K, and at low temperatures (lower than the Mott gap), the system effectively maps onto a spin-1/2 ladder. However, the ground states of the model in both cases cannot be smoothly connected (i.e., one is topologically trivial while the other is not), and therefore we anticipate that a gap-closing topological quantum phase transition (TQPT) must occur as a function of the ratio J^s_K/J^p_K. Indeed, in Ref. [21], it was proposed that while in the first case the (effective) spin-1/2 ladder forms *singlets* along the rungs, in the second case the Kondo coupling J^p_K favors the formation of *triplets* on the rungs, and therefore the system maps onto the Haldane spin-1 chain [38–41]. From this perspective, our toy-model Hamiltonian H allows to explore a transtition from topological to non-topological Kondo insulators, and to gain a valuable insight of the TQPT.

III. Field-theory analysis

The purpose of this section is to provide a simple and phenomenological understanding of the competition between the two Kondo interactions J^s_K and J^p_K, which is not evident in Eqs. (3) and (4). To that end, we introduce a field-theoretical representation of the model, valid at sufficiently low temperatures. We linearize the non-interacting spectrum $\epsilon_k = -2t\cos(ka)$ in the tight-binding chain H_1 around the Fermi energy $\mu = 0$, and take the continuum limit $a \to 0$, where a is the lattice constant. Then, the low-energy representation of the fermionic operators becomes [42–46]

$$\frac{c_{j,\sigma}}{\sqrt{a}} \sim e^{ik_F x_j} R_{1,\sigma}(x_j) + e^{-ik_F x_j} L_{1,\sigma}(x_j), \quad (7)$$

where $R_{1,\sigma}(x)$ and $L_{1,\sigma}(x)$ are right- and left-moving fermionic field operators, which vary slowly on the scale of a. Using this representation, the spin densities become [47]:

$$\frac{\mathbf{s}_j}{a} \to \left[\mathbf{J}_{1R}(x_j) + \mathbf{J}_{1L}(x_j) + (-1)^j \mathbf{N}_1(x_j)\right], \quad (8)$$

$$\frac{\boldsymbol{\pi}_j}{a} \to 2\left[\mathbf{J}_{1R}(x_j) + \mathbf{J}_{1L}(x_j) - (-1)^j \mathbf{N}_1(x_j)\right], \quad (9)$$

where we have defined slowly-varying spin densities for the *smooth* spin configurations:

$$\mathbf{J}_{1R}(x_j) = \sum_{\alpha,\beta} R^{\dagger}_{1,\alpha}(x_j)\left(\frac{\boldsymbol{\sigma}_{\alpha\beta}}{2}\right) R_{1,\beta}(x_j), \quad (10)$$

$$\mathbf{J}_{1L}(x_j) = \sum_{\alpha,\beta} L^{\dagger}_{1,\alpha}(x_j)\left(\frac{\boldsymbol{\sigma}_{\alpha\beta}}{2}\right) L_{1,\beta}(x_j), \quad (11)$$

and for the *staggered* spin density

$$\mathbf{N}_1(x_j) = \sum_{\alpha,\beta} R^{\dagger}_{1,\alpha}(x_j)\left(\frac{\boldsymbol{\sigma}_{\alpha\beta}}{2}\right) L_{1,\beta}(x_j) + \text{H.c.}. \quad (12)$$

Similarly, a continuum representation for the Heisenberg chain can be achieved, e.g., by fermionization of the S=1/2 spins by means of a Jordan-Wigner transformation. At low energies, the spin densities become [38–41, 47]

$$\frac{\mathbf{S}_j}{a} \to \mathbf{J}_{2R}(x_j) + \mathbf{J}_{2L}(x_j) + (-1)^j \mathbf{N}_2(x_j). \quad (13)$$

We now focus on the Kondo interaction and leave the analysis of H_1 and H_2 aside, as these terms are unimportant for the qualitative understading of the basic mechanism leading to the TQPT. Keeping the most relevant (in the RG sense) terms, we can write the Kondo interaction as:

$$H_K^{(s)} + H_K^{(p)} \to (J_K^s - 2J_K^p) \int dx\, \mathbf{N}_1(x) . \mathbf{N}_2(x) \quad (+\text{less relevant contributions}), \quad (14)$$

i.e., the Kondo interaction couples the staggered magnetization components in chains 1 and 2. In the above expression, note that while a large J_K^s favors a positive value of the effective coupling $(J_K^s - 2J_K^p)$, therefore promoting the formation of local singlets along the rungs, a large p-wave Kondo coupling J_K^p favors an *effective ferromagnetic* coupling which promotes the formation of local triplets with $S = 1$ (hence the connection to the $S = 1$ Haldane chain). The minus sign in front of J_K^p appears as a result of the p-wave nature of the orbitals in Eq. (6). From this qualitative analysis we can conclude that J_K^s and J_K^p will be competing interactions promoting different ground states, and since these grounstates cannot be adiabatically connected with each other, a TQPT must occur.

Strictly speaking, near the critical region where the bare coupling $(J_K^s - 2J_K^p)$ vanishes, the less relevant terms neglected in Eq. (14) should be taken into account. However, note that operators with conformal spin 1 (i.e., operators of the form $(\partial_x \mathbf{N}_1) . \mathbf{N}_2$ or $\mathbf{N}_1 . (\partial_x \mathbf{N}_2)$, see for instance Ref. [48]) are not allowed by the inversion symmetry of the Hamiltonian and the $p-$wave symmetry of the orbitals in Eq. (6), which demands that $c_{j+1,\alpha} - c_{j-1,\alpha} \to -\left(c_{-(j+1),\alpha} - c_{-(j-1),\alpha}\right)$ under the change $j \to -j$, and therefore forbids the occurrence of terms proportional to $\partial_x \mathbf{N}_1$. Therefore, only the marginal operators $\mathbf{J}_{1\nu} . \mathbf{J}_{2\nu'}$ (with $\nu = \{R, L\}$) and terms with conformal spin bigger than 1 are expected in the Hamiltonian. We do not expect these operators to change the physics qualitatively near the critical point, and we can ignore them for this simplified analysis. As we show below, our numerical DMRG results are in accordance with this qualitative picture.

IV. DMRG analysis

Before presenting the numerical results, it is worth providing technical details on the implementation of the DMRG method applied to the present model. This particular Hamiltonian contains two types of terms: (a) terms involving two local operators, as in most condensed-matter models with nearest-neighbor interactions, i.e., Eqs. (1)-(3), and (b) terms involving three local operators, which result from the expansion of Eq. (4). To make it easier to implement, we have found useful to define first a "supersite" representation of the system, where each supersite combines a spin $\mathbf{S}_j$ and the fermionic site along each rung (see Fig. 1), therefore spanning a new 8-dimensional local basis. The first kind of terms could be easily handled with standard DMRG implementations where the system is represented as $L(j)\otimes\bullet\otimes\bullet\otimes R(N-j-2)$, with $L(j)$ and $R(j)$ the left and right blocks with j supersites, respectively, and the two circles are the exactly-represented middle supersites $j+1$ and $j+2$. This comes at a price, however, since one is then forced to re-express the electron creation operator $c^\dagger_{j,\sigma}$, and the spin-1/2 operator $\mathbf{S}_j$ in this new basis in order to implement Eqs. (1) and (2). In this basis, note that the Hamiltonian $H_K^{(s)}$, Eq. (3), becomes an "on-site" term, which can be handled easily.

In the second type of contributions, the presence of three-operator terms in $H_K^{(p)}$, Eq. (4), must be properly treated in order to avoid extra truncation errors due to the tensor product of two (already truncated) operators inside each left and right blocks. Then, during each left-right DMRG sweep iteration, we save in the previous superblock configuration $L(j-1)\otimes\bullet\otimes\bullet\otimes R(N-j-3)$ the exact correlation matrices between spin and fermionic operators that involve positions $j-1$ (i.e., rightmost supersite of left block) and j (first single site), which will become the two rightmost supersites of the new $L(j)$ in the next sweep iteration step. These correlation matrices are

$$\left[A^{z\dagger}_{j-1,j,\sigma}\right]_{i;i'} = \rho_{i;i1i2}\left[S^z_{j-1}\right]_{i1;i1'}\left[c^\dagger_{j,\sigma}\right]_{i2;i2'}\rho^\dagger_{i1'i2';i'},$$

and

$$\left[A^{+\dagger}_{j-1,j,\bar\sigma}\right]_{i;i'} = \rho_{i;i1i2}\left[S^+_{j-1}\right]_{i1;i1'}\left[c^\dagger_{j,\bar\sigma}\right]_{i2;i2'}\rho^\dagger_{i1'i2';i'},$$

where $\rho_{i;i1i2}$ is the $m\times 8m$ reduced density matrix, with m the number of states kept, and where the indices i and $i1$ run over the truncated m-dimensional Hilbert space while $i2$ runs over the 8-dimensional supersite space. In addition, we have assumed summation over repeated internal indices. Similarly, correlations between spin and fermionic operators placed in the two leftmost supersites $j+3$ and $j+4$ of $R(N-j-2)$ should also be kept during the corresponding step of the rightf-left sweep. Therefore, we now deal with a standard "two-operator" interaction, for example the term $L(j)\otimes\bullet$ of $H_K^{(p)}$ is

$$-\frac{1}{4}J_K^p\sum_\sigma\left\{\left(\sigma A^{z\dagger}_{j-1,j,\sigma}+A^{+\dagger}_{j-1,j,\bar\sigma}\right)c_{l_{j+1},\sigma}+\text{h.c.}\right\}.$$

Finally, we mention that in our implementation we have kept up to a maximum of 800 states and we have swept 12 times, assuring truncation errors in the density matrix of the order 10^{-9} at worst.

We now turn to the results. We have used the tight-binding parameter t as our unit of energies, and in all of our calculations we have used the values $J_H/t=1$ and $J_K^p/t=2$. We have studied the evolution of the ground state upon the increase of J_K^s starting from the value $J_K^s=0$, where the system is in the Haldane phase. The nature of the topologically-ordered ground state and the precise detection of the TQPT are determined using, respectively: a) the analysis of the spin profile in the ground state, b) the value of the entanglement entropy and c) the analysis of the degeneracies in the full entanglement spectrum. As shown recently in seminal works [25, 26], the last two properties are useful *bona fide* indicators of symmetry-protected topological orders.

(a) Spin profile. One characteristic feature of a $S=1$ Haldane chain with open boundary conditions is the presence of topologically protected fractionalized spin-1/2 end-states, a consequence of the broken $Z_2\times Z_2$ symmetry of the ground state [24]. These states can be represented as: $|\uparrow_L\rangle\otimes|\uparrow_R\rangle$, $|\uparrow_L\rangle\otimes|\downarrow_R\rangle$, $|\downarrow_L\rangle\otimes|\uparrow_R\rangle$, $|\downarrow_L\rangle\otimes|\downarrow_R\rangle$, and correspond to the fourfold-degenerate ground state in the thermodynamic limit $L\to\infty$. In order to detect these fractionalized spin excitations in our system, we have defined the spin profile as $\langle T^z_j\rangle=\langle\psi_g^{M^z=1}|\,T^z_j\,|\psi_g^{M^z=1}\rangle$, where T^z_j is the $z-$projection of the total spin in the j-th rung $\mathbf{T}_j=\mathbf{S}_j+\mathbf{s}_j$, and $|\psi_g^{M^z=1}\rangle$ is the ground state of the system with total spin $M^z=1$ (where the visualization of the spin states at the ends is easier). In Fig. 2 we show $\langle T^z_j\rangle$ vs j for different values of J_K^s/J_K^p and for $L=80$. The presence of spin states localized at the edges can be clearly seen for $J_K^s/J_K^p=0$ and $J_K^s/J_K^p=0.85$, where

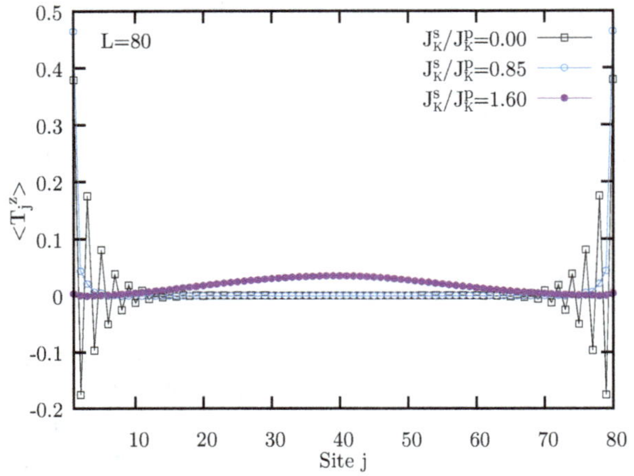

Figure 2: Spatial profile of $\langle T_j^z \rangle = \langle \psi_g^{M^z=1} | T_j^z | \psi_g^{M^z=1} \rangle$, i.e., the z component of the total spin in the supersite j, computed with the ground state of the subspace with total $M^z = 1$ for $L = 80$, and for parameters $J_H/t = 1$ and $J_K^p/t = 2$. The destruction of the topologically protected spin-1/2 states at the ends of the chain can be clearly seen as J_K^s/J_K^p is increased from $J_K^s/J_K^p = 0$ to $J_K^s/J_K^p = 1.6$.

the spin density is accumulated at the ends of the system. Note that since we are working in the subspace $M^z = 1$, this ground state corresponds to the state $| \uparrow_L \rangle \otimes | \uparrow_R \rangle$. For $J_K^s/J_K^p = 1.6$, the magnetic edge-states have already disappeared, indicating that the onset of the topologically trivial phase must occur at lower values of J_K^s/J_K^p. However, while this analysis is useful to understand the nature of the topologically-ordered ground state, it does not allow a precise determination of the TQPT. To that end, we have studied the entanglement entropy and entanglement spectrum (see below).

(b) Entanglement entropy. We have also calculated the entanglement entropy (i.e., the von Neumann entropy of the reduced density matrix), defined as [49]

$$S(L/2) = -\mathrm{Tr}\hat{\rho}_{L/2} \ln \hat{\rho}_{L/2} = -\sum_j \Lambda_j \ln \Lambda_j, \quad (15)$$

where $\hat{\rho}_{L/2}$ is the reduced density matrix obtained after tracing out half of the chain, and Λ_j the corresponding eigenvalues of $\hat{\rho}_{L/2}$, which are the squares of the Schmidt values. Recently, it has been clarified that the entanglement of a single quantum state is a crucial property not only from the perspective of quantum information, but also for condensed matter physics. In particular, the entanglement entropy has been shown to contain the *quantum dimension*, a property of topologically-ordered phases [50, 51]. Hirano and Hatsugai [52] have computed the entanglement entropy of the open-boundary spin-1 Haldane chain and obtained the lower-bound value $S(L/2) = \ln(4) = 2\ln(2)$ which, according to the edge-state picture in the thermodynamical limit $L \to \infty$, corresponds to the aforementioned 4 spin-1/2 edge states. In Fig. 3 we show the entanglement entropy of the system as a function of J_K^s/J_K^p, for different system sizes and in the subspace $M^z = 0$, where we expect to find the ground state (i.e., the ground state of an even-numbered antiferromagnetic chain is a global singlet [53]). Near the critical region, the entanglement entropy grows due to the contribution of the bulk, and exactly at the TQPT the entanglement entropy is predicted to show a logarithmic divergence $S(L/2) \sim \ln(L)$, characteristic of critical one-dimensional systems [49]. As the size of the system is increased, the logarithmic divergence becomes narrower and its position shifts to larger values. Using the fitting function $J_{K,c}^s(L) = J_{K,c}^s(\infty) + a/L^2$, we have obtained the extrapolated critical point $J_{K,c}^s(\infty) \approx 1.11\, J_K^p$ in the thermodynamic limit, see inset (a) in Fig. 3. Note that this value is smaller than the predicted value $J_{K,c}^s = 2\, J_K^p$ using the field-theory analysis of the previous section. We believe this to be the effect of the neglected marginal or irrelevant operators, which renormalize non-universal quantities such as the critical point.

We have also confirmed the logarithmic scaling of the entanglement entropy at the critical point, i.e., $S_{\max}(L/2) = \alpha \ln(L) + \text{constant}$, and we have obtained a prefactor $\alpha = 0.052$, see inset (b) in Fig. 3. A detailed analysis of this value and its connection to the corresponding central charge value of the conformal field theory is beyond the scope of the present work and is left to a subsequent publication.

In Fig. 3, note that for $J_K^s/J_K^p < J_{K,c}^s/J_K^p$, the value of the entanglement entropy roughly corresponds to $S(L/2) \sim 2\ln(2)$, consistent with the theoretical predictions in the Haldane phase. Values of $S(L/2)$ which are below the predicted lower-

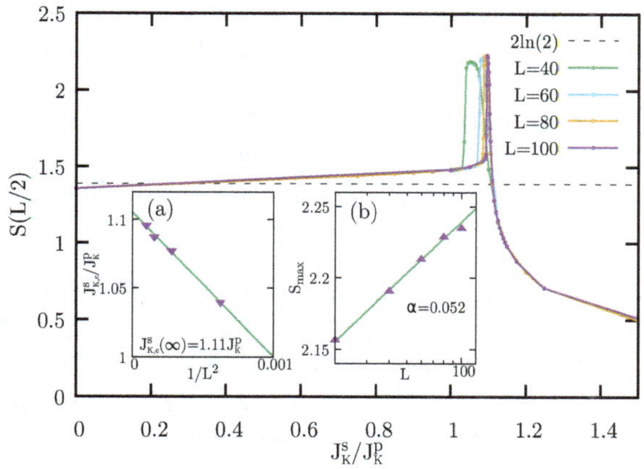

Figure 3: Entanglement entropy of the reduced density matrix, $S\,(L/2)$, as function of J^s_K/J^p_K for different lattice sizes. The entropy is computed within the subspace $M^z = 0$ and for parameters $J_H/t = 1$ and $J^p_K/t = 2$. The maximum of $S\,(L/2)$ indicates the position of the critical point. Inset (a): Finite-size scaling of the critical point, i.e., $J^s_{K,c}\,(L) = J^s_{K,c}\,(\infty) + a/L^2$, from where the value in the thermodynamic limit $J^s_{K,c}\,(\infty) \approx 1.11\,J^p_K$ is obtained. Inset (b): Maximum entropy $S_{\max}(L/2)$ vs L. The results reproduce the predicted logarithmic divergence $S_{\max}(L/2) = \alpha \ln(L) + \text{constant}$, with the fitting constant $\alpha = 0.052$.

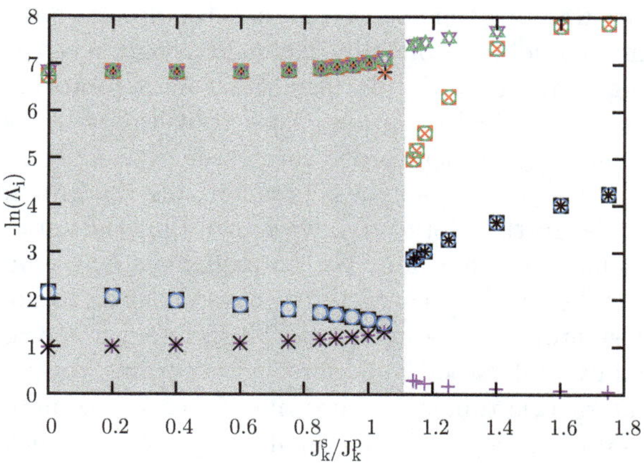

Figure 4: The largest eigenvalues of the entanglement spectrum (extrapolated to the thermodynamic limit $L \to \infty$), as a function of J^s_K. The results were obtained in the subspace $M^z = 0$. The gray zone corresponds to the Haldane phase, where the degeneracy in the spectrum of eigenvalues is even.

bound are presumably due to finite-size effects, which result in a decrease of the effective quantum dimension in small systems [52]. For $J^s_K/J^p_K > J^s_{K,c}/J^p_K$, $S(L/2)$ tends to zero as expected for a topologically trivial ground state. This result can be easily understood in the limit $J^s_K/J^p_K \to \infty$, where we expect the ground state to factorize as a product of local singlets (we recall that we are working in the subspace $M^z = 0$), for which $S(L/2) = 0$.

(c) Degeneracy of the entanglement spectrum. Finally, we focus on the full entanglement spectrum of the reduced density-matrix. Degeneracies in the entanglement spectrum are intimately related to the existence of discrete symmetries which protect the topological order. In particular, as shown in Ref. [25], an even degeneracy constitutes the most distinctive feature of the Haldane phase. This fact allows to make an interesting connection between the theory of symmetry-protected topological phases and the theory of topological Kondo insulators in one dimension, and might have important implications in the understanding of strongly interacting topological phases of fermions. In Fig. 4 we show the evolution of the largest eigenvalues extrapolated to the thermodynamic limit of the density matrix as a function of the parameter J^s_K/J^p_K, obtained in the subspace $M^z = 0$. Note that for $J^s_K/J^p_K < J^s_{K,c}/J^p_K \sim 1.11$ (i.e., gray-shaded area), the degeneracy of the eigenvalues is even, as expected for the Haldane phase. In contrast, for values $J^s_K/J^p_K > J^s_{K,c}/J^p_K$, the even-degeneracy breaks down, indicating the onset of the trivial phase. In particular, note the evolution of the largest eigenvalue of the density matrix $\Lambda_0 \approx 1$ (i.e., lowest "+" symbol) which becomes non-degenerate.

V. Conclusions

Using a combination of techniques, i.e., a field-theoretical analysis and the density-matrix renormalization group (DMRG), an essentially exact method in one dimension, we have studied the transition from a topological to a non-topological phase in a model for a one-dimensional strongly interacting topological Kondo insulator. As a prototypical topological quantum phase transition with no Landau-type local order-parameter, one must resort to global quantities characterizing the ground state. In this work, we have shown that the entanglement entropy and the entanglement spectrum can be used to characterize a topological Kondo insulator in one dimension. This system was originally understood and classified according to non-interacting topological invariants (i.e., Chern numbers), employing approximate large-N mean-field methods [20]. Here, by the means of the DMRG, we have shown that a more appropriate way to understand this system is by using the concepts developed for symmetry-protected topological phases [25–28]. In particular, for parameters $J_H/t = 1$ and $J^p_K/t = 2$, we have obtained the value of the critical point $J^s_{K,c}/J^p_K \simeq 1.11$ in the thermodynamic limit $L \to \infty$. This value is smaller than the expected from the qualitative field-theoretical estimation $J^s_{K,c}/J^p_K = 2$, a fact that is presumably originated in the effect of marginal or irrelevant operators, which were neglected in the qualitative analysis and which renormalize a non-universal quantity such as the critical point.

Acknowledgements - F.T.L, A.O.D. and C.J.G. acknowledge support from CONICET-PIP 11220120100389CO. A.M.L acknowledges support from PICT-2015-0217 of ANPCyT.

[1] B A Bernevig, T L Hughes, S C Zhang, *Quantum spin hall effect and topological phase transition in HgTe quantum wells*, Science **314**, 1757 (2006).

[2] L Fu, C L Kane, *Topological insulators with inversion symmetry*, Phys. Rev. B **76**, 045302 (2007).

[3] H Zhang, C X Liu, X L Dai, Z Fang, S C Zhang, *Topological insulators in Bi2Se3, Bi2Te3 and Sb2Te3 with a single Dirac cone on the surface*, Nat. Phys. **5**, 438 (2009).

[4] M König, S Wiedmann, C Brüne, A Roth, H Buhmann, L W Molenkamp, X L Qi, S C Zhang, *Quantum spin hall insulator state in HgTe quantum wells*, Science **318**, 766 (2007).

[5] D Hsieh, D Qian, L Wray, Y Xia, Y S Hor, R J Cava, M Z Hasan, *A topological dirac insulator in a quantum spin hall phase*, Nature **452**, 970 (2008).

[6] Y Xia, D Qian, L Hsieh, D Wray, A Pal, H Lin, A Bansil, D Grauer, Y S Hor, R J Cava, M Z Hasan, *Observation of a large-gap topological-insulator class with a single dirac cone on the surface*, Nat. Phys. **5**, 398 (2009).

[7] A Altland, M R Zirnbauer, *Nonstandard symmetry classes in mesoscopic normal-superconducting hybrid structures*, Phys. Rev. B **55**, 1142 (1997).

[8] A Kitaev, *Periodic table for topological insulators and superconductors*, AIP Conf. Proc. **1134**, 22 (2009).

[9] S Ryu, A P Schnyder, A Furusaki, A W W Ludwig, *Topological insulators and superconductors: tenfold way and dimensional hierarchy*, New J. Phys. **12**, 065010 (2010).

[10] M Dzero, K Sun, V Galitski, P Coleman, *Topological Kondo insulators*, Phys. Rev. Lett. **104**, 106408 (2010).

[11] M Dzero, K Sun, P Coleman, V Galitski, *Theory of topological Kondo insulators*, Phys. Rev. B **85**, 045130 (2012).

[12] M Dzero, J Xia, V Galitski, P Coleman, *Topological Kondo Insulators*, Annu. Rev. Condens. Matter Phys. **7**, 249 (2016).

[13] N Read, D M Newns, *On the solution of the Coqblin-Schreiffer Hamiltonian by the large-n expansion technique*, J. Phys. C **16**, 3273 (1983).

[14] P Coleman, *Mixed valence as an almost broken symmetry*, Phys. Rev. B **35**, 5072 (1987).

[15] D M Newns, N Read, *Mean-field theory of intermediate valence/heavy fermion systems*, Adv. Phys. **36**, 799 (1987).

[16] S Wolgast, C Kurdak, K Sun, J W Allen, D J Kim, Z Fisk, *Low-temperature surface conduction in the Kondo insulator SmB_6*, Phys. Rev. B **88**, 180405 (2013).

[17] X Zhang, N P Butch, P Syers, S Ziemak, R L Greene, J Paglione. *Hybridization, inter-ion correlation, and surface states in the Kondo insulator* SmB_6, Phys. Rev. X **3**, 011011 (2013).

[18] N Xu et al., *Direct observation of the spin texture in* SmB_6 *as evidence of the topological Kondo insulator*, Nat. Commun. **5**, 4566 (2014).

[19] D J Kim, J Xia, Z Fisk, *Topological surface state in the Kondo insulator samarium hexaboride*, Nat. Mater. **13**, 466 (2014).

[20] V Alexandrov, P Coleman, *End states in a one-dimensional topological Kondo insulator in large-n limit*, Phys. Rev. B **90**, 115147 (2014).

[21] A M Lobos, A O Dobry, V Galitski, *Magnetic end states in a strongly interacting one-dimensional topological Kondo insulator*, Phys. Rev. X **5**, 021017 (2015).

[22] I Affleck, T Kennedy, E H Lieb, H Tasaki, *Rigorous results on valence-bond ground states in antiferromagnets*, Phys. Rev. Lett. **59**, 799 (1987).

[23] I Affleck, T Kennedy, E H Lieb, H Tasaki, *Valence bond ground states in isotropic quantum antiferromagnets*, Commun. Math. Phys. **115**, 477 (1988).

[24] T Kennedy, H Tasaki, *Hidden* $z_2 \times z_2$ *symmetry breaking in haldane gap antiferromagnets*, Phys. Rev. B **45**, 304 (1992).

[25] F Pollmann, A M Turner, E Berg, M Oshikawa, *Entanglement spectrum of a topological phase in one dimension*, Phys. Rev. B **81**, 064439 (2010).

[26] F Pollmann, E Berg, A M Turner, M Oshikawa, *Symmetry protection of topological phases in one-dimensional quantum spin systems*, Phys. Rev. B **85**, 075125 (2012).

[27] Z C Gu, X G Wen, *Tensor-entanglement-filtering renormalization approach and symmetry-protected topological order*, Phys. Rev. B **80**, 155131 (2009).

[28] X Chen, Z C Gu, X G Wen, *Classification of gapped symmetric phases in one-dimensional spin systems*, Phys. Rev. B **83**, 035107 (2011).

[29] A Mezio, A M Lobos, A O Dobry, C J Gazza, *Haldane phase in one-dimensional topological Kondo insulators*, Phys. Rev. B **92**, 205128 (2015).

[30] I Hagymási, Ö Legeza, *Characterization of a correlated topological Kondo insulator in one dimension*, Phys. Rev. B **93**, 165104 (2016).

[31] O Zachar, S A Kivelson, V J Emery, *Exact results for a 1D Kondo lattice from bosonization*, Phys. Rev. Lett. **77**, 1342 (1996).

[32] A E Sikkema, I Affleck, S R White, *Spin gap in a doped Kondo chain*, Phys. Rev. Lett. **79**, 929 (1997).

[33] O Zachar, A M Tsvelik, *One dimensional electron gas interacting with a Heisenberg spin-1/2 chain*. Phys. Rev. B **64**, 033103 (2001).

[34] O Zachar, *Staggered liquid phases of the one-dimensional Kondo-Heisenberg lattice model*, Phys. Rev. B **63**, 205104 (2001).

[35] E Berg, E Fradkin, S A Kivelson, *Pair-density-wave correlations in the Kondo-Heisenberg model*, Phys. Rev. Lett. **105**, 146403 (2010).

[36] A Dobry, A Jaefari, E Fradkin, *Inhomogeneous superconducting phases in the frustrated Kondo-Heisenberg chain*, Phys. Rev. B **87**, 245102 (2013).

[37] G Y Cho, R Soto-Garrido, E Fradkin, *Topological Pair-Density-Wave Superconducting States*, Phys. Rev. Lett., **113**, 256405 (2014).

[38] D G Shelton, A A Nersesyan, A M Tsvelik, *Antiferromagnetic spin ladders: Crossover between spin S=1/2 and S=1 chains*, Phys. Rev. B **53**, 8521 (1996).

[39] A O Gogolin, A A Nersesyan, A M Tsvelik, L Yu, *Zero-modes and thermodynamics of disordered spin-1/2 ladders*, Nucl. Phys. B **540**, 705 (1999).

[40] P Lecheminant, E Orignac, *Magnetization and dimerization profiles of the cut two-leg spin ladder and spin-1 chain*, Phys. Rev. B **65**, 174406 (2002).

[41] N J Robinson, F H L Essler, E Jeckelmann, A M Tsvelik, *Finite wave vector pairing in doped two-leg ladders*, Phys. Rev. B, **85**, 195103 (2012).

[42] V J Emery, *Theory of the one-dimensional electron gas*, In: Highly conducting one-dimensional solids, Eds. J T Devreese, R P Evrard, V E van Doren, Pag. 247, Plenum, New York (1979).

[43] J Sólyom, *The fermi gas model of one-dimensional conductors*, Adv. in Phys. **28**, 201 (1979).

[44] J von Delft, H Schoeller, *Bosonization for beginners refermionization for experts*, Ann. Phys. (N. Y.) **7**, 225 (1998).

[45] T Giamarchi, *Quantum Physics in One Dimension*, Oxford University Press, Oxford (2004).

[46] A O Gogolin, A A Nersesyan, A M Tsvelik, *Bosonization and Strongly Correlated Systems*, Cambridge University Press, Cambridge (1999).

[47] I Affleck, In E. Brezin and J. Zinn-Justin, editors, *Fields, Strings and Critical Phenomena*, pag. 563, Elsevier Science, Amsterdam (1988).

[48] A A Nersesyan, A O Gogolin, Fabian H L Essler, *Incommensurate spin correlations in Heisenberg spin-1/2 zig-zag ladders*, Phys. Rev. Lett. **81**, 910 (1998).

[49] U Schollwöck, *The density-matrix renormalization group*, Rev. Mod. Phys. **77**, 259 (2005).

[50] A Kitaev, J Preskill, *Topological Entanglement Entropy*, Phys. Rev. Lett. **96**, 110404 (2006).

[51] M Levin, X G Wen, *Detecting Topological Order in a Ground State Wave Function*, Phys. Rev. Lett. **96**, 110405 (2006).

[52] T Hirano, Y Hatsugai, *Entanglement Entropy of One-dimensional Gapped Spin Chains*, J. Phys. Soc. Japan **76**, 074603 (2007).

[53] A Auerbach, *Interacting Electrons and Quantum Magnetism*, Springer, Berlin (1998).

10

An efficient density matrix renormalization group algorithm for chains with periodic boundary condition

Dayasindhu Dey,[1] Debasmita Maiti,[1] Manoranjan Kumar[1*]

The Density Matrix Renormalization Group (DMRG) is a state-of-the-art numerical technique for a one dimensional quantum many-body system; but calculating accurate results for a system with Periodic Boundary Condition (PBC) from the conventional DMRG has been a challenging job from the inception of DMRG. The recent development of the Matrix Product State (MPS) algorithm gives a new approach to find accurate results for the one dimensional PBC system. The most efficient implementation of the MPS algorithm can scale as $O(p \times m^3)$, where p can vary from 4 to m^2. In this paper, we propose a new DMRG algorithm, which is very similar to the conventional DMRG and gives comparable accuracy to that of MPS. The computation effort of the new algorithm goes as $O(m^3)$ and the conventional DMRG code can be easily modified for the new algorithm.

I. Introduction

The quantum many-body effect in the condensed matter gives rise to many exotic states such as superconductivity [1], multipolar phases [2,3], valence bond state [4], vector chiral phase [2,5] and topological superconductivity [6]. These effects are prominent in the one dimensional (1D) electronic systems due to the confinement of electrons. The confinement of electrons and the competition between the electron-electron repulsion and the kinetic energies of electrons produce many interesting phases like Spin Density Wave (SDW), dimer or the bond order wave phase and Charge Density Wave (CDW) phase in one dimensional systems [7–9]. Although these quantum many-body effects in the system are crucial for exotic phases, dealing with these systems is a challenging job because of the large degrees of freedom. The degrees of freedom increase as 2^N or 4^N for a spin-1/2 system or a fermionic system, respectively.

In most cases, the exact solutions for these systems with large degrees of freedom are almost impossible to reach. Therefore, during the last three decades many numerical techniques have been developed, e.g., Quantum Monte Carlo (QMC) [10], Density Functional Theory (DFT) [11], Renormalization Group (RG) [12] and Density Matrix Renormalization Group (DMRG) method [13, 14]. The DMRG is a state-of-the-art numerical technique for 1D systems with Open Boundary Condition (OBC). However, the numerical effort to maintain the accuracy for PBC systems becomes exponential [15, 16]. It is well known that the Periodic Boundary Condition (PBC) is essential to get rid of the boundary effect of a finite open chain and also to preserve the inversion symmetry in the systems [17].

The DMRG technique is based on the systematic truncation of irrelevant degrees of freedom and has been reviewed extensively in Ref. [15,16]. In a

*E-mail: manoranjan.kumar@bose.res.in

[1] S. N. Bose National Centre for Basic Sciences, Block JD, Sector III, Salt Lake, Kolkata 700106, India.

1D system with OBC, the number of relevant degrees of freedom is small [15, 16]. Let us consider that for a given accuracy of the OBC system, m_{obc} number of eigenvectors of the density matrix is required, then the conventional DMRG for the PBC system requires $O(m^2_{obc})$ [18]. In the conventional DMRG, computational effort for the OBC systems with sparse matrices goes as $O(m^3_{obc})$, whereas, it goes as $O(m^6_{obc})$ for the PBC system [19]. The accuracy of energies for the PBC systems calculated from the conventional DMRG decreases significantly, and there is a long bond in the system which connects both ends.

The accuracy of operators decreases with the number of renormalization, especially the raising/creation $S^+/a+$ and lowering/annihilation S^-/a^- operator of spin/fermionic systems. The conventional DMRG is solved in a S^z basis, therefore the exact S^z operator remains diagonal and, multiple times, renormalization deteriorates the accuracy slowly; but $S^+/a+$ and S^-/a^- are off diagonal in this exact basis and therefore, the multiple time renormalization of these operators decreases the accuracy of the operators. A similar type of accuracy problems occurs for multiple time renormalized a^+ and a^- in the fermionic systems. In fact, it has been noted that accuracy of energy of a system with PBC significantly increases if the superblock is constructed with very few times renormalized operators [9]. To avoid multiple renormalization, new sites are added at both ends of the chain in such a way that only second time renormalized operators are used to construct the superblock. In this algorithm, there is a connectivity between the old-old sites and their operators are renormalized; and this connectivity spoils the sparsity of the superblock Hamiltonian [9].

In this paper, a new DMRG algorithm is proposed, which can be implemented upon the existing conventional DMRG code in a few hours and gives accurate results which are comparable to those of MPS algorithm. In fact, this algorithm can be implemented for two-legged ladders without much effort [20]. We have studied the spectrum of density matrix of the system block, ground state energy and correlation functions of a Heisenberg antiferromagnetic Hamiltonian for spin-1/2 and spin-1 on a 1D chain with PBC.

This paper is divided into five sections. The model Hamiltonian is discussed in section II. The new algorithm and the comparative studies of algorithms are done in section III. The accuracy of various quantities is studied in the section IV. In section V, results and algorithm are discussed.

II. Model Hamiltonian

Let us consider a strongly correlated electronic system where Coulomb repulsion is dominant, therefore the charge degree of freedom gets localized, for example, Hubbard model in large U limit in a half filled band. In this limit, the system becomes insulating, but the electrons can still exchange their spin. The Heisenberg model is one of the most celebrated models in this limit, and only the spin degrees of freedom are active in the model. The Heisenberg model Hamiltonian can be written as

$$H = \sum_{\langle ij \rangle} J_{ij} \vec{S}_i \cdot \vec{S}_j \tag{1}$$

where, J_{ij} is the anti-ferromagnetic exchange interaction between the nearest neighbor spin. In the rest of the paper, J_{ij} is set to be 1.

III. Comparison of algorithms

A ground state wavefunction calculated from the conventional DMRG can be represented in terms of the Matrix Product State (MPS), as shown by Ostlund and Rommer [21]. The wavefunction can be written as

$$|\psi\rangle = \sum_{n_1, n_2, \ldots, n_L} \mathrm{tr}(A^1_{n_1} A^2_{n_2} \ldots A^L_{n_L}) |n_1, n_2, \ldots, n_L\rangle, \tag{2}$$

where $A^k_{n_k}$ are a set of matrices of dimension $m \times m$ for site k and with n_k degrees of freedom. The wave function $|\psi\rangle$ can be accurately found if m is sufficiently large. The expectation value of an operator O_k in the gs [19, 22] can be written as

$$\langle O_k \rangle = \mathrm{tr}\left(\sum_{n_k, n_k'} \langle n_k | O_k | n_k' \rangle A^k_{n_k} \otimes A^k_{n_k'} \right), \tag{3}$$

where n_k is the local degrees of freedom of site k. The matrix A can be evaluated by using the following equations

$$H_k|\psi_k\rangle = \lambda N_k|\psi_k\rangle, \tag{4}$$

$$N^k = E_1^{k+1}E_1^k \ldots E_1^L E_1^1 \ldots E_1^{k-1}, \tag{5}$$

where

$$E_1^k = \sum_{n_k,n_k^{'}} \langle n_k|1|n_k^{'}\rangle A_{n_k}^k \otimes A_{n_k^{'}}^k. \tag{6}$$

Here, H_k is the effective Hamiltonian of kth site and λ is the expectation value of energy. The A matrices are evaluated at this point and the matrices are rearranged to keep the algorithm stable. The H_k and N can be calculated recursively while evaluating A, one site at a time [19]. Here, N matrices are ill-conditioned and require storing, approximately L^2 matrices as well as multiplication of L^2 matrices of $m \times m$ size [19] at each step. The evaluation of A and N are done for all the sites and backward and forward sweeps for all the sites are executed similarly to the finite system DMRG. The mathematical operations of matrices of dimension $m^2 \times m^2$ represent the Hamiltonian cost $\sim (o)m^6$, but the special form of these matrices reduces the cost by a factor of m. Therefore, this algorithm scales as $\sim (o)m^5$ [19].

The above algorithm is extended by Verstraete et. al., for translational invariant systems [23]. Only two types of matrices, A^1 and A^2, are considered [23]. Product of the two matrices can be repeated to compute N. In this algorithm, only two matrices, A^1 and A^2, are updated and optimized to get the gs properties. This algorithm scales as $(o)m^3$, although it does not work for a finite system, or systems with impurity, etc. Pippan et. al. introduced another MPS based efficient algorithm for translational invariant PBC systems [19]. In the old version of MPS, most of the computation cost goes to constructing the product of $m^2 \times m^2$ matrices E defined in Eq. (6). The new MPS algorithm overcomes this problem by performing a SVD of the product of sufficiently large ($l \gg 1$) number of $m^2 \times m^2$ transfer matrices[19, 28]. The singular values, in general, decay very fast; therefore, only $p (\ll m^2)$ among m^2 singular values are significant [28]. Thus, the computational cost now is reduced to (o) $p \times m^3$ [19]. However, one requires $p \sim m$ to reach adequate numerical accuracy of physical measures, as pointed out in Ref. [28].

Although the above technique is efficient and accurate, there are various reasons for developing the new algorithm. First, the modified MPS works efficiently for a system where the singular values of products of matrices decay exponentially and this algorithm scales as (o) pm^3, where p can vary linearly with m. Second, the implementation of the MPS based numerical technique is quite different from the conventional DMRG, and the MPS algorithm should be written from scratch. Third, many conventional numerical techniques like the dynamical correction vector [24] or continued fraction [25], and implementation of symmetries like parity or inversion symmetries are difficult. In this paper, we will explain a new algorithm which is very similar to the conventional DMRG technique, and also show that the new algorithm can give accuracy comparable to that of MPS based techniques. This algorithm is applied for $S = 1/2$ and $S = 1$ chains with PBC. But first, let us try to understand the algorithm before discussing the results.

In this algorithm, we will try to avoid the multiple renormalization of operators, whereas the other parts of the algorithm remain the same as the conventional DMRG. Before going to the new algorithm, let us recap the conventional DMRG.

1. Start with a superblock of four sites consisting of one site for both the left and the right block and two new sites.

2. Get desired eigenvalues and eigenvectors of the superblock and construct the density matrix ρ of the system which consists of the left or the right block and a new site.

3. Now, construct an effective $\tilde{\rho}$ with m number of eigenvectors of ρ, corresponding to the m largest eigenvalues. The effective system Hamiltonian and all operators in the truncated basis are constructed using the following equations:

$$H = \tilde{\rho}^{\dagger} H \tilde{\rho}; \qquad O = \tilde{\rho}^{\dagger} H \tilde{\rho} \tag{7}$$

4. Superblock is constructed using the effective Hamiltonian and operators of the system block and two new sites.

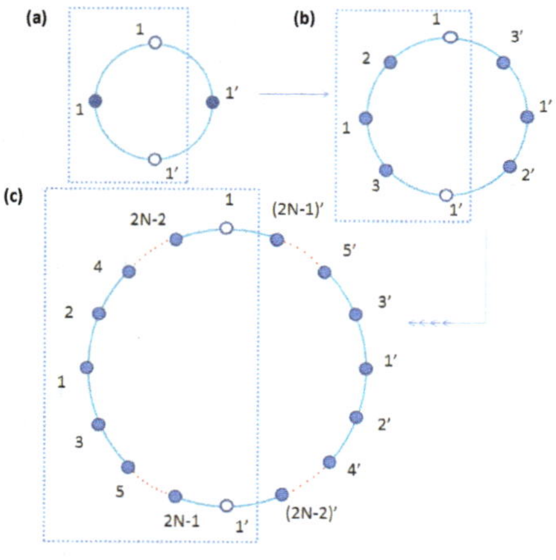

Figure 1: Pictorial representation of the new DMRG algorithm with only one site in the new block. (a) One starts with two blocks left and right represented by filled circles and two new sites blocks represented as open circles. The dotted box represents the system block for the next step. (b) Superblock of the next DMRG step is shown. (c) The final step of infinite DMRG of $N = 4N$ system size is shown with $2N - 1$ number of sites in each of the left and the right blocks and two new sites.

5. Repeat all the steps from 2 until the desired system size is reached. The full process is called infinite DMRG.

As mentioned earlier, the conventional algorithm is excellent for a 1D open chain as superblock is constructed with only one time renormalized operators. However, for a PBC system, one needs a long bond; therefore, at least two operators of superblocks are renormalized multiple times. In the new algorithm, the multiple time renormalization of operator is avoided and the algorithm goes as:

1. Start with a superblock with four blocks consisting of a left and a right block and two new site blocks. The blocks are shown in Fig. 1 as filled circles and may have more than one site. Here, we have considered only one site in each block. New blocks may also have more than one site and are shown as open circles. In this paper, new blocks have one site in a chain or two for a ladder, like a structure with PBC [17].

2. Get the eigenvalues and eigenvectors of the superblock and construct the density matrix ρ of the system which consists of the left or the right block and two new blocks. The left system block is shown inside the box in Fig. 1a.

3. Now, construct an effective $\tilde{\rho}$ with m number of eigenvectors of ρ corresponding to the m largest eigenvalues of the density matrix. The effective system Hamiltonian and operators in the truncated basis are calculated using Eq. (7).

4. The superblock is constructed using the effective Hamiltonian and operators of system blocks and two new sites.

5. Now, go to step 2 and the processes 2 – 5 are repeated till the desired system size is reached.

We notice that the superblock Hamiltonian is constructed using the effective Hamiltonian of blocks and operator which are renormalized once. Therefore, the massive truncation because of long bond is avoided in this algorithm. Bonds in the superblock are only between new-new sites or new-old sites. For the construction of a Hamiltonian matrix of old-new site bond, a new site operator is highly sparse. However, old sites renormalized operators are highly dense. The old-old sites interaction in the conventional algorithm generates a large number of non-zero matrix elements in the superblock Hamiltonian matrix and the diagonalization of dense matrix goes as m^4. But, in the new algorithm, old-new sites interaction in the superblock generates only a sparse Hamiltonian matrix, and its digonalization scales as m^3.

IV. Results

We consider spin-1/2 and spin-1 chains with PBC of length up to $N = 500$ to check the accuracy of results for the Heisenberg Hamiltonian. In this part, we study the truncation error of density matrix T, error in relative ground state energy $\frac{\Delta E}{|E_0|}$ and dependence of correlation function $C(r)$ on m. The correlation function $C(r)$ is defined as

$$C(r) = \vec{S}_0 \cdot \vec{S}_r, \tag{8}$$

where $\vec{S}_0$ corresponds to the reference spin and $\vec{S}_r$ is the spin at a distance r from the reference spin.

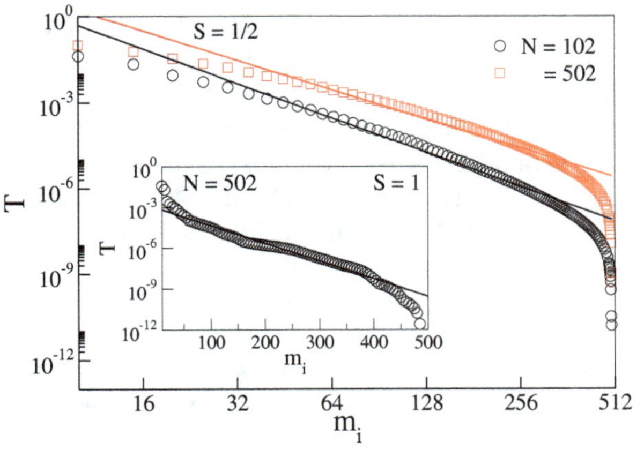

Figure 2: Truncation error T of the density matrix for spin-1/2 chain (main). The inset shows the truncation error for the $S = 1$ chain. For $S = 1/2$, the truncation error follows power law decay whereas it follows exponential decay for a $S = 1$ system.

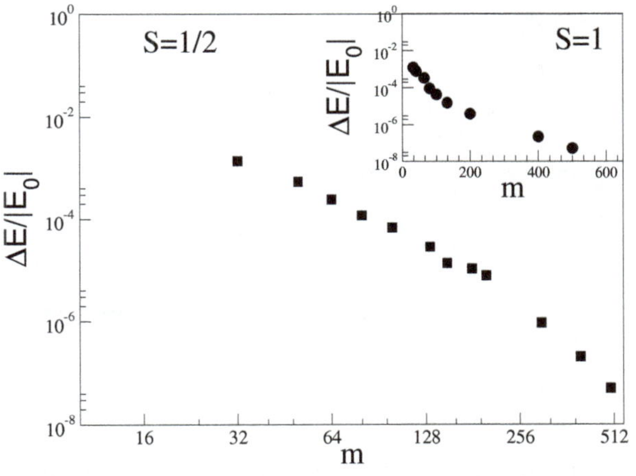

Figure 3: Energy accuracy $\Delta E/|E_0|$ for spin half chain with PBC (main) which shows a power law behavior with m. The inset shows the energy accuracy for spin one chain with PBC which shows exponential behavior with m.

The relative ground state energy can be defined as $\Delta E/|E_0|$, where $\Delta E = E(m) - E_0$ with E_0 is the most accurate value for $S = 1$ chain [19] and $E_0 = E(m = 1200)$ for $S = 1/2$ chain.

As discussed earlier, the DMRG is based on the systematic truncation of the irrelevant degrees of freedom and the eigenvalues of the density matrix represent the importance of the respective states. In the DMRG, only selected states corresponding to the highest eigenvalues are kept and the rest of the other states are thrown. We define truncation error T as

$$T = 1 - \sum_{i=1}^{m} \lambda_i \qquad (9)$$

where λ_i are eigenvalues of density matrix of the system block arranged in descending order. In the main Fig. 2, T is shown as function of m for $S = 1/2$ and the inset shows the same for a $S = 1$ system with N=102 and 502. We notice that $m \sim 350$ for $S = 1/2$ and $m \sim 300$ for $S = 1$ are sufficient to achieve $T = 10^{-9}$. In the main Fig. 2, T vs m in a log-log plot shows a linear behavior, i.e., T for both system sizes of $N = 102$ and 502 for $S = 1/2$ follows a power law $T \propto m_i^{\alpha}$ with $\alpha = 4.0$ and 3.4, respectively. The m dependence of T for $S = 1$ ring is shown in the inset of Fig. 2. The truncation error T in this case decays exponentially, i.e., $T \propto \exp(-\beta m_i)$ with $\beta = 0.03$ for both $N = 102$ and 502.

The relative error in energies $\Delta E/|E_0|$ for $S = 1/2$ and 1 with $N = 102$ spins are shown in Fig. 3, main and inset respectively. The exact energies of a spin-1 system is $E_0/N \sim 1.4014840386$ and this value is obtained by using conventional DMRG with $m = 2000$ and up to $N = 100$ site chain [19]. Extrapolation of energies with m is done to obtain the above value [19]. We notice that $\frac{\Delta E}{E_0}$ for $S = 1/2$ systems goes to 10^{-6} for $m = 256$ whereas it goes to 10^{-8} for $m \approx 500$, as shown in the main Fig. 3. Although similar accuracy of the energy can be achieved with $m = 100$ in MPS approach, the scaling is $\sim m^4$, rather than $\sim m^3$ in our algorithm. For $S = 1$ accuracy of 10^{-8} can be reached with $m \sim 450$, as shown in the inset of Fig. 3.

The dependence of accuracy of correlation function $|C(r)|$ of $S = 1/2$ for $N = 130$ as a function of m is shown in the main Fig. 4. We notice that $m = 256$ is sufficient to achieve an accuracy of $\sim 10^{-4}$. We also notice that $|C(r)| \propto r^{-1}$ with the well known logarithmic correction specially for smaller r [26]. We have fitted the correlations with the well known form $|C(r)| = Ar^{-1}\ln^{1/2}(r/r_0)$ with $A = 0.22$ and $r_0 = 0.08$ [29]. Deviation in function for large r is the artifact of finite sys-

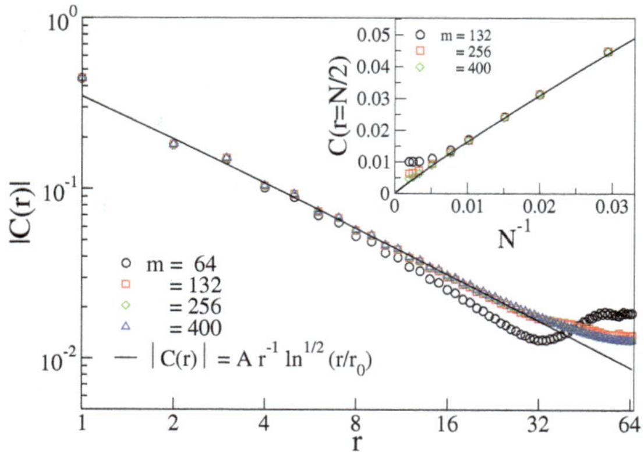

Figure 4: The main figure shows the variation of $|C(r)|$ (defined in Eq. (8)) as a function of r for different m. The solid curve in the main figure is the logarithmic correction formula: $Ar^{-1}\ln^{1/2}(r/r_0)$ with $A = 0.22$ and $r_0 = 0.08$. [26, 29] The inset shows the variation of $C(N/2)$ with the inverse of N for different m. The solid curve is the form: $A(N/2)^{-1}\ln^{1/2}(N/2r_0)$ with $A = 0.323$ and $r_0 = 0.08$.

tems. In our algorithm, two sites are added symmetrically and the new sites are added $N/2$ sites apart, consequently the distance between two new sites is $N/2$. Therefore, the new-new sites correlation function is $C(N/2)$ and is plotted with N^{-1} in the inset of Fig 4. We observe that $m = 256$ is sufficient for $N \sim 200$ to achieve sufficient numerical accuracy. The curve behaves almost linearly with the logarithmic correction: $|C(N/2)| = 2AN^{-1}\ln^{1/2}(N/2r_0)$ with $A = 0.323$ and $r_0 = 0.08$.

V. Summary

The DMRG is a state-of-the-art numerical technique for solving the 1D quantum many-body systems with open boundary condition. However, the accuracy of the 1D PBC system is rather poor. The MPS approach gives very accurate results but the computational cost goes as (o) m^5 [23]. Later, this algorithm was modified and the computational cost of the modified algorithm goes as (o) $p \times m^3$, where p in general varies linearly with m [28], but p can go as m^2 in case of long range order in the system. The computational cost of the algorithm presented in this paper scales as (O) m^3 because of the sparse superblock Hamiltonian and is very similar to the conventional DMRG. To achieve this goal, we avoid the multiple times renormalization of the operators which are used to construct the superblock. This algorithm can readily be used to solve general 1D and quasi-1D systems, e.g., J_1–J_2 model, two-legged ladder. The new algorithm can be implemented with ease using the conventional DMRG code.

Our calculation suggests that most of the quantities, e.g., ground state energies, energy gaps and correlation function, can accurately be calculated by keeping $m \sim 400$. The superfluity stiffness [27] and dynamical structure factors using the correction vector technique [24] or continued fraction method [25] can be calculated with this algorithm. The symmetries, e.g., spin parity, electron-hole, inversion, can easily be implemented in this algorithm [24]. This algorithm is used by us in calculating accurate static structure factors and correlation function for $J_1 - J_2$ model for a spin-1/2 ring geometry [17].

Acknowledgements - We thank Z. G. Soos for his valuable comments. M. K. thanks DST for Ramanujan fellowship and computation facility provided under the DST project SNB/MK/14-15/137.

[1] M Tinkham, *Introduction to Superconductivity: Second Edition* (Dover Books on Physics) (Vol I), Dover Publications INC, New York (2004).

[2] A V Chubukov, *Chiral, nematic, and dimer states in quantum spin chains*, Phys. Rev. B **44**, 4693 (1991).

[3] L Kecke, T Momoi, A Furusaki, *Multimagnon bound states in the frustrated ferromagnetic one-dimensional chain*, Phys. Rev. B **76**, 060407 (2007).

[4] P W Anderson, *The resonating valence bond state in la2cuo4 and superconductivity*, Science **235**, 1196 (1987).

[5] A Parvej, M Kumar, *Degeneracies and exotic phases in an isotropic frustrated spin-1/2 chain*, J. Magn. Magn. Mater. **401**, 96 (2016).

[6] X L Qi, S C Zhang, *Topological insulators and superconductors*, Rev. Mod. Phys. **83**, 1057 (2011).

[7] M Nakamura, *Tricritical behavior in the extended hubbard chains*, Phys. Rev. B **61**, 16377 (2000).

[8] P Sengupta, A W Sandvik, D K Campbell, *Bond-order-wave phase and quantum phase transitions in the one-dimensional extended hubbard model*, Phys. Rev. B **65**, 155113 (2002).

[9] M Kumar, S Ramasesha, Z G Soos, *Tuning the bond-order wave phase in the half-filled extended hubbard model*, Phys. Rev. B **79**, 035102 (2009).

[10] M Suzuki (Ed.), *Quantum Monte Carlo Methods in Condensed Matter Physics*, World Scientific, Singapore (1993).

[11] W Kohn, L J Sham, *Self-consistent equations including exchange and correlation effects*, Phys. Rev. **140**, A1133 (1965).

[12] K G Wilson, *The renormalization group and critical phenomena*, Rev. Mod. Phys. **55**, 583 (1983).

[13] S R White, *Density matrix formulation for quantum renormalization groups*, Phys. Rev. Lett. **69**, 2863 (1992).

[14] S R White, *Density-matrix algorithms for quantum renormalization groups*, Phys. Rev. B **48**, 10345 (1993).

[15] U Schollwöck, *The density-matrix renormalization group*, Rev. Mod. Phys. **77**, 259 (2005).

[16] K A Hallberg, *New trends in density matrix renormalization*, Adv. Phys. **55**, 477 (2006).

[17] Z G Soos, A Parvej, M Kumar, *Numerical study of incommensurate and decoupled phases of spin-1/2 chains with isotropic exchange J_1 , J_2 between first and second neighbors*, J. Phys. Condens. Mat. **28**, 175603 (2016).

[18] U Schollwöck, *The density-matrix renormalization group in the age of matrix product states*, Ann. Phys. **326**, 96 (2011).

[19] P Pippan, S R White, H G Evertz, *Efficient matrix-product state method for periodic boundary conditions*, Phys. Rev. B **81**, 081103 (2011).

[20] E Dagotto, T M Rice, *Surprises on the way from one- to two-dimensional quantum magnets: The ladder materials*, Science **271**, 618 (1996).

[21] S Östlund, S Rommer, *Thermodynamic limit of density matrix renormalization*, Phys. Rev. Lett. **75**, 3537 (1995).

[22] F Verstraete, D Porras, J I Cirac, *Density matrix renormalization group and periodic boundary conditions: A quantum information perspective*, Phys. Rev. Lett. **93**, 227205 (2004).

[23] F Verstraete, J I Cirac, J I Latorre, E Rico, M M Wolf, *Renormalization-group transformations on quantum states*, Phys. Rev. Lett. **94**, 140601 (2005).

[24] S K Pati, S Ramasesha, Z Shuai, J L Brédas, *Dynamical nonlinear optical coefficients from the symmetrized density-matrix renormalization-group method*, Phys. Rev. B **59**, 14827 (1999).

[25] K A Hallberg, *Density-matrix algorithm for the calculation of dynamical properties of low-dimensional systems*, Phys. Rev. B **52**, R9827 (1995).

[26] I Affleck, D Gepner, H J Schulz, T Ziman, *Critical behaviour of spin-s Heisenberg antiferromagnetic chains: analytic and numerical results*, J. Phys. A - Math. Gen. **22**, 511 (1989).

[27] D Rossini, V Giovannetti, R Fazio, *Spin-supersolid phase in Heisenberg chains: A characterization via matrix product states with periodic boundary conditions*, Phys. Rev. B **83**, 140411 (2011).

[28] D Rossini, V Giovannetti, R Fazio, *Stiffness in 1D matrix product states with periodic boundary conditions*, J. Stat. Mech. **2011**, P05021 (2011).

[29] A W Sandvik, *Computational Studies of Quantum Spin Systems*, AIP Conf. Proc. **1297**, 135 (2010).

11

Influence of bottleneck lengths and position on simulated pedestrian egress

D. R. Parisi,[1,2*] G. A. Patterson[1†]

In this paper, the problem of pedestrian egress under different geometries is studied by means of two numerical models. The length of the bottleneck after the exit and the distance of the exit to the lateral wall of a squared room are investigated. Both models show that an increase in the bottleneck length increases the evacuation time by more than 20%, for any exit position. Hence, a bottleneck length tending to zero is the best choice. On the contrary, the results of moving the exit closer to the lateral wall are different in both models and, thus, its convenience cannot be stated. To unveil whether this layout modification is favorable, experimental data are required. Moreover, the discrepancy between models indicates that they should be validated considering several scenarios.

I. Introduction

The characterization of pedestrian flow through doors and bottlenecks is a key feature for designing and dimensioning pedestrian facilities. These observables were extensively studied with field observations, experiments and simulations. However, a systematic study of the influence of bottleneck length and position is missing. Besides, the terms "door" and "bottleneck" are used sometimes as synonyms and should be better specified.

As a starting point, we will define a bottleneck. Considering the flow of pedestrians with a preferred direction of motion (for instance, in an open space, a room or a corridor), we will call "bottleneck" to any geometrical reduction that impedes the pedestrians' flow and that it is orientated in the perpendicular direction of this flow, as shown in Fig.1.

The geometry of a simple bottleneck can be characterized by its width (b) and its length (L_B). From this definition, it is natural to see that a door (or an opening) is a particular case of a bottleneck whose length tends to zero ($L_B \rightarrow 0$).

The pedestrian flow rate, J, is defined as the number of persons ΔN going out during a certain time lapse Δt. For the egress from a facility where N pedestrians are initially inside, the time lapse considered could be the evacuation time (T_E). A derived quantity, the specific flow rate J_s, is defined as the flow per unit length of the perpendicular cross section of the way out, $J_s = J/b = \frac{\Delta N}{b \cdot \Delta t}$.

Legal codes and engineering tables provide values for the specific flow rate, usually ranging between $J_s = 1.1$ $(\mathrm{m} \cdot \mathrm{s})^{-1}$ and $J_s = 1.8$ $(\mathrm{m} \cdot \mathrm{s})^{-1}$ (see, for instance, [1–5]). Usually, these specifications only consider the width of the exit, but do not specify L_B, which is assumed to be as small as possible.

*E-mail: dparisi@itba.edu.ar
†E-mail: gpatters@itba.edu.ar

[1] Instituto Tecnológico de Buenos Aires, Lavardén 389, C1437FBG Ciudad Autónoma de Buenos Aires, Argentina.

[2] Consejo Nacional de Investigaciones Científicas y Técnicas, Godoy Cruz 2290, 1425 Ciudad Autónoma de Buenos Aires, Argentina.

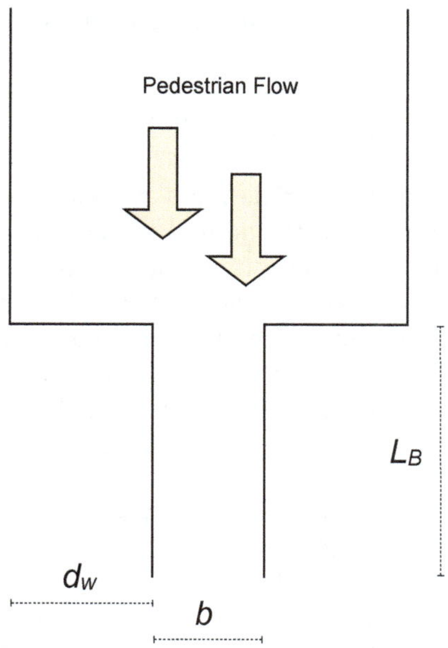

Figure 1: Schematic representation of a bottleneck for a pedestrian flow. The main geometrical quantities are shown: the bottleneck length (L_B); the bottleneck narrow width (b); and the distance to the nearest lateral wall (d_w).

In contrast, some experimental papers specify the value of the bottleneck length for which the flow rate is obtained. Hoogendoorn and Daamen [6] studied the influence of the bottleneck width (b) on the pedestrian flow and found values ranging from $J_s = 0.8$ $(\mathrm{m}\cdot\mathrm{s})^{-1}$ for $b = 1$ m to $J_s = 1.4$ $(\mathrm{m}\cdot\mathrm{s})^{-1}$ for $b = 2.25$ m. In all cases, the bottleneck length was kept constant at some value $L_B \geq 5$ m.

Kretz et al. [7] reported specific flows ranging between 1.5 and 2.2 $(\mathrm{m}\cdot\mathrm{s})^{-1}$ for $N = 100$ persons exiting through doors of widths between 0.40 and 1.40 m, keeping the value of the bottleneck length at $L_B = 0.4$ m.

Seyfried and collaborators [8] found $J_s = 1.6$ $(\mathrm{m}\cdot\mathrm{s})^{-1}$ for $b = 0.8$ m and $J_s = 1.97$ $(\mathrm{m}\cdot\mathrm{s})^{-1}$ for $b = 1.2$ m, both of them for $N = 60$ pedestrians with $L_B = 2.8$ m.

In the work of Liao et al. [9], the flow for different bottleneck widths (b) was also studied. They remarked that the flow rate grows linearly with b (thus, J_s is constant). In that paper, the specific flow $J_s = 2.3$ $(\mathrm{m}\cdot\mathrm{s})^{-1}$ was computed considering the total number of pedestrians and the total evacuation time. Alternatively, the steady-state specific flow was calculated only during an intermediate period of time in which the flow was nearly steady, finding $J_s = 2.5$ $(\mathrm{m}\cdot\mathrm{s})^{-1}$. Furthermore, the bottleneck length was kept constant at $L_B = 1$ m.

In the above-referenced papers, the experiments were done by varying the width b and reporting the length (L_B), which was kept constant. However, there are few papers where the influence of the bottleneck length is studied.

Liao et al. [10] presented experimental results for $L_B = 0.1$; 1.0; and 4.0 m, obtaining $J_s = 2.0$; 2.5; and 1.8 $(\mathrm{m}\cdot\mathrm{s})^{-1}$, respectively, and for steady-state conditions, i.e., measured only for the stationary part of the evacuation.

The first experiments that considered different bottleneck lengths keeping the other parameters constant were the ones presented in Refs. [11] and [12]. Here, the results of three evacuation drills with $N = 180$ soldiers, $b = 1.2$ m and $L_B = 0.06; 2.00; 4.00$ m were reported. The corresponding values of the pedestrian flow rate J were $J = 3.05$; 2.54; 2.56 s^{-1}. The flow rate, in this case, was calculated from the pedestrian 1 to 120 (non-stationary state). From these data, it can be seen that a small bottleneck length leads to an improvement of about 20% with respect to large bottlenecks and that, moreover, it seems to be the same for $L_B = 2$ m and $L_B = 4$ m.

The improvement of the flow for a short bottleneck length is expected because doors do not impose any restriction on the movement once the door is passed, generating a reduction of the density and thus gaining speed. On the other hand, bottlenecks reduce the available space, increasing the density and reducing the velocity, which could result in some limitations to the motion upstream the constriction.

In order to confirm these results and get a deeper understanding of how L_B influences the evacuation time, in the present paper we propose studying this variation systematically by means of numerical simulations. Moreover, we point out that another geometrical parameter can influence the evacuation time, that is, the distance of the bottleneck from its closest wall (d_w), as shown in Fig. 1.

In this regard, we can refer to the work of Nagai et al. [13] who performed evacuation drills in a configuration where the door was located in the corner of a classroom. In this experiment, the boundaries were built with tables and thus, the shoulders of the people were above these boundaries (walls and bottleneck). However, this study does not compare the evacuation performance for different positions of the door.

As first empirical evidence, preliminary experiments of mice egress under stressed conditions suggest that proximity of the door to the wall improves the evacuation process [14].

Moreover, two papers have compared the pedestrian flow at two different positions of the door: at the center and at the corner of the room, by means of simulations with cellular automata (C.A.) models, in particular the floor field model [15,16]. They found that the exits at the corner produce a larger flow than the ones at the center because, in this C.A. model, conflicts are reduced.

On the contrary, Wu et al. [17] presented results of evacuations that were simulated with a C.A.: model based on game theory by using preferential direction, and found that the best location for the door is far from the walls.

Considering that how the door position influences the evacuation time is an open question, we are also going to study the evacuation performance as a function of the distance between the bottleneck and the wall (d_w).

II. Models of pedestrian dynamics

i. Social force model with a respect area

The physical model implemented is the one described in Ref. [18], which is a modification of the social force model (SFM) [19]. This modification allows a better approximation to the fundamental diagram of Ref. [1], commonly used in the design of pedestrian facilities.

The SFM is a continuous-space and force-based model that describes the dynamics considering the forces exerted over each particle (p_i). The Newton equation for each particle reads

$$m_i \mathbf{a}_i = \mathbf{F}_{Di} + \mathbf{F}_{Si} + \mathbf{F}_{Ci}, \tag{1}$$

where $\mathbf{a}_i$ is the acceleration of particle p_i. The equations are solved using standard molecular dynamics techniques. The three forces are: 'Driving Force' ($\mathbf{F}_{Di}$), 'Social Force' ($\mathbf{F}_{Si}$) and 'Contact Force'($\mathbf{F}_{Ci}$). The corresponding expressions are as follows:

$$\mathbf{F}_{Di} = m_i \, \frac{(v_{di} \, \mathbf{e}_i - \mathbf{v}_i)}{\tau}, \tag{2}$$

where m_i is the particle mass, $\mathbf{v}_i$ and v_{di} are the actual velocity and the desired velocity magnitude, respectively, $\mathbf{e}_i$ is the unit vector pointing to the desired target (particles inside the corridors or rooms have their targets located at the closest position over the line of the exit door), τ is a constant related to the time needed for the particle to achieve v_d.

$$\mathbf{F}_{Si} = \sum_{j=1, j\neq i}^{N} A \, \exp\left(\frac{-\epsilon_{ij}}{B}\right) \mathbf{e}_{ij}^{n}, \tag{3}$$

where N is the total number of pedestrians in the system, A and B are constants that determine the strength and range of the social interaction, $\mathbf{e}_{ij}^{n}$ is the unit vector pointing from particle p_j to p_i; this direction is the 'normal' direction between two particles, and ϵ_{ij} is defined as

$$\epsilon_{ij} = d_{ij} - (r_i + r_j), \tag{4}$$

where d_{ij} is the distance between the centers of p_i and p_j, and r is their corresponding particle radius.

$$\mathbf{F}_{Ci} = \tag{5}$$
$$\sum_{j=1, j\neq i}^{N} \left[(-\epsilon_{ij} \, k_n) \, \mathbf{e}_{ij}^{n} + (v_{ij}^{t} \, \epsilon_{ij} \, k_t) \, \mathbf{e}_{ij}^{t}\right] \, g(\epsilon_{ij}),$$

where the tangential unit vector ($\mathbf{e}_{ij}^{t}$) indicates the perpendicular direction of $\mathbf{e}_{ij}^{n}$, k_n and k_t are the normal and tangential elastic restorative constants, v_{ij}^{t} is the tangential projection of the relative velocity seen from $p_j(\mathbf{v}_{ij} = \mathbf{v}_i - \mathbf{v}_j)$, and the function $g(\epsilon_{ij})$ is: $g = 1$ if $\epsilon_{ij} < 0$ (if particles overlap) or $g = 0$ otherwise.

Because this version of the SFM does not provide any self-stopping mechanism for the particles, it cannot reproduce the fundamental diagram of pedestrian traffic as shown in Ref. [18]. Consequently, the modification consists on providing virtual pedestrians with a way to stop pushing other pedestrians. This is achieved by incorporating a circular respect area close to and ahead of the particle (p_i). The center of this circular area is located at the *respect distance* (D_{Ri}) from the center of the particle p_i, also having D_{Ri} as its radius, and in the direction of the desired velocity. The respect distance is parametrized for each particle depending on its own radius, $D_{Ri} = R_F \times r_i$, R_F being the *respect factor*. While any other pedestrian is inside this respect area, the desired velocity of pedestrians (p_i) is set equal to zero ($v_{di} = 0$). For further details and benefits of this modification to the SFM, we refer the reader to Ref. [18].

ii. Contractile particle model

This model was proposed by Baglietto and Parisi [20] and it is useful for modeling pedestrian flows in normal conditions. It consists of a set of rules defining automata particles that can move on the real plane. It has the benefit of being computationally faster than force-based models and it can reproduce very well specific flow rates and fundamental diagrams of different experiments reported in the bibliography by only changing the parameters related to the particle radii.

The model consists of particles of variable radii. Each radius can range between r_{min} and r_{max} in a continuous way. The minimum radii are associated with the maximum physical compressibility of the pedestrian and the maximum radii are related to the necessary personal space needed for taking normal steps under low density conditions. The particle has a desired velocity ($\mathbf{v}_{di}$) pointing at the desired target location and its magnitude (v_{di}) is a function of the particle radius (r_i). When a particle (p_i) enters in contact with a boundary, obstacle or another particle, an escape velocity ($\mathbf{v}_{ei}$) appears, having a fixed magnitude (v_e) and opposite direction to the interaction. Then, the particle radius instantaneously collapses to r_{min}.

The relation between the radius and the desired speed must fulfill $v_d(r_{min}) = 0$ and $v_d(r_{max}) = v_{dmax}$ with a functional form given by Eq. (6):

$$v_d = v_{dmax}\left[\frac{(r - r_{min})}{(r_{max} - r_{min})}\right]^{\beta}, \tag{6}$$

where v_{dmax} is the desired speed at which a pedestrian would walk in an open and free space and β is a constant.

When a particle is free of contact with any other particle or boundary, its radius increases in each time step according to

$$\Delta r = \frac{r_{max}}{(\frac{\tau}{\Delta t})}, \tag{7}$$

where the constant τ is set to $\tau = 0.5\ s$ and Δt is the time step for computing the evolution of the system. Once r reaches r_{max}, it stops increasing.

Finally, in each iteration, the positions of particles are updated by the equation

$$\mathbf{x}^i(t + \Delta t) = \mathbf{x}^i(t) + \mathbf{v}^i \Delta t. \tag{8}$$

More details of this model can be found in Ref. [20]

III. Simulation Results

We present the results of simulations with two different models analyzing variations of the geometrical parameters L_B and d_w (see Fig. 1) and looking at the total evacuation time and the specific flow patterns inside the room.

In the subsections i., ii. and iii., the exit width (b) is fixed at $b = 1.2$ m, the initial number of pedestrians is $N = 200$, which are uniformly distributed inside a squared room of 20×20 m^2; this configuration generates an initial density $\rho = 0.5$ particles/m^2.

The pedestrians' behavior corresponds to normal conditions (competitive behavior with pushing and shoving are excluded in the present paper). The simulated pedestrian's plan simply consists of moving toward the nearest point of a segment placed centered and parallel to the exit line (with a size 0.2 m shorter than the exit width), and then toward a second target placed at a long distance location beyond the end of the bottleneck.

In Fig. 2, the geometry and typical configurations of simulated pedestrians are shown for the two models used.

i. Bottleneck length

First, we consider simulations with the social force model with a respect area. The parameters used are uniformly distributed in the following ranges: pedestrian mass $m \in [70\ \text{kg},\ 90\ \text{kg}]$; shoulder width $2 \times r \in [48\ \text{cm}, 56\ \text{cm}]$; desired velocity $v_d \in [1.15\ \text{m/s}, 1.45\ \text{m/s}]$. The constant parameters are: $\tau = 0.5$ s, $A = 2000$ N, $B = 0.08$ m,

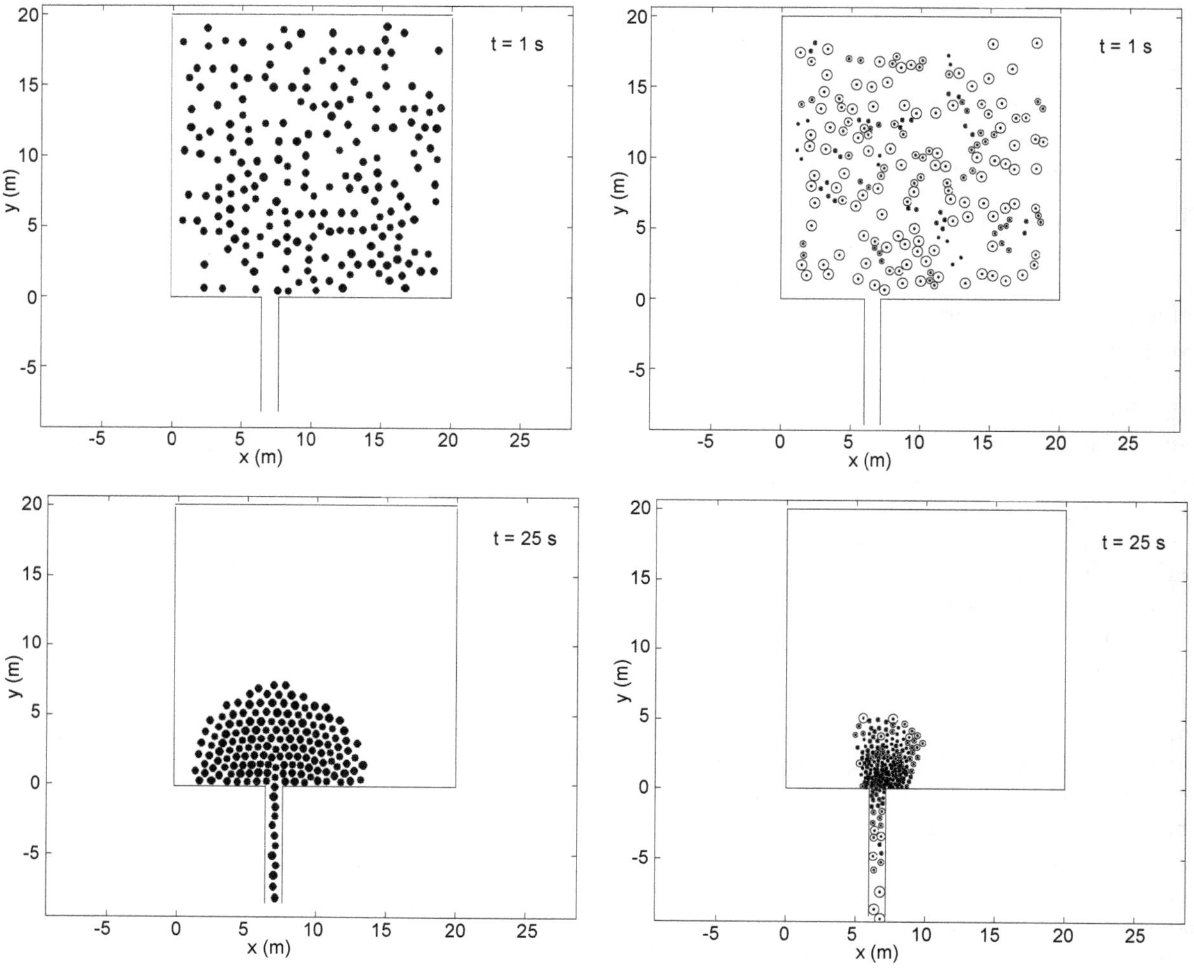

Figure 2: Snapshots of simulations. The left column displays two configurations of the system at different times for the simulations performed with the social force model with respect area. For the same times, the corresponding configurations of the system simulated with the contractile particle model are shown on the right column.

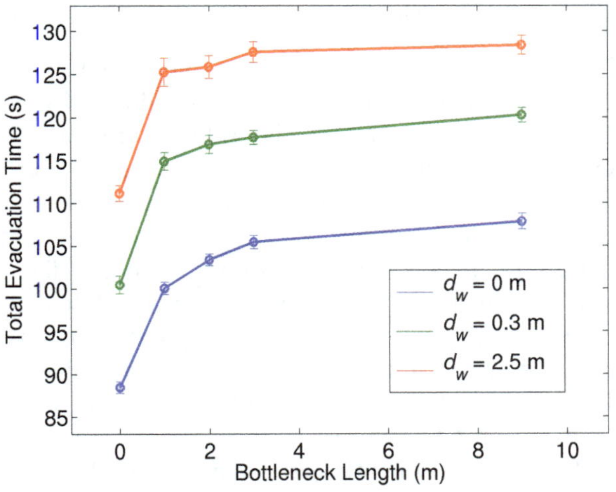

Figure 3: Mean total evacuation times as a function of bottleneck length for the social force model with a respect area. Error bars indicate one standard deviation.

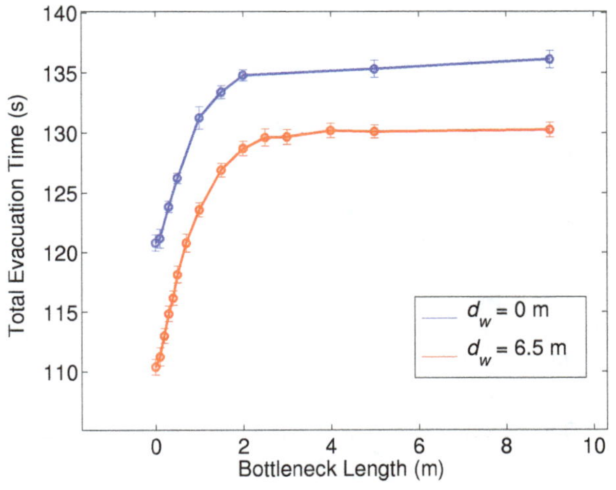

Figure 4: Mean total evacuation times as a function of bottleneck length for the contractile particle model. Error bars indicate one standard deviation.

$k_n = 1.2\ 10^5$ N/m, $k_t = 2.4\ 10^5$ kg/m/s, and $R_F = 0.7$.

It is worth noting that because we are simulating normal conditions, the low values of the desired velocity (and the respect mechanism) lead to the absence of contact interaction [Eq. (5)] as can be seen in Fig. 2, bottom-left panel. This may suggest that under the present conditions the contact term of the SFM-respect has no influence on the results.

The total evacuation times as a function of bottleneck length are displayed in Fig. 3 for three different values of d_w. Each condition (each point in the figure) was simulated ten times with initial positions uniformly distributed inside the room ($\rho = 0.5$ particles/m^2).

As expected, the evacuation time increases with increasing values of L_B, confirming that imposing long boundaries after the exit has a negative influence on the evacuation performance. This influence grows rapidly over the first 1 or 2 meters and reaches an asymptotic value of about 20% higher evacuation time with respect to the one of a clean exit. Both facts are in agreement with experimental studies [11,12] with a similar number of pedestrians ($N = 180$) and the same exit width ($b = 1.2$ m). In particular for this model, the results for different d_w reveal that the proximity of the exit to the wall is favorable.

Now, we consider the simulation results of the contractile particle model (CPM). The model parameters used are: $r_{min} = 0.10$ m; $r_{max} = 0.37$ m; $\Delta t = 0.05$ s; $v_{dmax} = 0.95$ m/s; and $\beta = 0.9$. Figure 4 shows the total evacuation time as a function of the bottleneck length. The same tendency as in the previous model can be observed. It presents two main facts: (a) increasing evacuation time for increasing L_B, and (b) the asymptotic upper bound is reached at similar values of L_B. Furthermore, for the reference scenario of the exit far from the lateral wall, a similar percentage increase of the evacuation times is observed. However, the contractile particle model shows a reversed trend when the exit is near the lateral wall, increasing the evacuation time. This issue will be further discussed in next sections.

ii. Distance of the exit from the lateral wall

In this section, we present the results of mean evacuation times as a function of the distance of the exit from the nearest wall (d_w). Figure 5 shows the related information for three different values of L_B. In all cases, it can be observed that it is favorable to bring the exit near the sidewall. Also for this

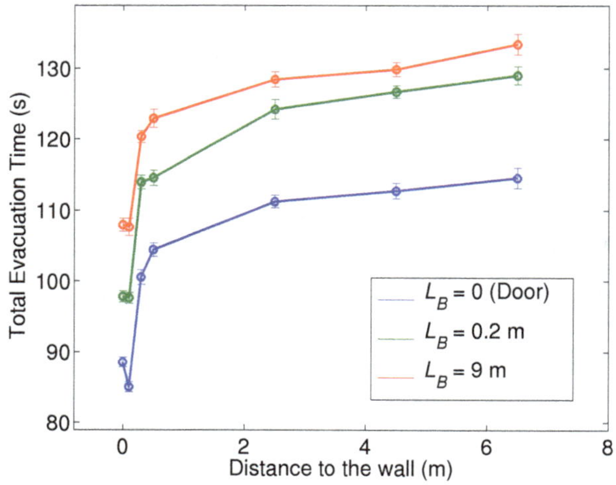

Figure 5: Mean total evacuation times as a function of the distance of the exit from the closest lateral wall for the social force model with a respect area. Error bars indicate one standard deviation.

geometrical modification, the maximum reduction of the evacuation time is roughly 20%.

It must be noted that the maximum evacuation performance does not occur at $d_w = 0$ m but at a slightly greater value, in the case of the door ($L_B = 0$ m), at $d_w \sim 0.1$ m. This improvement in evacuation time for a location of the exit near the sidewalls is in accordance with mice experiments under competitive conditions [14].

On the contrary, the contractile particle model displays a slightly higher value of the evacuation time for the exit located near the wall, as indicated by the two curves in Fig. 4, which in the worst case (the door: $d_w = 0$) is only 10%. The tendency to worsening the evacuation performance for lower d_w is in accordance with another cellular automaton model [17]; however, in this case, the value is much higher, i.e., an increase of about 70% of the evacuation time.

It is worth emphasizing that, although both models studied here (SFM-respect and CPM) were validated by reproducing experimental flow rates at the exit for a geometry with a central door, when the geometry is changed the responses of the models are different. Thus, this standard way of validating pedestrian models may not be enough.

iii. Specific Flow Maps

In the present section, we will get some insight into the different evacuation times when different distances of the exit from the lateral wall are considered. To this end, specific flow maps will be computed and, again, we compare results using the two models described above.

Given a velocity field $v(x, y)$ and a density field $\rho(x, y)$, the specific flow field can be calculated as $J_s(x, y) = v(x, y)\ \rho(x, y)$ at any point (x, y) inside the area of interest. This can be done at any time step, and then the average specific flow map $\overline{J_s}(x, y)$ can be obtained.

This calculation was performed using the tool *JPSreport*, which is the analysis module of *JuPedSim* developed at the Forschungszentrum Jülich in Germany [21, 22]. This tool implements an established protocol for calculating the density using Voronoi cells as described in Ref. [23].

We considered data with a frequency of one frame per second. Averages were computed between 30% and 70% of the total evacuation time for each run. Furthermore, the spatial grid (x_m, y_n), over which the different fields were computed, was a square grid of side 0.2 m.

In Fig. 6, the specific flow maps for both models are presented for the case of doors ($L_B = 0$).

The specific flow patterns reveal important differences between the examined models.

The upper panels, corresponding to the social force model with a respect area, show that the specific flow is greater along the walls. Thus, a wall orientated in the normal direction of the door will increase the output flow and hence, the evacuation time will be shorter.

On the other hand, the three lower panels show that the walls do not lead to any improvement in the flow. Instead, the higher values of J_s occur away from the walls, at the middle of the cluster of simulated pedestrians. So, having a wall perpendicular to and near the exit reduces the effective flow and increases the evacuation time.

iv. Size of the System

In order to check whether the improvement of the flow rate for shorter bottleneck length holds for larger systems, we also study the egress of $N = 400$ pedestrian from a room of 28.3×28.3 m^2 and

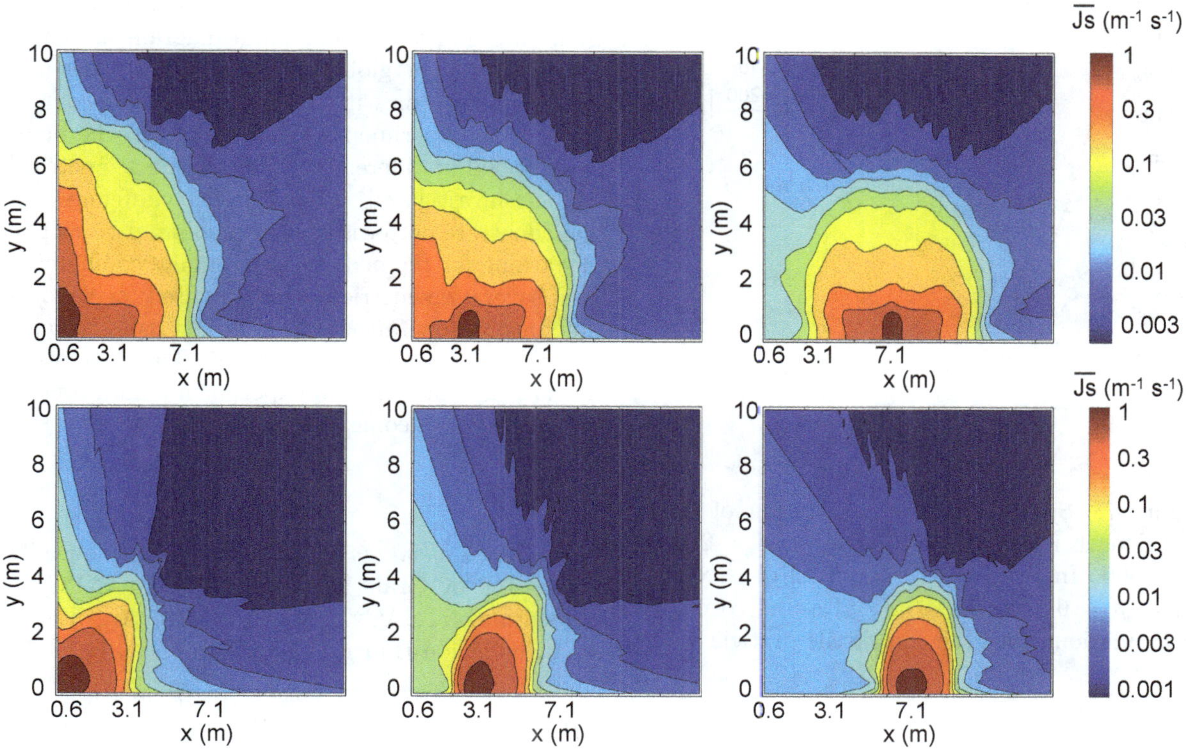

Figure 6: Specific flow maps for both models investigated. The three upper panels (first row) correspond to the social force model with a respect area [18], the three lower panels (second row) correspond to the contractile particle model [20]. Each column corresponds, from left to right, to the door center at 0.6; 3.1; and 7.1 m ($d_w = 0$; 2.5; and 6.5 m). In all cases, $L_B = 0$.

with the same door width ($b = 1.2$ m) centered at $x = 14$ m. The new dimension of the room allows us to keep the same initial density ($\rho = 0.5$ particles/m^2) as in the smaller system studied above. We simulate this larger system by means of the contractile particle model.

Because we want to compare systems with different N, it is more suitable to look at the specific flow that is shown in Fig. 7.

Both curves of specific flow (J_s) as a function of L_B look similar, showing a decrease of J_s as L_B increases, thus confirming the original result presented in section i. For the model used (CPM), the specific flow is a little bit higher in the case of $N = 400$ particles, and the enhancement due to a shorter bottleneck length is more pronounced. In the system of $N = 200$ particles, the percentage difference is about 18% when comparing the maximum and minimum specific flows, but for the larger system ($N = 400$) this difference is 28%.

These results suggest that the door (or very short bottleneck) is the best option for any size of the system. We think this can be explained as it was already outlined in the introduction: a bottleneck reduces the available space after the exit, which produces a higher density that can decrease the flow (exceeding the capacity).

IV. Conclusions

In this paper, we studied the influence of two geometrical factors on the evacuation time: the bot-

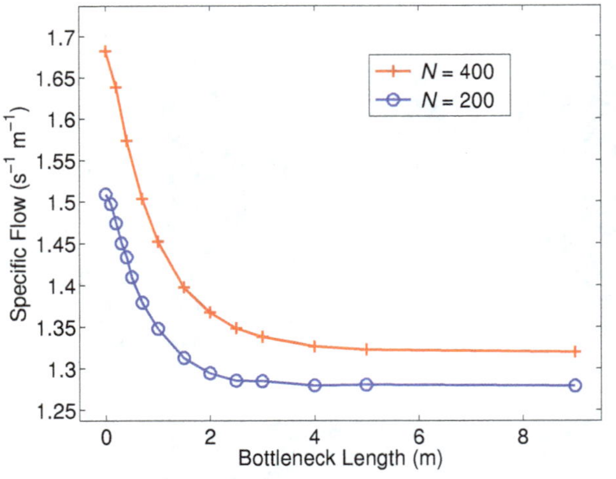

Figure 7: Specific flow as a function of bottleneck length for two system sizes of $N = 200$ and $N = 400$. In both cases, the initial density is the same ($\rho = 0.5$ particles/m^2). Results correspond to simulations with the contractile particle model.

tleneck length and the distance of the exit to the lateral wall in a squared room. Two pedestrian dynamics models were considered: the social force model with a respect area [18] and the contractile particle model [20].

For evacuation at normal desired velocities ($1-1.3$ m/s), both models agree in that the longer the bottleneck length after the exit, the longer the evacuation time and, in particular, a door (which is a special case of bottleneck of length $L_B \to 0$ m) is the best option. These results also agree with experiments in very similar conditions [11, 12]. Furthermore, the convenience of using short bottlenecks seems to be independent of the number of simulated particles.

Regarding the convenience of moving the exit closer to the lateral wall, the two models used in this work do not agree, and pedestrian experiments are not reported in the literature. Furthermore, other models [15–17] provide contradictory results and thus, the optimal door position is an open question. Considering the possibility that placing the door near a sidewall could improve the evacuation performance, it would be worth exploring this geometrical modification experimentally in the future.

It is important to remark that both models used in the present study were validated considering the flow rate and the fundamental diagram in simple geometries. The global flow rate values measured in a configuration with a centered door are comparable with experimental ones. However, the models respond in different ways under the geometrical change investigated, presenting very different specific flow rate patterns inside the room. This can be a warning for the pedestrian dynamics community, and several geometries should be used for validating output flow rates, instead of just using a single door far away from lateral walls. This is another reason for expecting new experimental data considering different geometries.

Acknowledgements - This work was financially supported by Grant PICT2011 - 1238 (ANPCyT, Argentina); ITBACyT2015-33 (ITBA, Argentina); and Fundación Bunge & Born (Argentina).

[1] P J Di Nenno (Ed.), *SFPE Handbook of fire protection engineering*, Society of Fire Protection Engineers and National Fire Protection Association, Quincy, Massachusetts (2002).

[2] J Fruin, *Pedestrian planning and design*. The Metropolitan Association of Urban Designers and Environmental Planners, New York (1971).

[3] IMO Correspondence Group, *Interim guidelines for evacuation analyses for new and existing passenger ships*, Technical Report MSC/Circ.1033, International Maritime Organization (2002).

[4] SGSA, *The green guide guide to safety at sports grounds*, HSMO, London (2008). Available at: http://www.safetyatsportsgrounds.org.uk/sites/default/files/publications/green-guide.pdf

[5] Department of The Environment and The Welsh Office, *The building regulations*, Approved Document B-Section B1, HMSO, London (1992).

[6] S P Hoogendoorn, W Daamen, *Pedestrian behavior at bottlenecks*, Transport. Sci. **39**, 147 (2005).

[7] T Kretz, A Grünebohm, M Schreckenberg, *Experimental study of pedestrian flow through a bottleneck*, J. Stat. Mech. P10014 (2006).

[8] A Seyfried, O Passon, B Steffen, M Boltes, T Rupprecht, W Klingsch, *New insights into pedestrian flow through bottlenecks*, Transport. Sci. **43**, 395 (2009).

[9] W Liao, A Seyfried, J Zhang, M Boltes, X Zheng, Y Zhao, *Experimental study on pedestrian flow through wide bottleneck*, Transport. Res. Procedia **2**, 26 (2014).

[10] W Liao, A Tordeux, A Seyfried, M Chraibi, K Drzycimski, X Zheng, Y Zhao, *Measuring the steady state of pedestrian flow in bottleneck experiments*, Physica A **461**, 248 (2016).

[11] J Liddle, A Seyfried, W Klingsch, T Rupprecht, A Schadschneider, A Winkens, *An experimental study of pedestrian congestions: Influence of bottleneck width and length*, arXiv:0911.4350 (2009).

[12] T Rupprecht, W Klingsch, A Seyfried, *Influence of geometry parameters on pedestrian flow through bottleneck*, In: Pedestrian and Evacuation Dynamics, Eds. R D Peacock, E D Kuligowski, J D Averill, Pag. 71, Springer US (2011).

[13] R Nagai, M Fukamachi, T Nagatani, *Evacuation of crawlers and walkers from corridor through an exit*, Physica A **367**, 449 (2006).

[14] P Lin, J Ma, Personal communication (2016).

[15] D Yanagisawa, K Nishinari, *Mean-field theory for pedestrian outflow through an exit*, Phys. Rev. E **76**, 061117 (2007).

[16] T Ezaki, D Yanagisawa, K Nishinari, *Pedestrian flow through multiple bottlenecks*, Phys. Rev. E **86**, 026118 (2012).

[17] J Wu, X Wang, J Chen, G Shu, Y Li, *The position of a door can significantly impact on pedestrians evacuation time in an emergency*, Appl. Math. Comput. **258**, 29 (2015).

[18] D R Parisi, B M Gilman, H Moldovan, *A modification of the social force model can reproduce experimental data of pedestrian flows in normal conditions*, Physica A **388**, 3600 (2009).

[19] D Helbing, I Farkas, T Vicsek, *Simulating dynamical features of escape panic*, Nature **407**, 487 (2000).

[20] G Baglietto, D R Parisi, *Continuous-space automaton model for pedestrian dynamics*, Phys. Rev. E. **83**, 056117 (2011).

[21] http://www.fz-juelich.de/ias/jsc/EN/Research/ModellingSimulation/CivilSecurityTraffic/PedestrianDynamics/Activities/JuPedSim/jupedsimNode.html

[22] https://github.com/JuPedSim/JuPedSim

[23] B Steffen, A Seyfried, *Methods for measuring pedestrian density, flow, speed and direction with minimal scatter*, Physica A **389**, 1902 (2010).

12

Green's functions technique for calculating the emission spectrum in a quantum dot-cavity system

Edgar A. Gómez,[1*] J. D. Hernández-Rivero,[2] Herbert Vinck-Posada[3]

We introduce the Green's functions technique as an alternative theory to the quantum regression theorem formalism for calculating the two-time correlation functions in open quantum systems at the steady state. In order to investigate the potential of this theoretical approach, we consider a dissipative system composed of a single quantum dot inside a semiconductor cavity and the emission spectrum is computed due to the quantum dot as well as the cavity. We propose an algorithm based on the Green's functions technique for computing the emission spectrum that can easily be adapted to more complex open quantum systems. We found that the numerical results based on the Green's functions technique are in perfect agreement with the quantum regression theorem formalism. Moreover, it allows overcoming the inherent theoretical difficulties associated with the direct application of the quantum regression theorem in open quantum systems.

I. Introduction

The measurement and control of light produced by quantum systems have been the focus of interest of the cavity quantum electrodynamics [1, 2]. Specially, the emission of light powered by solid-state devices coupled to nanocavities is an extensive area of research due to its promising technological applications, such as infrared and low-threshold lasers [3, 4], single and entangled photon sources [5, 6], as well as various applications in quantum cryptography [7] and quantum information theory [8]. Experiments with semiconductor quantum dots (QDs) embedded in microcavities have revealed a plethora of quantum effects and offer desirable properties for harnessing coherent quantum phenomena at the single photon level. For example, the Purcell enhancement [9], photon antibunching [10], vacuum Rabi splitting [11] and strong light matter coupling [12]. These and many other quantum phenomena are being confirmed experimentally by observing the power spectral density of the light (PSD) emitted by the quantum-dot cavity systems (QD-cavity). Thus, the PSD, or the so-called emission spectrum of the system, becomes the only relevant information that allows to study the properties of the light via measurements of correlation functions, as it is stated by the Wiener-Khintchine theorem [13]. In order to compute the emission spectrum of the QD-cavity systems in the framework of open quantum systems, different approaches have been elaborated from the theoretical point of view. For example, the method of thermodynamic Green's functions has been applied to the determination of the susceptibilities and absorption spectrum of

*E-mail: eagomez@uniquindio.edu.co

[1] Universidad del Quindío, A.A. 2639, 630004 Armenia, Colombia.

[2] Departamento de Física, Universidade Federal de Minas Gerais, A.A. 486, 31270-901 Pampulha, Belo Horizonte, Minas Gerais, Brazil.

[3] Universidad Nacional de Colombia, A.A. 055051, 111321 Bogotá, Colombia.

atomic systems embedded in nanocavities [14], the time-resolved photo-luminescence approach whose application allows to determine the emission spectrum when an additional subsystem is considered, the so-called the photon reservoir [15]. These theoretical approaches are based on several approximations and therefore, they have their own limitations when they are considered in more general scenarios. In consequence, these methods are not used extensively.

Frequently, the emission spectrum of QD-cavity systems is computed through the Quantum Regression Theorem (QRT) [16–18], since it relates the evolution of mean values of observables and the two-time correlation functions. It is worth mentioning that the QRT approach can be difficult to implement in a computer program because computational complexity increases significantly as the number of QDs or modes inside the cavity are being considered; more precisely, the dimensionality associated with the Hilbert space is large. In general, the QRT approach is time-consuming because it is required to solve a large system of coupled differential equations and numerical instabilities that can arise. Moreover, theoretical complications related to dynamics of the operators involved can appear, as we will point out in the next section. In spite of this, the QRT approach is widely used in theoretical works, for example, in studies about photoluminescence spectra of coupled light-matter systems in microcavities in the presence of a continuous and incoherent pumping [19, 20]. Also, in studies considering the relation between dynamical regimes and entanglement in QD-cavity systems [21, 22]. In the past, the Green's functions technique (GFT) was successfully applied for calculating the emission spectrum for a very simple quantum system, e.g., the micro-maser [23]. Nevertheless, this approach has not been widely noticed in many significant situations in open quantum systems. The purpose of this work is to provide a simple and efficient numerical method based on the GFT in order to overcome the inherent difficulties associated with the direct application of the QRT approach by solving the dynamics of the system in the frequency domain directly. This paper is structured as follows: the theoretical background of the QRT as well as the GFT are presented in section II. A concrete application of our proposed methodology for calculating the emission spectrum of the QD-cavity system is considered in section III. Moreover, for comparison purposes with the GFT, we discuss in some detail the methodology of the QRT for calculating the emission spectrum of the cavity. The numerical results for the emission spectrum of the quantum dot, as well as of the cavity obtained from both the GFT and the QRT, are shown in section IV. A discussion about our findings is summarized in section V.

II. Theoretical background

i. Quantum regression theorem

One of the most important measurements when the light excites resonantly a QD-cavity system is the emission spectrum of the system. From a theoretical point of view, it is assumed that it corresponds to a stationary and ergodic process which can be calculated as a PSD of light by using the well-known Wiener-Khintchine theorem [13]. This theorem states that the emission spectrum is given by the Fourier Transform of the two-time correlation function of the operator field $\hat{a}$; explicitly, it is

$$S(\omega) = \mathfrak{Re} \lim_{t\to\infty} \int_0^\infty \langle \hat{a}^\dagger(t+\tau)\hat{a}(t)\rangle e^{i\omega\tau} d\tau. \quad (1)$$

In order to calculate the two-time correlation function involved in Eq. (1), a theoretical approach based on the QRT is frequently considered. It states that if a set of operators $\{\hat{O}_i(t+\tau)\}$ satisfy the dynamical equations $\frac{d}{d\tau}\langle\hat{O}_i(t+\tau)\rangle = \sum_j L_{ij}\langle\hat{O}_j(t+\tau)\rangle$ then $\frac{d}{d\tau}\langle\hat{O}_i(t+\tau)\hat{O}(t)\rangle = \sum_j L_{ij}\langle\hat{O}_j(t+\tau)\hat{O}(t)\rangle$ is valid for any operator $\hat{O}(t)$ at an arbitrary time t. Here, L_{ij} represents the matrix of coefficients associated with the coupled linear equations of motion. It is worth mentioning that validity of this theorem holds whenever a closed set of operators is associated with the dynamics. In general, to obtain the closed set of operators can be difficult or an impossible task, since there must be added as many operators as necessary in order to close the dynamics of the system. For example, to calculate the emission spectrum in a model of QD-cavity system [20, 21], two new operators are required because the field operators in the interaction picture do not lead to a complete set.

ii. Green's functions technique

Let us consider a QD-cavity system and an operator $\hat{A}$ which does not operate on the reservoirs, then its single-time expectation value in the Heisenberg representation is given by

$$\langle\tilde{\hat{A}}(t+\tau)\rangle = Tr_{S\otimes R}[\tilde{\hat{A}}(t+\tau)\tilde{\hat{\rho}}_{S\otimes R}(t)]. \quad (2)$$

The density operator system-reservoir can be evolved from an initial state at time 0 to an arbitrary time t via $\tilde{\hat{\rho}}_{S\otimes R}(t) = \hat{U}^{\dagger}(t,0)\hat{\rho}_{S\otimes R}(0)\hat{U}(t,0)$, with $\hat{U}(t,0)$ being a unitary time-evolution operator involving the Hamiltonian terms of the system and reservoirs. Moreover, the operator $\tilde{\hat{\rho}}_{S\otimes R}(t) = \tilde{\hat{\rho}}_{S}(t)\otimes\tilde{\hat{\rho}}_{R}(t)$ depicts the composite density operator of the system and reservoir. It is worth pointing out that tilde means that the operator has been transformed to the Heisenberg representation and that the dynamics of the system depend directly on $\tilde{\hat{\rho}}_{S\otimes R}(t)$ for all times. The validity of the Markovian approximation requires that the state of the system is sufficiently well described when it is considered that $\tilde{\hat{\rho}}_{S}(t) = Tr_{R}(\tilde{\hat{\rho}}_{S\otimes R}(t))$. Therefore, it is sufficient to write $\tilde{\hat{\rho}}_{S\otimes R}(t) = \tilde{\hat{\rho}}_{S}(t)\otimes\tilde{\hat{\rho}}_{R}(t)$ for all times. If we assume that at $t=0$ the initial state of the system is the steady state, then $\tilde{\hat{\rho}}_{S\otimes R}(0) = \hat{\rho}^{(ss)}_{S\otimes R}$. Here, the superscript "(ss)" should be understood to be the steady state of the system-reservoir. After tracing over degrees of freedom of the reservoirs, we have that the Eq. (2) takes the form

$$\langle\tilde{\hat{A}}(\tau)\rangle = Tr_{S}[\tilde{\hat{A}}(0)\tilde{\hat{\rho}}_{S}(\tau)], \quad (3)$$

where the reduced density operator for the system is given by $\tilde{\hat{\rho}}_{S}(\tau) = Tr_{R}[\hat{U}(\tau,0)\tilde{\hat{\rho}}_{S\otimes R}(0)\hat{U}^{\dagger}(\tau,0)]$ and $\tilde{\hat{A}}(0)=\hat{A}$. If $\tilde{\hat{\rho}}_{S}(\tau)$ satisfies the Lindblad master equation $d\tilde{\hat{\rho}}_{S}(\tau)/d\tau = \mathcal{L}\tilde{\hat{\rho}}_{S}(\tau)$ with $\mathcal{L}$ the superoperator defined as $\mathcal{L}\tilde{\hat{\rho}}_{S}(\tau) = -i[\hat{H}_S,\tilde{\hat{\rho}}_{S}(\tau)] + \sum_j \frac{\Gamma_j}{2}(2\hat{X}_j\tilde{\hat{\rho}}_{S}(\tau)\hat{X}_j^{\dagger} - \hat{X}_j^{\dagger}\hat{X}_j\tilde{\hat{\rho}}_{S}(\tau) - \tilde{\hat{\rho}}_{S}(\tau)\hat{X}_j^{\dagger}\hat{X}_j)$ for an operator $\hat{X}_j$, then the expectation value $\langle\tilde{\hat{A}}(\tau)\rangle$ can be computed by solving the dynamics associated to the Lindblad master equation. It is worth mentioning that the Hamiltonian operator $\hat{H}_S$ describes the QD-cavity system and Γ_j corresponds to the damping (pumping) rate associated to the operator $\hat{X}_j$.

In order to calculate the two-time correlation function $\langle\tilde{\hat{A}}(t+\tau)\tilde{\hat{B}}(t)\rangle$ where $\tilde{\hat{A}}(t+\tau) = \hat{U}^{\dagger}(t+\tau,t)\tilde{\hat{A}}(t)\hat{U}(t+\tau,t)$ and $\tilde{\hat{B}}(t) = \hat{U}^{\dagger}(t,0)\hat{B}\hat{U}(t,0)$ are arbitrary Heisenberg operators which do not operate on the reservoirs, we proceed similarly to the case of the single-time expectation value, it is

$$\begin{aligned}\langle\tilde{\hat{A}}(\tau)\tilde{\hat{B}}(0)\rangle &= Tr_{S\otimes R}[\tilde{\hat{A}}(\tau)\tilde{\hat{B}}(0)\tilde{\hat{\rho}}_{S\otimes R}(0)],\\ &= Tr_S[\tilde{\hat{A}}(0)\tilde{\hat{G}}(\tau)], \quad (4)\end{aligned}$$

where we have used the well-known properties of the unitary time-evolution operator and the fact that the system at time $t=0$ is in the steady state. We have defined the operator

$$\tilde{\hat{G}}(\tau) = Tr_R[\hat{U}(\tau,0)\tilde{\hat{B}}(0)\tilde{\hat{\rho}}_{S\otimes R}(0)\hat{U}^{\dagger}(\tau,0)] \quad (5)$$

where the trace operation is performed on the reservoirs only. By performing the time derivation of Eq. (4), we have that

$$\frac{d}{d\tau}\langle\tilde{\hat{A}}(\tau)\tilde{\hat{B}}(0)\rangle = Tr_S[\tilde{\hat{A}}(0)\frac{d\tilde{\hat{G}}(\tau)}{d\tau}]. \quad (6)$$

where

$$\begin{aligned}\frac{d\tilde{\hat{G}}(\tau)}{d\tau} &= \frac{d}{d\tau}Tr_R[\hat{U}(\tau,0)\tilde{\hat{B}}(0)\tilde{\hat{\rho}}_{S\otimes R}(0)\hat{U}^{\dagger}(\tau,0)],\\ &= \frac{d}{d\tau}Tr_R[\tilde{\hat{B}}(\tau)\tilde{\hat{\rho}}_{S\otimes R}(\tau)],\\ &= Tr_R[\tilde{\hat{B}}(\tau)\frac{d\tilde{\hat{\rho}}_{S\otimes R}(\tau)}{d\tau} + \tilde{\hat{\rho}}_{S\otimes R}(\tau)\frac{d\tilde{\hat{B}}(\tau)}{d\tau}],\\ &= Tr_R[\tilde{\hat{B}}(\tau)\frac{d\tilde{\hat{\rho}}_{S\otimes R}(\tau)}{d\tau}],\\ &+ Tr_R[\frac{d\tilde{\hat{B}}(\tau)}{d\tau}\tilde{\hat{\rho}}_{S\otimes R}(\tau)]. \quad (7)\end{aligned}$$

Notice that the last term vanishes since $Tr_R[\frac{d\tilde{\hat{B}}(\tau)}{d\tau}\tilde{\hat{\rho}}_{S\otimes R}(\tau)] = \frac{d}{d\tau}Tr_R[\tilde{\hat{B}}(0)\tilde{\hat{\rho}}_{S\otimes R}(0)]$ is independent of time τ. Thus, the Eq. (7) can be reduced to the form

$$\begin{aligned}\frac{d\tilde{\hat{G}}(\tau)}{d\tau} &= Tr_R[\frac{d\tilde{\hat{\rho}}_{S\otimes R}(\tau)}{d\tau}\tilde{\hat{B}}(\tau)],\\ &= Tr_R[\mathcal{L}\tilde{\hat{\rho}}_{S\otimes R}(\tau)\tilde{\hat{B}}(\tau)],\\ &= \mathcal{L}Tr_R[\tilde{\hat{\rho}}_{S\otimes R}(\tau)\tilde{\hat{B}}(\tau)],\\ &= \mathcal{L}Tr_R[\hat{U}(\tau,0)\tilde{\hat{B}}(0)\tilde{\hat{\rho}}_{S\otimes R}(0)\hat{U}^{\dagger}(\tau,0)],\\ &= \mathcal{L}\tilde{\hat{G}}(\tau) \quad (8)\end{aligned}$$

where we have taken into account that the superoperator $\mathcal{L}$ acts only on the system Hilbert space and not on the reservoir. It is straightforward to conclude that $\hat{\hat{G}}(\tau)$ is an operator that obeys the same dynamical equations as $\hat{\rho}_S(\tau)$. More precisely, $d\hat{\hat{G}}(\tau)/d\tau = \mathcal{L}\hat{\hat{G}}(\tau)$ with the boundary condition $\hat{\hat{G}}(0) = \hat{\hat{B}}(0)\hat{\rho}_S(0)$ at the steady state of the system.
We also conclude that the two-time correlation function in the long-time limit can be written as

$$\lim_{t\to\infty} \langle \hat{\hat{A}}(t+\tau)\hat{\hat{B}}(t)\rangle = Tr_S[\hat{A}\hat{\hat{G}}(\tau)], \qquad (9)$$

where $\hat{\hat{G}}(\tau) = Tr_R[\hat{U}(\tau)\hat{B}\hat{\rho}^{(ss)}_{S\otimes R}\hat{U}^\dagger(\tau)]$ is defined as the Green's functions operator and the operators $\hat{A}$, $\hat{B}$ and $\hat{\rho}^{(ss)}_{S\otimes R}$ are in the Schrödinger representation. $\hat{A} = \hat{\hat{A}}(0)$ and $\hat{B} = \hat{\hat{B}}(0)$ are operators considered at the steady state of the system. In the remainder of the paper, we assume that $\hat{U}(\tau) \equiv \hat{U}(\tau, 0)$. Particularly, the Eq. (9) takes the form of the Eq. (1) after performing the integral transformation. More precisely, by taking the real part of the Laplace transform to the Eq. (9), we obtain an expression in terms of the Green's functions operator in the frequency domain as follows

$$S(\omega) = \mathfrak{Re}\; Tr_S[\hat{A}\hat{\tilde{G}}(i\omega)]. \qquad (10)$$

Notice that the operators $\hat{A}$ and $\hat{B}$ should be defined appropriately for describing the emission spectrum due to the cavity or the quantum dot. Moreover, the wide tilde is used to indicate that the Laplace transform was taken. The superscript "(ss)" should be understood to be the steady state of the reduced density operator of the system. After taking the Laplace transform of the Eq. (9), we obtain an expression for the emission spectrum of the system in terms of the Green's functions operator in the frequency domain as follows

$$S(\omega) = \frac{1}{\pi n_c}\mathfrak{Re}\, Tr_S[\hat{A}\hat{\tilde{\hat{G}}}(i\omega)]. \qquad (11)$$

Prior to leaving this section, we mention that this result will be the starting point for calculating the emission spectrum due to the cavity as well as the quantum dot by considering the photon and fermionic operators in a separated way.

III. Application to the QD-cavity system

i. Model

In order to illustrate the potential of the Green's function technique for calculating the emission spectrum in a QD-cavity system, we will consider an open quantum system composed of a quantum dot interacting with a confined mode of the electromagnetic field inside a semiconductor cavity. This quantum system is well described by the Jaynes-Cummings Hamiltonian [24]

$$\hat{H}_S = \omega_X\hat{\sigma}^\dagger\hat{\sigma} + (\omega_X - \Delta)\hat{a}^\dagger\hat{a} + g(\hat{\sigma}\hat{a}^\dagger + \hat{a}\hat{\sigma}^\dagger), \qquad (12)$$

where the quantum dot is described as a fermionic system with only two possible states, e.g., $|G\rangle$ and $|X\rangle$ are the ground and excited state. $\hat{\sigma} = |G\rangle\langle X|$ and $\hat{a}$ ($\hat{\sigma}^\dagger = |X\rangle\langle G|$ and $\hat{a}^\dagger$) are the annihilation (creation) operators for the fermionic system and the cavity mode, respectively. The parameter g is the light-matter coupling constant. Moreover, note that we have set $\hbar = 1$. We also define the detuning between frequencies of the quantum dot and the cavity mode as $\Delta = \omega_X - \omega_a$, where ω_X and ω_a are the energies associated to an exciton and the photons inside the cavity, respectively. This Hamiltonian system is far from describing any real physical situation since it is completely integrable [25] and no measurements could be done since the light remains always inside the cavity.
In order to incorporate the effects of the environment on the dynamics of the system, we consider the usual approach to model an open quantum system by considering a whole system-reservoir Hamiltonian which is frequently split in three parts. Namely, $\hat{H} = \hat{H}_S + \hat{H}_{SR} + \hat{H}_R$, where $\hat{H}_S$ defines the Hamiltonian term of the QD-cavity system as it is defined in the Eq. (12). The Hamiltonian terms $\hat{H}_{SR}$ and $\hat{H}_R$ corresponding to a bilinear coupling between the system-reservoir and its respective reservoirs $\hat{H}_R$ have been discussed in detail by Perea et al. in [26]. The reader can find a detailed discussion of the Markovian master equation in [27, 28]. In the framework of open quantum systems, different reservoirs have been proposed in order to describe the dissipation, decoherence or decays. Particularly, for QD-cavity systems, a reservoir is considered for describing the

physical situation where the photons are absorbed in a semiconductor and electron-hole pairs (excitons) can be produced which can be associated to either electrical injection or the capture of excitons optically created at frequencies larger than the typical ones of our system. This process corresponds to the so-called continuous and incoherent pumping of the QD. Also, when the excitons are coupled to the leaky modes of the cavity with energy different than the cavity mode, there is a residual density of states inside the cavity and this process is responsible for the spontaneous emission (radiative recombination) to an independent reservoir of photons. Another physical process is known as the coherent emission and it is due to the direct dissipation of the cavity mode, more precisely, the cavity mode is coupled to the continuum of photonic modes out of the cavity. For obtaining the master equation for the QD-cavity system, it is convenient to consider the interaction picture with respect to $\hat{H}_S + \hat{H}_{SR}$ and assume the validity of the Born-Markov approximation. After tracing out the degrees of freedom of all the reservoirs, one arrives to the Lindblad master equation for the reduced density matrix of the system

$$\begin{aligned}\frac{d\hat{\rho}_S}{d\tau} &= -i\left[\hat{H}_S, \hat{\rho}_S\right] \\ &+ \frac{\kappa}{2}(2\hat{a}\hat{\rho}_S\hat{a}^\dagger - \hat{a}^\dagger\hat{a}\hat{\rho}_S - \hat{\rho}_S\hat{a}^\dagger\hat{a}) \\ &+ \frac{\gamma}{2}(2\hat{\sigma}\hat{\rho}_S\hat{\sigma}^\dagger - \hat{\sigma}^\dagger\hat{\sigma}\hat{\rho}_S - \hat{\rho}_S\hat{\sigma}^\dagger\hat{\sigma}) \\ &+ \frac{P}{2}(2\hat{\sigma}^\dagger\hat{\rho}_S\hat{\sigma} - \hat{\sigma}\hat{\sigma}^\dagger\hat{\rho}_S - \hat{\rho}_S\hat{\sigma}\hat{\sigma}^\dagger). \quad (13)\end{aligned}$$

The parameter γ is the decay rate due to the spontaneous emission, κ is the decay rate of the cavity photons across the cavity mirrors, and P is the rate at which the excitons are being pumped. Figure 1 shows a scheme of the simplified model of the QD-cavity system showing the processes of continuous pumping P and cavity loses κ. The physical process begins when the light from the pumping laser enters into the cavity and excites one of the quantum dots in the QD layer. Thus, light from this source couples to the cavity and a fraction of photons escapes through the partly transparent mirror from the cavity and goes to the spectrometer for measurements of the emission spectrum.

A general approach for solving the dynamics of the system consists in writing the corresponding Bloch equations for the reduced density matrix of the system in the bared basis. It is an extended Hilbert space formed by taking the tensor product of the state vectors for each of the system components, $\{|G\rangle, |X\rangle\} \otimes \{|n\rangle\}_{n=0}^{\infty}$. In this basis, the reduced density matrix $\hat{\rho}_S$ can be written in terms of its matrix elements as $\tilde{\rho}_{S\alpha n,\beta m} \equiv \langle\alpha n|\hat{\rho}_S(\tau)|\beta m\rangle$. Hence, the Eq. (13) explicitly reads

$$\begin{aligned}\frac{d\tilde{\rho}_{S\alpha n,\beta m}}{d\tau} &= i\Big[(\omega_X - \Delta)(m-n)\tilde{\rho}_{S\alpha n,\beta m} \\ &+ \omega_X(\delta_{\beta X}\tilde{\rho}_{S\alpha n,Xm} - \delta_{\alpha X}\tilde{\rho}_{SXn,\beta m})\Big] \\ &+ ig\Big[\Big(\sqrt{m+1}\delta_{\beta X}\tilde{\rho}_{S\alpha n,Gm+1} \\ &+ \sqrt{m}\delta_{\beta G}\tilde{\rho}_{S\alpha n,Xm-1}\Big) \\ &- \Big(\sqrt{n}\delta_{\alpha G}\tilde{\rho}_{SXn-1,\beta m} \\ &+ \sqrt{n+1}\delta_{\alpha X}\tilde{\rho}_{SGn+1,\beta m}\Big)\Big] \\ &+ \frac{\kappa}{2}\Big(2\sqrt{(m+1)(n+1)}\tilde{\rho}_{S\alpha n+1,\beta m+1} \\ &- (n+m)\tilde{\rho}_{S\alpha n,\beta m}\Big) - \frac{\gamma}{2}\Big(\delta_{\alpha X}\tilde{\rho}_{SXn,\beta m} \\ &- 2\delta_{\alpha G}\delta_{\beta G}\tilde{\rho}_{SXn,Xm} + \delta_{\beta X}\tilde{\rho}_{S\alpha n,Xm}\Big) \\ &+ \frac{P}{2}\Big(2\delta_{\alpha X}\delta_{\beta X}\tilde{\rho}_{SGn,Gm} - \delta_{\alpha G}\tilde{\rho}_{SGn,\beta m} \\ &- \delta_{\beta G}\tilde{\rho}_{S\alpha n,Gm}\Big). \quad (14)\end{aligned}$$

Note that we use the convention that all indices written in Greek alphabet are used for the fermionic states and take only two possible values $|G\rangle$, $|X\rangle$. The indices written in Latin alphabet are used for the Fock states which take the possible values $0, 1, 2, 3\ldots$ Additionally, it is worth mentioning that our proposed method does not require to solve a system of coupled differential equations, instead of it, we solve a reduced set of algebraic equations that speed up the numerical solution.

Prior to leaving this section, we point out that the number of excitations of the system is defined by the operator $\hat{N} = \hat{a}^\dagger\hat{a} + \hat{\sigma}^\dagger\hat{\sigma}$. The closed system and the number of excitations of the system is conserved, i.e., $[\hat{H}_S, \hat{N}] = 0$. It allows us to organize the states of the system through the number of excitations criterion such that the density matrix elements $\tilde{\rho}_{Gn,Gn}$, $\tilde{\rho}_{Xn-1,Xn-1}$, $\tilde{\rho}_{Gn,Xn-1}$ and $\tilde{\rho}_{Xn-1,Gn}$ are related by having the same number of

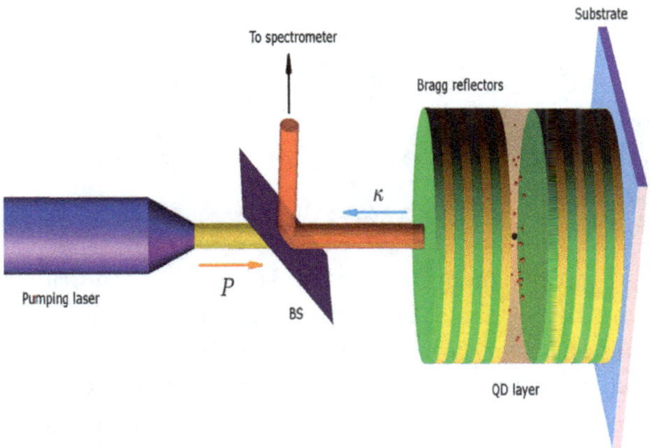

Figure 1: The picture represents a QD-cavity system showing the processes of continuous pumping P and cavity loses κ.

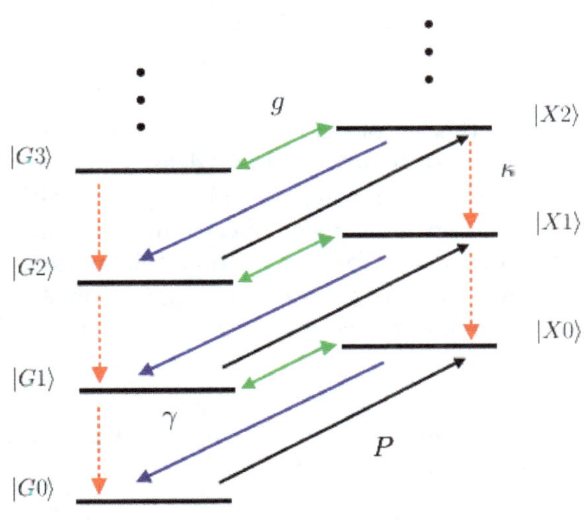

Figure 2: Ladder of bared states for a two-level quantum dot coupled to a single cavity mode. The double headed green arrow depicts the matter coupling constant g, dashed red lines the emission of the cavity mode κ, solid black lines the exciton pumping rate P and solid blue lines the spontaneous emission rate γ.

quanta; sub-spaces of a fixed number of excitation evolve independently from each other. The Fig. 2 shows a schematic representation of the action of the dissipative processes involved in the dynamics of the system according to the excitation number (N_{exc}).

ii. Emission spectrum of the cavity based on the GFT

In order to compute the emission spectrum of the cavity, we will consider the two-time correlation function accordingly with the Eq. (9) for the field operator as follows

$$\lim_{t\to\infty} \langle \hat{\tilde{a}}^\dagger(t+\tau)\hat{\tilde{a}}(t)\rangle = Tr_S[\hat{a}^\dagger \hat{\tilde{G}}(\tau)]. \quad (15)$$

Where we have considered that the field operator is given by $\hat{\tilde{a}}(0) = \hat{a}$ at the steady state. After performing the partial trace over the degrees of freedom of the system, we have that

$$\begin{aligned} Tr_S[\hat{a}^\dagger \hat{\tilde{G}}(\tau)] &= \sum_{\alpha,\beta,\gamma,l,m,n} \sqrt{(l+1)(m+1)} \\ &\times \; Tr_R[U_{\alpha l,\beta m}(\tau) \\ &\times \; \langle \beta m+1|\hat{\rho}^{(ss)}_{S\otimes R}|\gamma n\rangle U^\dagger_{\gamma n,\alpha l+1}(\tau)], \end{aligned} \quad (16)$$

where the matrix elements for the time evolution operator are given by $U_{\alpha l,\beta m}(\tau) = \langle \alpha l|\hat{U}(\tau)|\beta m\rangle$ and $U^\dagger_{\gamma n,\alpha l+1}(\tau) = \langle \gamma n|\hat{U}^\dagger(\tau)|\alpha l+1\rangle$. In what follows, we assume the validity of the Markovian approximation, it means that the correlations between the system and the reservoir must be unimportant even at the steady state. Thus, the density operator system-reservoir can be written as $\hat{\rho}^{(ss)}_{S\otimes R} = \hat{\rho}^{(ss)}_S \otimes \hat{\rho}^{(ss)}_R$ which implies that

$$\langle \beta m+1|\hat{\rho}^{(ss)}_{S\otimes R}|\gamma n\rangle = \hat{\rho}^{(ss)}_R \langle \beta m+1|\hat{\rho}^{(ss)}_S|\gamma n\rangle. \quad (17)$$

Replacing the previous expression in Eq. (16), it is straightforward to show that the two-time correlation function reads

$$Tr_S[\hat{a}^\dagger \hat{\tilde{G}}(\tau)] = \sum_{\alpha l} \sqrt{l+1}\langle \alpha l|\hat{\tilde{G}}(\tau)|\alpha l+1\rangle, \quad (18)$$

where the Green's functions operator $\hat{\tilde{G}}(\tau)$ is given by

$$\begin{aligned} \hat{\tilde{G}}(\tau) &= Tr_R\Big[\hat{U}(\tau)\hat{\rho}^{(ss)}_R \sum_{\beta\gamma mn} \Big(\sqrt{m+1}\,|\beta m\rangle\,\langle\gamma n| \\ &\times \; \langle \beta m+1|\,\hat{\rho}^{(ss)}_S\,|\gamma n\rangle\Big)\hat{U}^\dagger(\tau)\Big]. \end{aligned} \quad (19)$$

As we pointed out in section II, this operator must obey the same master equation as the reduced density operator of the system. In fact, the terms that only contribute in the Eq. (18) are given by the matrix elements $\tilde{G}_{\beta m,\gamma n}(\tau) \equiv \langle \beta m| \hat{\tilde{G}}(\tau) |\gamma n\rangle$ of the Green's functions operator. This is due to the fact that the projection operator $|\beta m\rangle\langle\gamma n|$ enters into $\hat{\tilde{G}}(\tau)$ in the same way as into the reduced density operator of the system.

In order to identify these matrix elements, it should be considered that for the QD-cavity system, the dynamics of all coherences asymptotically vanish and there only remains the reduced density matrix elements which are ruled by the number of excitations criterion, i.e., $\rho_{Gn,Gn}$, $\rho_{Xn-1,Xn-1}$, $\rho_{Gn,Xn-1}$, $\rho_{Xn-1,Gn}$. Thus, the Eq. (17) can be written as follows

$$
\begin{aligned}
\langle \beta m+1|\hat{\rho}^{(ss)}_{S\otimes R}|\gamma n\rangle &= \hat{\rho}^{(ss)}_R \Big(\delta_{\beta G}\delta_{\gamma G}\delta_{m+1,n} \\
&+ \delta_{\beta X}\delta_{\gamma X}\delta_{m,n-1} \\
&+ \delta_{\beta G}\delta_{\gamma X}\delta_{m,n} \\
&+ \delta_{\beta X}\delta_{\gamma G}\delta_{m+1,n-1}\Big) \\
&\times \rho^{(ss)}_{S\beta m+1,\gamma n}.
\end{aligned} \tag{20}
$$

By replacing the Eq. (20) into Eq. (19), we find that the Green's functions operator explicitly reads

$$
\begin{aligned}
\hat{\tilde{G}}(\tau) &= Tr_R\Big[\hat{U}(\tau)\hat{\rho}^{(ss)}_R \sum_m \sqrt{m+1} \\
&\times \Big(|Gm\rangle\langle Gm+1|\,\rho^{(ss)}_{SGm+1,Gm+1} \\
&+ |Xm\rangle\langle Xm+1|\,\rho^{(ss)}_{SXm+1,Xm+1} \\
&+ |Gm\rangle\langle Xm|\,\rho^{(ss)}_{SGm+1,Xm} \\
&+ |Xm\rangle\langle Gm+2| \\
&\times \rho^{(ss)}_{SXm+1,Gm+2}\Big)\hat{U}^{\dagger}(\tau)\Big].
\end{aligned} \tag{21}
$$

Note that from this expression, it is easy to identify the nonzero matrix elements of the Green's functions operator that contribute to the emission spectrum. Finally, after performing the Laplace transform to the Eq. (18), we have that the emission spectrum of the cavity is given by

$$
\begin{aligned}
S(\omega) &= \mathfrak{Re}\sum_l \sqrt{l+1}\Big(\widetilde{G}_{Gl,Gl+1}(i\omega) \\
&+ \widetilde{G}_{Xl,Xl+1}(i\omega) + \widetilde{G}_{Gl,Xl}(i\omega) \\
&+ \widetilde{G}_{Xl,Gl+2}(i\omega)\Big).
\end{aligned} \tag{22}
$$

It is worth mentioning that the initial conditions may be obtained by evaluating the Green's function operator at $\tau = 0$. Moreover, by using the fact that the time evolution operators become the identity and $Tr_R[\hat{\rho}^{(ss)}_R] = 1$. We obtain a set of initial conditions given by

$$
\begin{aligned}
\tilde{G}_{Gl,Gl+1}(0) &= \sqrt{l+1}\rho^{(ss)}_{SGl+1,Gl+1}, \\
\tilde{G}_{Xl,Xl+1}(0) &= \sqrt{l+1}\rho^{(ss)}_{SXl+1,Xl+1}, \\
\tilde{G}_{Gl,Xl}(0) &= \sqrt{l+1}\rho^{(ss)}_{SGl+1,Xl}, \\
\tilde{G}_{Xl,Gl+2}(0) &= \sqrt{l+1}\rho^{(ss)}_{SXl+1,Gl+2}.
\end{aligned} \tag{23}
$$

Note that this set of initial conditions corresponds to the steady state of the reduced density matrix of the system. A general algorithm based on the GFT for computing the emission spectrum is presented in the appendix. We mention that this approach can be adapted easily for calculating the emission spectrum due to the cavity as well as the quantum dot.

iii. Emission spectrum of the quantum dot based on the GFT

In order to compute the emission spectrum of the quantum dot, we will consider the two-time correlation function given by Eq. (9), but for the case of the matter operator

$$
\lim_{t\to\infty}\langle \hat{\tilde{\sigma}}^{\dagger}(t+\tau)\hat{\tilde{\sigma}}(t)\rangle = Tr_S[\hat{\sigma}^{\dagger}\hat{\tilde{G}}(\tau)] \tag{24}
$$

where we have considered that the matter operator is given by $\hat{\tilde{\sigma}}(0) = \hat{\sigma}$ at the steady state. It is straightforward to show, after performing the partial trace over the degrees of freedom of the system, that the two-time correlation function reads

$$
Tr_S[\hat{\sigma}^{\dagger}\hat{\tilde{G}}(\tau)] = \sum_{\alpha l}\delta_{\alpha X}\langle Gl|\hat{\tilde{G}}(\tau)|\alpha l\rangle, \tag{25}
$$

where the Green's functions operator $\hat{\tilde{G}}(\tau)$ is given by

$$\hat{\tilde{G}}(\tau) = Tr_R\Big[\hat{U}(\tau)\sum_{\beta\gamma mn}\Big(\delta_{\beta X}\left|Gm\right\rangle\left\langle\gamma m\right| \times \left\langle\beta m\right|\hat{\rho}^{(ss)}_{S\otimes R}\left|\gamma n\right\rangle\Big)\hat{U}^\dagger(\tau)\Big]. \quad (26)$$

Assuming again the validity of the Markovian approximation and taking into account the number of excitations criterion, we have that the density operator system-reservoir can be written as

$$\begin{aligned}\langle\beta m|\hat{\rho}^{(ss)}_{S\otimes R}|\gamma n\rangle &= \hat{\rho}^{(ss)}_R\Big(\delta_{\beta G}\delta_{\gamma G}\delta_{m,n} \\ &+ \delta_{\beta X}\delta_{\gamma X}\delta_{m,n} + \delta_{\beta G}\delta_{\gamma X}\delta_{m,n+1} \\ &+ \delta_{\beta X}\delta_{\gamma G}\delta_{m,n-1}\Big)\rho^{(ss)}_{S\beta m,\gamma n}. \quad (27)\end{aligned}$$

By inserting the Eq. (27) into Eq. (26), we find that the Green's functions operator explicitly reads

$$\begin{aligned}\hat{\tilde{G}}(\tau) &= Tr_R\Big[\hat{U}(\tau)\hat{\rho}^{(ss)}_R \\ &\times \sum_m\Big(\left|Gm\right\rangle\left\langle Xm\right| rho^{(ss)}_{SXm,Xm} \\ &+ \left|Gm\right\rangle\left\langle Gm+1\right|\rho^{(ss)}_{SXm,Gm+1}\Big) \\ &\times \hat{U}^\dagger(\tau)\Big]. \quad (28)\end{aligned}$$

Analogously as in section ii, we identify the nonzero matrix elements of the Green's functions operator that contribute to the emission spectrum. After performing the Laplace transform to Eq. (25), the emission spectrum of the quantum dot is given by

$$S(\omega) = \mathfrak{Re}\sum_l\Big(\tilde{G}_{Gl,Xl}(i\omega)+\tilde{G}_{Gl,Gl+1}(i\omega)\Big). \quad (29)$$

Taking into account that the initial conditions are obtained by evaluating the Green's function operator at $\tau = 0$, we have the time evolution operators become the identity and $Tr_R[\hat{\rho}^{(ss)}_R] = 1$, thus, we obtain a set of initial conditions given by

$$\begin{aligned}\tilde{G}_{Gl,Xl}(0) &= \rho^{(ss)}_{SXl,Xl}, \\ \tilde{G}_{Gl,Gl+1}(0) &= \rho^{(ss)}_{SXl,Gl+1}, \\ \tilde{G}_{Xl,Xl+1}(0) &= 0, \\ \tilde{G}_{Gl+2,Xl}(0) &= 0. \quad (30)\end{aligned}$$

at the steady-state.

iv. Emission spectrum of the cavity based on the QRT

In what follows, we apply the QRT approach for the model of QD-cavity system described in section III. In order to compute the emission spectrum of the cavity, the knowledge of the two-time correlation function for the field operator is required, it is $\langle\hat{a}^\dagger(\tau)\hat{a}(0)\rangle$ in concordance with the Eq. (1). Moreover, by following the approach presented in Ref. [26], it is straightforward to show that the two-time correlation function is given by

$$\begin{aligned}\langle\hat{a}^\dagger(\tau)\hat{a}(0)\rangle &= \sum_n\sqrt{n+1}\Big(\langle\hat{a}^\dagger_{Gn}(\tau)\hat{a}(0)\rangle \\ &+ \langle\hat{a}^\dagger_{Xn}(\tau)\hat{a}(0)\rangle\Big), \quad (31)\end{aligned}$$

where the following definitions have been used

$$\begin{aligned}\hat{a}^\dagger_{Gn} &= \left|Gn+1\right\rangle\left\langle Gn\right|, \\ \hat{a}^\dagger_{Xn} &= \left|Xn+1\right\rangle\left\langle Xn\right|, \\ \hat{\sigma}^\dagger_n &= \left|Xn\right\rangle\left\langle Gn\right|, \\ \hat{\zeta}_n &= \left|Gn+1\right\rangle\left\langle Xn-1\right|. \quad (32)\end{aligned}$$

It is worth mentioning that the last two operators should be added in order to close the dynamics of the system accordingly to the QRT as pointed out in section II. More precisely, we are interested in solving the dynamical equations associated to the expectation values $\langle\hat{a}^\dagger_{Gn}(\tau)\hat{a}(0)\rangle$ and $\langle\hat{a}^\dagger_{Xn}(\tau)\hat{a}(0)\rangle$ as a function of τ. Therefore, we need to solve a set of coupled differential equations given by

$$\begin{aligned}\frac{d}{d\tau}\langle\hat{a}^\dagger_{Gn}(\tau)\hat{a}(0)\rangle &= \sum_j L_{ij}\langle\hat{a}^\dagger_{Gn}(\tau)\hat{a}(0)\rangle, \\ \frac{d}{d\tau}\langle\hat{a}^\dagger_{Xn}(\tau)\hat{a}(0)\rangle &= \sum_j L_{ij}\langle\hat{a}^\dagger_{Xn}(\tau)\hat{a}(0)\rangle, \\ \frac{d}{d\tau}\langle\hat{\sigma}^\dagger_n(\tau)\hat{a}(0)\rangle &= \sum_j L_{ij}\langle\hat{\sigma}^\dagger_n(\tau)\hat{a}(0)\rangle, \\ \frac{d}{d\tau}\langle\hat{\zeta}_n(\tau)\hat{a}(0)\rangle &= \sum_j L_{ij}\langle\hat{\zeta}_n(\tau)\hat{a}(0)\rangle. \quad (33)\end{aligned}$$

In order to find explicitly the set of dynamical equations and its corresponding initial conditions, we obtain first the set of differential equations for the single-time expectation values for the operators given by Eq. (32), it is explicitly

$$\begin{aligned}
\frac{d}{d\tau}\langle \hat{a}^{\dagger}_{Gn}(\tau)\rangle &= \left(-P - i\Delta - n\kappa - \frac{\kappa}{2} + i\omega_X\right) \\
&\times \langle \hat{a}^{\dagger}_{Gn}(\tau)\rangle + \kappa\sqrt{(n+1)(n+2)} \\
&\times \langle \hat{a}^{\dagger}_{Gn+1}(\tau)\rangle + \gamma\langle \hat{a}^{\dagger}_{Xn}(\tau)\rangle \\
&- ig\sqrt{n}\langle \hat{\zeta}_n(\tau)\rangle + ig\sqrt{n+1} \\
&\times \langle \hat{\sigma}^{\dagger}_n(\tau)\rangle, \\
\frac{d}{d\tau}\langle \hat{\sigma}^{\dagger}_n(\tau)\rangle &= ig\sqrt{n+1}\langle \hat{a}^{\dagger}_{Gn}(\tau)\rangle \\
&- ig\sqrt{n}\langle \hat{a}^{\dagger}_{Xn-1}(\tau)\rangle \\
&+ \frac{1}{2}(-P - \gamma - 2n\kappa + 2i\omega_X) \\
&\times \langle \hat{\sigma}^{\dagger}_n(\tau)\rangle \\
&+ (n+1)\kappa\langle \hat{\sigma}^{\dagger}_{n+1}(\tau)\rangle, \\
\frac{d}{d\tau}\langle \hat{a}^{\dagger}_{Xn-1}(\tau)\rangle &= P\langle \hat{a}^{\dagger}_{Gn-1}(\tau)\rangle + \kappa\sqrt{n(n+1)} \\
&\times \langle \hat{a}^{\dagger}_{Xn}(\tau)\rangle \\
&+ (-\gamma - i\Delta - n\kappa + \frac{\kappa}{2} + i\omega_X) \\
&\times \langle \hat{a}^{\dagger}_{Xn-1}(\tau)\rangle + ig\sqrt{n+1}\langle \hat{\zeta}_n(\tau)\rangle \\
&- ig\sqrt{n}\langle \hat{\sigma}^{\dagger}_n(\tau)\rangle, \\
\frac{d}{d\tau}\langle \hat{\zeta}_n(\tau)\rangle &= -ig\sqrt{n}\langle \hat{a}^{\dagger}_{Gn}(\tau)\rangle + ig\sqrt{n+1} \\
&\times \langle \hat{a}^{\dagger}_{Xn-1}(\tau)\rangle \\
&+ (-\frac{P}{2} - \frac{\gamma}{2} - 2i\Delta - n\kappa + i\omega_X) \\
&\times \langle \hat{\zeta}_n(\tau)\rangle \\
&+ \sqrt{n(n+2)}\kappa\langle \hat{\zeta}_{n+1}(\tau)\rangle. \qquad (34)
\end{aligned}$$

The QRT approach implies that the following two-time correlation functions $\langle \hat{a}^{\dagger}_{Gn}(\tau)\hat{a}(0)\rangle$, $\langle \hat{a}^{\dagger}_{Xn}(\tau)\hat{a}(0)\rangle$, $\langle \hat{\sigma}^{\dagger}_n(\tau)\hat{a}(0)\rangle$ and $\langle \hat{\zeta}_n(\tau)\hat{a}(0)\rangle$ satisfy the same dynamical equations given by Eq. (34), subject to the initial conditions

$$\begin{aligned}
\langle \hat{a}^{\dagger}_{Gn}(0)\hat{a}(0)\rangle &= \sqrt{n+1}\rho_{Gn+1,Gn+1}(0), \\
\langle \hat{a}^{\dagger}_{Xn}(0)\hat{a}(0)\rangle &= \sqrt{n+1}\rho_{Xn+1,Xn+1}(0), \\
\langle \hat{\sigma}^{\dagger}_n(0)\hat{a}(0)\rangle &= \sqrt{n+1}\rho_{Gn+1,Xn}(0), \\
\langle \hat{\zeta}_n(0)\hat{a}(0)\rangle &= \sqrt{n}\rho_{Xn,Gn+1}(0). \qquad (35)
\end{aligned}$$

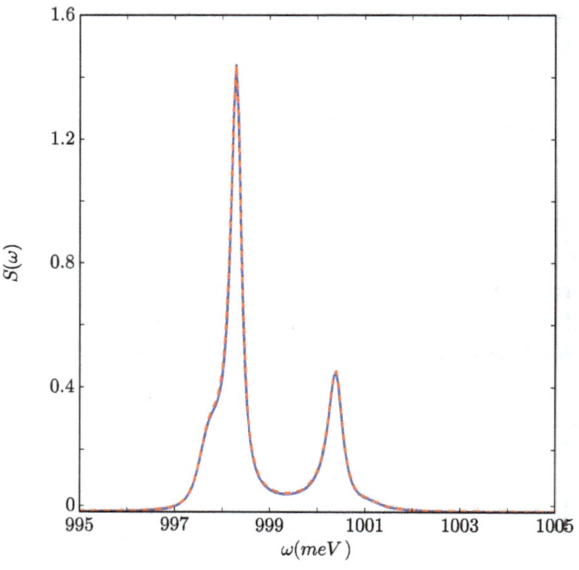

Figure 3: Emission spectrum of the cavity based on the GFT as a solid blue line and the corresponding numerical calculation based on the QRT as a dashed red line. The parameters values are $g = 1$ meV, $\gamma = 0.005$ meV, $\kappa = 0.2$ meV, $P = 0.3$ meV, $\Delta = 2$ meV, $\omega_a = 1000$ meV.

More precisely, it is done explicitly by performing the following replacements: $\langle \hat{a}^{\dagger}_{Gn}(\tau)\rangle \to \langle \hat{a}^{\dagger}_{Gn}(\tau)\hat{a}(0)\rangle$, $\langle \hat{a}^{\dagger}_{Xn}(\tau)\rangle \to \langle \hat{a}^{\dagger}_{Xn}(\tau)\hat{a}(0)\rangle$, $\langle \hat{\sigma}^{\dagger}_n(\tau)\rangle \to \langle \hat{\sigma}^{\dagger}_n(\tau)\hat{a}(0)\rangle$ and $\langle \hat{\zeta}_n(\tau)\rangle \to \langle \hat{\zeta}_n(\tau)\hat{a}(0)\rangle$. The parameters of the system $\omega_X, \Delta, g, \kappa, \gamma$ determine the dynamics of the two-time correlation function, as well as setting the initial conditions that will be propagated according to the dynamical equations given by Eq. (34). Since we are interested in the light that the quantum system emits, we have considered the steady state of the system as the initial state into equation Eq. (35).

IV. Results and discussion

In this section, we compare the numerical calculations based on the GFT and the QRT approach for the emission spectrum of the cavity as well as the quantum dot. In particular, the QD-cavity system can display two different dynamical regimes by changing the parameters of the system and two regimes can be achieved when the loss and pump rates are modified. In fact, the relation $g > |\kappa - \gamma|/4$ holds for the strong coupling regime and the relation $g < |\kappa - \gamma|/4$ remains valid for the weak coupling regime. Figure 3 shows the numerical results for the emission spectrum associated to the cavity in the strong coupling regime, where the emission spectrum of the cavity based on the GFT is shown as a solid blue line and the emission spectrum based on the QRT approach as a dashed red line. The parameters of the system are $g = 1$ meV, $\gamma = 0.005$ meV, $\kappa = 0.2$ meV, $P = 0.3$ meV, $\Delta = 2$ meV and $\omega_a = 1000$ meV. Particularly, for this set of parameters values, we can identify two different peaks which are associated to the energy of the cavity and the quantum dot, they are $\omega_a \approx 998.3$ meV and $\omega_X \approx 1000.3$ meV. We have considered the relative error as a quantitative measure of the discrepancy between the GFT and the QRT approaches. More precisely, by monitoring the numerical computations of the emission spectrum, we have estimated that the maximum relative error is on the order of 10^{-3} in all numerical calculations that we have performed. Similarly, the emission spectrum of the cavity based on the GFT (solid blue line) and the QRT approach (dashed red line) for the strong coupling regime are shown in Fig. 4. Here, the parameters values of the system are given by $g = 1$ meV, $\gamma = 0.005$ meV, $\kappa = 2$ meV, $P = 0.005$ meV, $\Delta = 0.0$ meV and $\omega_a = 1000$ meV. Note that we have considered the resonant case, more precisely, the same energy values for the cavity and the quantum dot. Here, the emission spectrums do not match but repel each other, resulting in a structure of two separate peaks for a distance of approximately two times the coupling constant, i.e., $2g \approx 2$ meV. It is worth mentioning that this quantum effect is well-known as Rabi splitting in QD-cavity systems. The emission spectrum of the quantum dot in the weak coupling regime is shown in Fig. 5. The numerical result for the emission spectrum of the quantum dot based on the GFT is shown as a solid blue line and the corresponding numerical result for QRT approach is shown as a dashed red line. We set the weak coupling regime by considering high values of the decay and pump rates $\kappa = 5$ meV and $P = 1$ meV, respectively. The rest of the parameters values are $g = 1$ meV, $\gamma = 0.1$ meV, $\Delta = 5$ meV and $\omega_a = 1000$ meV. We conclude that the method based on the GFT is in perfect agreement with the QRT approach and reproduces very well the emission spectrum of the QD-cavity system.

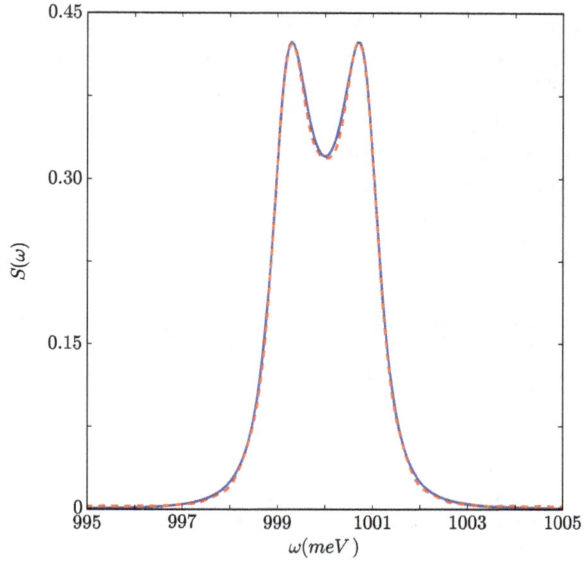

Figure 4: Emission spectrum of the cavity based on the GFT as a solid blue line and the corresponding numerical calculation based on the QRT approach as a dashed red line. The parameters values are $g = 1$ meV, $\gamma = 0.005$ meV, $\kappa = 2$ meV, $P = 0.005$ meV, $\Delta = 0$ meV, $\omega_a = 1000$ meV.

For comparison purposes with our GFT approach, we have also implemented the numerical method based on the QRT for the QD-cavity system (see details in section iv). In the conducted simulations, we have considered the same truncation level in the bare-state basis, e.g., $N_{exc} = 10$. Moreover, we have solved numerically the dynamical equations of the system given by the Eq. (34) until time $t_{max} = 2^{17}$ ps for obtaining an acceptable resolution in frequency domain, it is $\Delta\omega \approx 0.048$ meV. In order to test the performance of the GFT approach in terms of efficiency, we have compared the computational time involved on the numerical calculation of the emission spectrum of the cavity at four different excitation numbers N_{exc}. Table 1 shows in first column the excitation number. Second and third columns show the elapsed time (CPU time) in seconds during the simulations for the GFT and the QRT approach, respectively. It is worth mentioning that we have considered, for this comparison, exactly the same

Table 1: Comparison of computational time (CPU time) between the Green's Functions Technique (GFT) and the Quantum Regression Theorem (QRT) approach for the numerical calculation of the emission spectrum of the cavity. The computational times were made using a commercial Intel(R) Core(TM) $i7-4770$ processor of 3.4 GHz $\times 8$, and 12 GB RAM.

Excitation number (N_{exc})	CPU time(s) for the GFT	CPU time(s) for the QRT
5	0.4	92.5
10	2.0	273.4
20	14.0	390.2
40	100.2	673.8

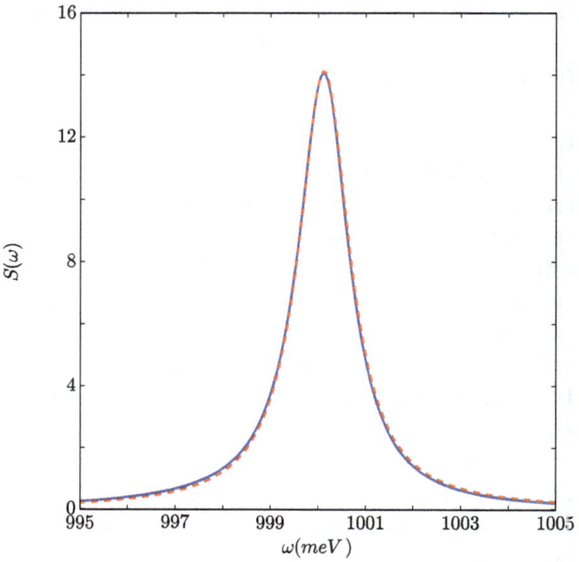

Figure 5: Emission spectrum of the quantum dot based on the GFT as a solid blue line and the corresponding numerical calculation based on the QRT approach as a dashed red line. The parameters values are $\kappa = 5$ meV, $P = 1$ meV, $g = 1$ meV, $\gamma = 0.1$ meV, $\Delta = 5$ meV and $\omega_a = 1000$ meV

resolution in the frequency domain and the numerical calculations were carried out with the same parameters values as in Fig. 4 for both methods. It is straightforward to observe that the QRT approach is time-consuming compared with the GFT approach when the excitation number is increased. From the computational point of view, it is due to the fact that the QRT approach requires solving a large number of coupled differential equations in contrast to the GFT approach which requires a relatively small system of algebraic equations.

V. Conclusions

We have presented the GFT as an alternative methodology to the QRT approach for calculating the two-time correlation functions in open quantum systems. We have applied the GFT and the QRT approach for calculating the emission spectrum in a QD-cavity system. In particular, the performance of the GFT in terms of accuracy and efficiency by comparison of the emission spectrum of the cavity and the quantum dot is demonstrated, as well as by comparison of the computational times involved during the numerical simulations. In fact, we have shown that the GFT offers a computational advantage, namely, the speeding up numerical calculations. We conclude that the GFT allows to overcome the inherent theoretical difficulties presented in the QRT approach, i.e., to find a closure condition on the set of operators involved in the dynamical equations. We mention that our methodology based on the GFT can be extended for calculating the emission spectrum in significant situations where the quantum dots are in biexcitonic regime or when the quantum dots are coupled to photonic cavities.

Acknowledgements - EAG acknowledges the finacial support from Vicerrectoría de Investigaciones at Universidad del Quindío through research grant No. 752. HVP acknowledges the financial support from Colciencias, within the project with code 11017249692, and HERMES code 31361.

Appendix

The emission spectrum in a QD-cavity system can easily be computed by taking into account that the dynamics of the operators $\hat{G}(\tau)$ and $\hat{\rho}_S(\tau)$ are governed by the same Linblad master equation, i.e., $d\hat{G}(\tau)/d\tau = \mathcal{L}\hat{G}(\tau)$ with $\mathcal{L}$ the Liouvillian superoperator discussed in section II. Moreover, it has effectively a larger tensor rank than the reduced density operator of the system. Thus, we can write the dynamical equations for the Green's functions operator in a component form

$$\frac{dG_{\tilde{\alpha}}(\tau)}{d\tau} = \sum_{\tilde{\beta}} \mathcal{L}_{\tilde{\alpha}\tilde{\beta}} G_{\tilde{\beta}}(\tau), \qquad (36)$$

together with the initial condition $G_{\tilde{\beta}}(0)$. Here, the symbol $\tilde{\alpha}$ corresponds to a composite index for labeling the states of the reduced density operator of the system, e.g., for indexing both matter and photon states in the QD-cavity system, see section III. for details. It is worth mentioning that $G_{\tilde{\beta}}$ and $\mathcal{L}_{\tilde{\alpha}\tilde{\beta}}$ act as a column vector and a matrix in this notation. In order to obtain the solution to the Eq. (36) in frequency domain, we perform a Laplace Transform and it explicitly takes the form $-\tilde{G}_{\tilde{\alpha}}(0) = \sum_{\tilde{\beta}}(\mathcal{L}_{\tilde{\alpha}\tilde{\beta}} - i\omega\delta_{\tilde{\alpha}\tilde{\beta}})\tilde{G}_{\tilde{\beta}}(i\omega)$. It is straightforward to obtain the solution by performing the matrix inversion to $\mathcal{M}_{\tilde{\alpha}\tilde{\beta}} = (i\omega\delta_{\tilde{\alpha}\tilde{\beta}} - \mathcal{L}_{\tilde{\alpha}\tilde{\beta}})$ and the emission spectrum is computed easily in terms of the initial conditions given by

$$\tilde{G}_{\tilde{\beta}}(i\omega) = \sum_{\tilde{\alpha}} \mathcal{M}^{-1}_{\tilde{\beta}\tilde{\alpha}} \tilde{G}_{\tilde{\alpha}}(0). \qquad (37)$$

The initial conditions are obtained by evaluating the Green's function operator at $\tau = 0$.

---

[1] H Walther *et al.*, *Cavity quantum electrodynamics*, Rep. Prog. Phys. **69**, 395603 (2006).

[2] A Kavokin, J J Baumberg, G Malpuech, F P Laussy, *Microcavities*, Oxford University Press (2007).

[3] H Altug, D Englund, J Vuckovic, *Ultrafast photonic crystal nanocavity laser*, Nat. Phys. **2**, 484 (2006).

[4] Y Mu, C M Savage, *One-atom lasers*, Phys. Rev. A **46**, 5944 (1992).

[5] R M Stevenson, R J Young, P Atkinson, K Cooper, D A Ritchie, A J Shields, *A semiconductor source of triggered entangled photon pairs*, Nature **439**, 179 (2006).

[6] T M Stace, G J Milburn, C H W Barnes, *Entangled two-photon source using biexciton emission of an asymmetric quantum dot in a cavity*, Phys. Rev. B **67**, 085317 (2003).

[7] N Gisin, G Ribordy, W Tittel, H Zbinden, *Quantum cryptography*, Rev. Mod. Phys. **74**, 145 (2002).

[8] C Monroe, *Quantum information processing with atoms and photons*, Nature **416**, 238 (2002).

[9] Y Todorov, I Sagnes, I Abram, C Minot, *Purcell enhancement of spontaneous emission from quantum cascades inside mirror-grating metal cavities at THz frequencies*, Phys. Rev. Lett. **99**, 223603 (2007).

[10] J Wiersig *et al.*, *Direct observation of correlations between individual photon emission events of a microcavity laser*, Nature **460**, 245 (2009).

[11] G Khitrova *et al.*, *Vacuum Rabi splitting in semiconductors*, Nat. Phys. **2**, 81 (2006).

[12] J P Reithmaier *et al.*, *Strong coupling in a single quantum dot-semiconductor microcavity system*, Nature **432**, 197 (2004).

[13] L Mandel, E Wolf, *Optical Coherence and Quantum Optics*, Cambridge University Press, (1997).

[14] O Jedrkiewicz, R Loudon, *Atomic dynamics in microcavities: absorption spectra by Green function method*, J. Opt. B. - Quantum S. O. **2**, R47 (2000).

[15] N V Hieu, N B Ha, *Time-resolved luminescence of the coupled quantum dotmicrocavity system: general theory*, Adv. Nat. Sci.: Nanosci. Nanotechnol. **1**, 045001 (2010).

[16] D F Walls, G J Milburn, *Quantum Optics*, Springer-Verlag, Berlin (1994).

[17] M Lax, *Quantum Noise. IV. Quantum Theory of Noise Sources*, Phys. Rev. **145**, 110 (1966).

[18] S Swain, *Master equation derivation of quantum regression theorem*, J. Phys. A. - Math. Gen. **14**, 2577 (1981).

[19] E del Valle, F P Laussy, C Tejedor, *Luminescence spectra of quantum dots in microcavities. II. Fermions*, Phys. Rev. B **79**, 235326 (2009).

[20] N Quesada, H Vinck-Posada, B A Rodríguez, *Density operator of a system pumped with polaritons: a Jaynes-Cummings-like approach*, J. Phys. Condens. Matter **23**, 025301 (2011).

[21] C A Vera, N Quesada, H Vinck-Posada, B A Rodríguez, *Characterization of dynamical regimes and entanglement sudden death in a microcavity quantum dot system*, J. Phys. Condens. Matter **21**, 395603 (2009).

[22] N Ishida, T Byrnes, F Nori, Y Yamamoto, *Photoluminescence of a microcavity quantum dot system in the quantum strong-coupling regimes*, Sci. Rep. **3**, 1180 (2013).

[23] T Quang, G S Agarwal, J Bergou, M O Scully, H Walther, K Vogel, W P Schleich, *Calculation of the micromaser spectrum. I. Greens-function approach and approximate analytical techniques*, Phys. Rev. A **48**, 803 (1993).

[24] E T Jaynes, F W Cummings, *Comparison of quantum and semiclassical radiation theories with application to the beam maser*, Proc. IEEE **51**, 89 (1963).

[25] M O Scully, M S Zubairy, *Quantum Optics*, Cambridge, Cambridge University Press (1996).

[26] J I Perea, D Porras, C Tejedor, *Dynamics of the excitations of a quantum dot in a microcavity*, Phys. Rev. B **70**, 115304 (2004).

[27] H P Breuer, F Petruccione, *The theory of open quantum systems*, Oxford, University Press (2002).

[28] A Rivas, S F Huelga, *Open Quantum Systems*, Springer (2012).

13

Self-sustained oscillations with delayed velocity feedback

D. H. Zanette[1*]

We study a model for a nonlinear mechanical oscillator, relevant to the dynamics of micro- and nanomechanical time-keeping devices, where periodic motion is sustained by a feedback force proportional to the oscillation velocity. Specifically, we focus our attention on the effect of a time delay in the feedback loop, assumed to originate in the electric circuit that creates and injects the self-sustaining force. Stationary oscillating solutions to the equation of motion, whose stability is insured by the crucial role of nonlinearity, are analytically obtained through suitable approximations. We show that a delay within the order of the oscillation period can suppress self-sustained oscillations. Numerical solutions are used to validate the analytical approximations.

I. Introduction

Inside any modern time-keeping device, the principal component is an oscillator which autonomously generates a stationary periodic signal with a well-defined frequency. The only external input to the system is the power needed to sustain the oscillations. In devices based on mechanical oscillators –which comprise essentially all present-day clocks, with the exception of the atomic kind–, sustained periodic motion is achieved by forcing the oscillator with a conditioned version of the signal generated by the oscillator itself [1]. If this reinjected force is in-phase with the oscillation velocity, the resonant response is maximal, thus optimizing power consumption during the process of signal conditioning.

Purely mechanical devices employed this feedback scheme already in the Middle Ages. In fact, the escapement mechanism was routinely used in clock building since the 13th century. In modern clocks, oscillators are built from synthetic quartz crystals and feedback is implemented electronically. At the micro- and nanoscale, quartz crystals are expected to be replaced by simpler mechanical oscillators such as tiny vibrating silica beams [2, 3], which are easily built during circuit printing and can be actuated by very small electric fields [4, 5].

In a series of recent experiments on micromechanical oscillators, it has been shown that self-sustained oscillations can be achieved with a feedback force *proportional* to the oscillation velocity [6]. Since a purely *linear* mechanical system cannot display stable periodic motion, this kind of feedback force must necessarily be compensated by some *nonlinear* contribution from the oscillator dynamics itself. In the experiments, this balance turned out to come from the damping force, which was proportional to the velocity but showed a nonlinear dependence on the oscillation amplitude.

In the present paper, we analyze a model for self-sustained periodic motion in a mechanical oscillator subjected to a feedback force proportional to the velocity, and nonlinear amplitude-dependent terms both in damping and in the restoring force. Emphasis is put on the effects of time delays in the

*E-mail: zanette@cab.cnea.gov.ar

[1] Centro Atómico Bariloche and Instituto Balseiro (Comisión Nacional de Energía Atómica, Universidad Nacional de Cuyo), Consejo Nacional de Investigaciones Científicas y Técnicas, Avda. E. Bustillo 9500, 8400 San Carlos de Bariloche, Río Negro, Argentina.

feedback circuit, which modify the phase shift between the feedback force and the velocity. It is shown that these delays affect the response of the oscillator, to the point that stable periodic motion can even be suppressed. The model is studied analytically within suitable approximations, and the results are compared with numerical solutions to the equation of motion.

II. Mechanical model for self-sustained oscillations

Our model is based on an equation of motion for a one-dimensional variable $x(t)$, which represents the departure from equilibrium of a mechanical oscillator:

$$\ddot{x} + \mu(1 + \alpha x^2)\dot{x} + (1 + \beta x^2)x = g\dot{x}(t - \tau), \quad (1)$$

where μ is the damping coefficient per unit mass. The coefficients α and β weight the amplitude-dependent nonlinear corrections to the damping force and to the elastic force, respectively. The former is a Van der Pol-like nonlinearity, while the cubic contribution to the restoring force defines a Duffing oscillator [7]. Both kinds of nonlinearity, with $\alpha, \beta > 0$, have been experimentally verified to occur in micromechanical oscillators formed by silica beams clamped at their two ends (clamped-clamped, or c-c beams [4, 6]). In the main oscillation mode, c-c beams vibrate much like a plucked string, so that $x(t)$ can be associated with the displacement of the middle point of the beam with respect to its rest position.

The right-hand side of Eq. (1) represents the feedback force per unit mass, which is proportional to the velocity at the delayed time $t - \tau$. As stated in the introduction, this delay is expected to originate in the electric circuit that reads, conditions, and reinjects the oscillator signal, due to the time elapsed during signal processing. Although τ should be a very short time, it is not necessarily negligible as compared with other time scales in the system, in particular with the oscillation period. In fact, micro- and nano-oscillators vibrate with frequencies from the order of 100 kHz [4] to 1 GHz [8].

Time units in Eq. (1) have been chosen in such a way that the natural frequency of the undamped ($\mu = 0$), linear ($\beta = 0$), unforced ($g = 0$) oscillator equals unity. Meanwhile, the units of x can be fixed in such a way that the coefficient α adopts any prescribed value. Hence, without generality loss and for future convenience, we fix $\alpha = 4$.

i. Approximate stationary solutions

Nonlinearity in Eq. (1) prevents obtaining an exact solution. Approximate stationary solutions can be found by the standard procedure of neglecting higher-harmonic contributions to the oscillations [7]. Within this approximation, and proposing $x(t) = \frac{1}{2}A\exp(i\omega t) + \text{c.c.}$, we get a complex algebraic equation whose real and imaginary parts read

$$1 - \omega^2 + bA^2 = g\omega \sin \omega\tau, \quad (2)$$

with $b = 3\beta/4$, and

$$\mu(1 + A^2) = g \cos \omega\tau. \quad (3)$$

These are equations for the unknowns A and ω, which have been obtained assuming $A, \omega \neq 0$.

Note that, to have a non-negative solution for A^2, Eq. (3) requires that $g \geq \mu$. Physically, this amounts to require that the energy input from the feedback force is not less than the energy dissipated by damping. Otherwise, stationary periodic motion cannot be sustained. In the following, thus, we work under the assumption that such condition holds.

Squaring and summing up Eqs. (2) and (3), we obtain

$$(1 - \omega^2 + bA^2)^2 + \mu^2\omega^2(1 + A^2)^2 = g^2\omega^2, \quad (4)$$

namely, a relation between the oscillation amplitude A and the frequency ω which does not involve explicitly the delay τ. Black (full and dotted) lines in Fig. 1 represent this relation for the parameters indicated in the figure.

The graph of the relation between amplitude and frequency expressed by Eq. (4) can be interpreted as a resonance curve, in the sense that it characterizes the response of the oscillator to the feedback force. In fact, its shape resembles the upper part

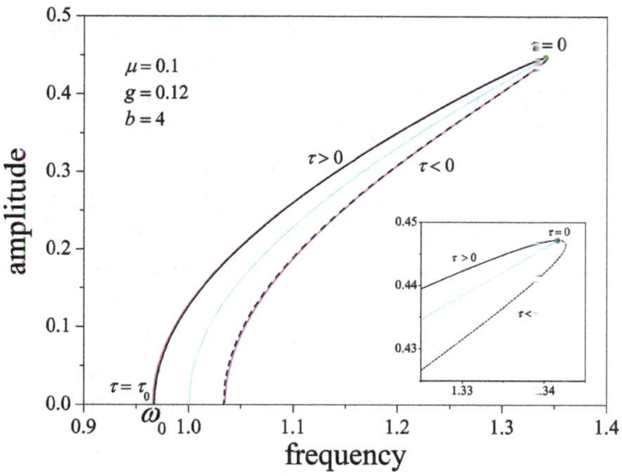

Figure 1: Amplitude-frequency relation, Eq. (4), for $\mu = 0.1$, $g = 0.12$, and $b = 4$. Full and dotted black lines show the branches where τ is positive and negative, respectively. The green dot stands at the point of maximal amplitude, corresponding to delay $\tau = 0$. For positive delays, oscillations become suppressed for $\tau = \tau_0$, at frequency ω_0, where the amplitude vanishes. The magenta line is the short-delay approximation, and the cyan line is the backbone curve. The inset shows a close-up around the peak.

of the Duffing resonance curve [7], and its mathematical origin is similar. From the physical point of view, however, it is important to stress that ω is not a control parameter –as in the standard case of an externally forced system– but emerges as an autonomous dynamical property of the self-sustained oscillator. The parameter whose variation defines the resonance curve of Fig. 1 is, on the other hand, the time delay τ. In the figure, we have highlighted the point corresponding to $\tau = 0$ which, as can be seen from Eq. (3), corresponds to the maximum of the amplitude,

$$A_{\max} = \sqrt{\frac{g}{\mu} - 1}. \tag{5}$$

Note that, due to the smooth profile of the resonance curve at its peak, the frequency at $\tau = 0$, given by $\sqrt{1 + bA_{\max}^2}$, does not equal the maximum frequency attainable by the oscillator, but lies slightly below. This is better appreciated in the figure inset.

Different line types (full and dotted) in Fig. 1 represent the branches of positive and negative τ. Naturally, in an experiment, only the branch of positive time delays can be observed. However, stationary oscillatory solutions exist –and, as we discuss later, are stable– irrespectively of the sign of τ.

ii. Oscillation suppression

The most relevant dynamical feature emerging from Eqs. (2) and (3) is that, as τ varies from zero to both positive and negative values, the oscillation amplitude decreases and, eventually, vanishes for sufficiently long delays. This situation is reached at the points where the resonance curve in Fig. 1 intersects the horizontal axis. Focusing on the branch of positive delays, oscillations are suppressed when their frequency reaches

$$\omega_0 = \sqrt{1 + \Delta - \sqrt{(1+\Delta)^2 - 1}}, \tag{6}$$

with $\Delta = (g^2 - \mu^2)/2$, corresponding to a critical delay

$$\tau_0 = \omega_0^{-1} \left| \arccos \frac{\mu}{g} \right|. \tag{7}$$

Oscillation suppression can be understood in qualitative terms as a consequence of the increasing phase shift between the feedback force and the velocity when τ grows. As the phase shift becomes larger, the "timing" of energy supply by feedback fails to counteract energy dissipation by damping. The energy balance –implicit in Eq. (3)– cannot be further maintained, and periodic motion dies out.

Figure 2 shows the critical delay τ_0 at which oscillations are suppressed, as a function of the feedback amplitude g, and for various values of the damping μ. For g just above μ, τ_0 grows from zero and –if μ is small enough– attains a plateau around $\pi/2$. Note that, since the oscillation frequency is always near unity, in this plateau the phase shift between feedback and velocity is also close to $\pi/2$. In other words, in this zone the feedback force is in-phase with the displacement $x(t)$. For larger values of g, τ_0 leaves the plateau and grows further.

We point out that very small values of the damping μ, as considered in Fig. 2, are realistic when

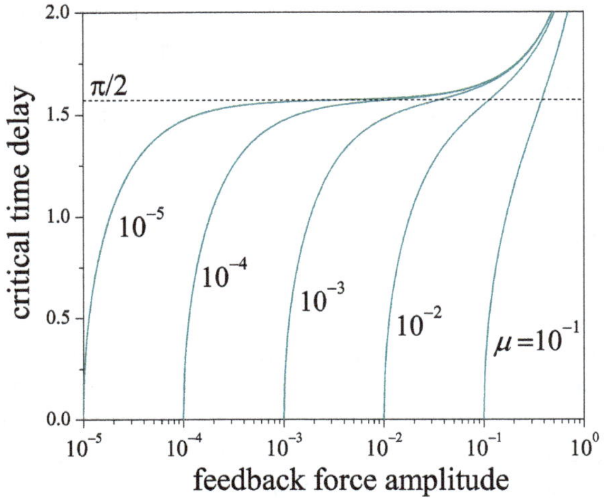

Figure 2: Critical time delay for oscillation suppression, τ_0 as given by Eq. (7), vs. the feedback force amplitude g, for various values of the damping coefficient μ.

working with micro- and nanomechanical oscillators. Indeed, these devices have quality factors Q ($\sim \mu^{-1}$) typically above 10^4 [4]. In Fig. 1, on the other hand, we have chosen a relatively large value of μ for clarity in the graphical representation. For very small values of μ, in fact, the resonance curve becomes too narrow for its features to be clearly discerned in a plot.

iii. Stability of self-sustained oscillations

Stability of the stationary oscillatory solutions presented above can be assessed by the method of multiple time scales [7], which provides equations of motion for the relatively slow dynamics of the oscillation amplitude and phase. The method works under the assumption that there is a clear separation between the oscillation period and other time scales involved in the system. In the present case, this condition is fulfilled for small damping, $\mu \ll 1$, as it is energy dissipation which controls the relaxation of amplitude and frequency to their asymptotic values.

Application of the method of multiple scales to Eq. (1) straightforwardly shows that when oscillatory solutions of non-vanishing amplitude do exist, they are stable under arbitrary perturbations to their amplitude and phase (or frequency), irrespectively of the sign of the delay. Instead of giving the details of this standard calculation, we resort here to a simple physical argument to explain stability, which sheds useful light on the role of nonlinearity.

Assume that the oscillator is in stationary periodic motion. This implies, in particular, that the energy input from feedback is exactly balanced by dissipation. If motion is now perturbed in such a way that the oscillation amplitude increases, the velocity increases accordingly, thus making feedback and damping to grow in the same proportion (since both are proportional to the velocity). However, due to the amplitude-dependent nonlinear correction in the damping coefficient, there is an extra growth in damping which enhances dissipation and therefore counteracts the perturbation, as the amplitude will tend to decrease. A symmetric argument applies if, on the contrary, the amplitude is perturbed to lower values.

If, on the other hand, the perturbation makes the frequency to increase, the velocity also increases as so do feedback and damping. Through the cubic nonlinearity in the restoring force, however, such change in the frequency leads the amplitude to grow. This growth, in turn, implies an enhancement in dissipation, as explained in the preceding paragraph. Consequently, the perturbation is counteracted by the oscillator response. The case where the frequency decreases is analogous. Under perturbations of both amplitude and frequency, therefore, nonlinearity plays a key role in insuring the stability of stationary oscillations.

III. Further approximations

i. Backbone approximation

As mentioned in section II.ii., a realistic limit when working with micro- and nanomechanical oscillators consists in considering very small values of the damping coefficient μ. In order to insure the existence of physically meaningful solutions for the amplitude and the frequency, however, Eq. (3) requires that the limit of $\mu \to 0$ is taken along with the limit $g \to 0$, maintaining a finite ratio $r = g/\mu$. In physical terms, this joint limit is justified by the fact that the smaller the rate of energy dissipation, the smaller the feedback force necessary to maintain oscillatory motion.

In the crudest approximation, the resonance

curve collapses to a single-valued curve with equation

$$A = \sqrt{\frac{1-\omega^2}{b}}, \tag{8}$$

plotted as a cyan line in Fig. 1. This is the so-called backbone approximation to the resonance curve [7, 9]. Within this approximation, the time delay as a function of the frequency along the backbone curve is

$$\tau = \omega^{-1} \arccos\left(\frac{1+b-\omega^2}{rb}\right). \tag{9}$$

Although, for the relatively large value of μ considered in Fig. 1, the backbone curve gives a poor approximation to the resonance curve; as μ becomes smaller the two branches of the latter collapse to the backbone. A measure of this collapse is provided by computing the frequency of oscillation suppression, ω_0, in the limit of small damping. Equation (6) yields

$$\omega_0 \approx 1 - \frac{\mu}{2}\sqrt{r^2-1}, \tag{10}$$

showing that the width of the resonance curve at its base ($A = 0$) is proportional to μ. In this limit, the time delay for oscillation suppression is still given by Eq. (7).

ii. Short-delay approximation

Another relevant approximation, which drastically simplifies the problem of solving Eqs. (2) and (3), is obtained for small values of τ. In fact, approximating $\sin\omega\tau \approx \omega\tau$ and $\cos\omega\tau \approx 1-\omega^2\tau^2/2$, we obtain linear equations for A^2 and ω^2, whose solutions yield

$$A = \sqrt{\frac{2(g-\mu)(1+g\tau)-g\tau^2}{2\mu(1+g\tau)+bg\tau^2}} \tag{11}$$

and

$$\omega = \sqrt{2\frac{b(g-\mu)+\mu}{2\mu(1+g\tau)+bg\tau^2}}. \tag{12}$$

The magenta line in Fig. 1 represents these results. For the parameters of the figure, thus, this is an excellent approximation along the entire resonance curve.

IV. Numerical results

As a validation of the analytical results obtained by means of the approximation considered in section II., i.e., neglecting the higher-harmonic contributions to oscillatory motion, we have numerically solved Eq. (1) for various parameter sets. We have used a fourth-order Runge-Kutta scheme, with the only non-standard feature that, due to the delay in the feedback force, it is necessary to specify the solution $x(t)$ for all times $-\tau \le t \le 0$, instead of the initial condition at $t = 0$. In all cases, we have considered a constant value of $x(t)$, hence with $\dot{x}(t) = 0$, in that interval.

For all the parameter sets taken into account in the numerical calculations, we have found that, after a transient time which scaled as μ^{-1}, the oscillator reached periodic motion. This observation is in qualitative agreement with our argument of section II.iii., which predicts stability of periodic motion under very general conditions.

A more quantitative comparison between theoretical and numerical results, illustrated for the parameter set of Fig. 1, is given in Fig. 3. The main panel shows, as green dots, numerical measurements of the frequency and amplitude of long-time periodic motion for various (positive) values of the delay in the interval $0 \le \tau \le 0.4$. The black full line stands for the theoretical result. The inset shows the same data in the amplitude-delay plane. We see that the agreement is generally very good. Not unexpectedly, the theoretical approximation improves towards smaller amplitudes, where nonlinear effects are weaker and periodic motion is better described by a pure harmonic oscillation, as assumed in section II. As a matter of fact, Fig. 3 is limited to amplitudes above 0.25 since, in the plot, theoretical and numerical results are indistinguishable for smaller values.

The agreement between our theoretical and numerical results rises the question to what extent higher-harmonic components, which are a direct product of nonlinearity but have been neglected in our analytical approximation, constitute a substan-

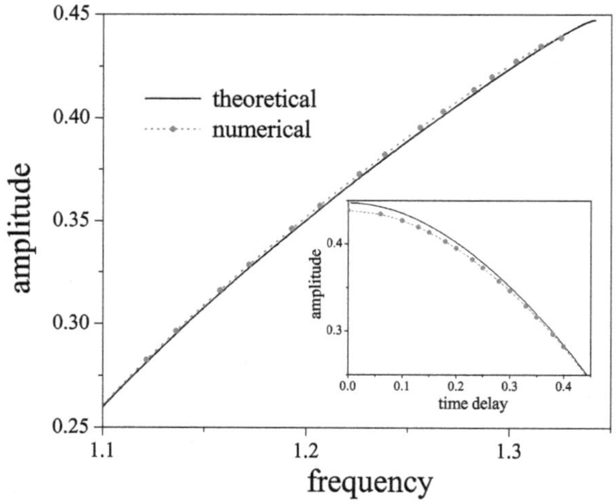

Figure 3: Theoretical (black line) and numerical (green dots) results for the amplitude-frequency relation of periodic oscillatory motion, with $\mu = 0.1$, $g = 0.12$, and $b = 4$, and for various values of the delay in the interval $[0, 0.4]$. The inset shows the same data in the amplitude-delay domain. Dotted green lines have been plotted as a guide to the eye.

tial contribution to oscillatory motion. To evaluate this, we calculate the amplitudes of different Fourier components in the long-time numerical solution for $x(t)$. With the parameters of Fig. 3 and $\tau = 0.1$, for which the oscillation amplitude and frequency are $A = 0.428$ and $\omega = 1.303$, the first Fourier amplitude (corresponding to frequency ω) is $A_1 = 0.42$. Due to the cubic nonlinearity in the restoring force, in turn, the next significant amplitude corresponds to the third-harmonic component (frequency 3ω), $A_3 = 0.0076$. Higher-harmonic amplitudes are even smaller. The first correction to harmonic motion, thus, is almost two orders of magnitude weaker than the main contribution, which reasonably justifies our analytical approximation. On the other hand, this modest contribution of higher-harmonic components to the overall motion markedly contrasts with the sizable phenomenology studied in section II., which is also a direct consequence of nonlinearity.

V. Conclusions

We have analyzed the dynamics of a mechanical oscillator whose periodic motion is sustained by a feedback force proportional to the oscillation velocity. The key ingredient that makes oscillations stable is a nonlinear dependence of damping with the oscillation amplitude, such that energy dissipation increases or decreases when the amplitude respectively grows or drops. This kind of nonlinearity, together with a cubic component in the restoring force, has been experimentally observed to occur in c-c beam micromechanical oscillators [6], which can therefore exhibit self-sustained motion under the action of linear velocity feedback.

Our emphasis was put on the effect of a time delay in the feedback force, assumed to originate in the electric circuit that reads, conditions, and reinjects the oscillation signal. The most significant consequence of this ingredient is that oscillation can be suppressed if the delay is large enough. It is well known that differential equations with time-delayed terms can exhibit either stationary oscillatory solutions or fixed rest points, depending on the delay [10]. Also, oscillation suppression (or "death") due to delays has been reported to occur in a variety of dynamical systems, such as synchronized limit-cycle oscillators [11,12]. Here, this same occurrence has been characterized for a nonlinear mechanical system that can become relevant for the design of micromechanical time-keeping devices.

In connection with possible applications of this phenomenology, it is worth mentioning that classical-mechanical models such as Eq. (1) provide a suitable starting point for the description of micro- and nanomechanical machines. As length scales become smaller, however, the effects of thermal fluctuations and electrical noise cannot be further ignored [13, 14]. From the viewpoint of our model, therefore, an important step forward would be to include noise in the theoretical analysis.

Acknowledgements - This project has been supported by Agencia Nacional de Promoción Científica y Tecnológica, Argentina, through grant PICT2014-1611. Fruitful collaboration with D. Antonio, S. Arroyo, Changyao Chen, D. Czaplewski, J. Guest, D. López, and F. Mangussi, as well as discussions with S. Risau Gusmn, are gratefully acknowledged.

---

[1] B Yurke, D S Greywall, A N Pargellis, P A Busch, *Theory of amplifier-noise evasion in an oscillator employing a nonlinear resonator*, Phys. Rev. A **51**, 4211 (1995).

[2] D Bishop, P Gammel, R Giles, *The little machines that are making it big*, Phys. Today **54**, 38 (2001).

[3] K L Ekinci, M L Roukes, *Nanoelectromechanical systems*, Rev. Sci. Instrum. **76**, 061101 (2005).

[4] D Antonio, D H Zanette, D López, *Frequency stabilization in nonlinear micromechanical oscillators*, Nat. Commun. **3**, 806 (2012).

[5] D Antonio, D A Czaplewski, J R Guest, D López, S I Arroyo, D H Zanette, *Nonlinearity-induced synchronization enhancement in micromechanical oscillators*, Phys. Rev. Lett. **114**, 034103 (2015).

[6] Changyao Chen, D H Zanette, J R Guest, D A Czaplewski, D López, *Self-sustained micromechanical oscillator with linear feedback*, Phys. Rev. Lett. **117**, 017203 (2016).

[7] A H Nayfeh, D T Mook, *Nonlinear Oscillations*, Wiley, New York (2008).

[8] H B Peng, C W Chang, S Aloni, T D Yuzvinsky, A Zettl, *Ultrahigh frequency nanotube resonators*, Phys. Rev. Lett. **97**, 087203 (2006).

[9] S I Arroyo, D H Zanette, *Duffing revisited: Phase-shift control and internal resonance in self-sustained oscillators*, Eur. Phys. J. B **89**, 12 (2016).

[10] T Erneux, *Applied Delay Differential Equations*, Springer, New York (2009).

[11] D V Ramana Reddy, A Sen, G L Johnston, *Time delay induced death in coupled limit cycle oscillators*, Phys. Rev. Lett. **80**, 5109 (1998).

[12] S H Strogatz, *Death by delay*, Nature **394**, 316 (1998).

[13] A N Cleland, M L Roukes, *Noise processes in nanomechanical resonators*, J. Appl. Phys. **92**, 2758 (2002).

[14] P Ward, A Duwel, *Oscillator phase noise: systematic construction of an analytical mode encompassing nonlinearity*, IEEE Trans. Ultrason. Ferroelectr. Freq. Control **58**, 195 (2011).

14

Photoionization of water molecules by high energy photons

L. Martini,[1*] D. I. R. Boll,[1] O. A. Fojón[1†]

We theoretically study the photoionization of water molecules by high energy photon impact. We develop a model in which the final state wavefunction is given by a Coulomb continuum wavefunction with effective charges and the water molecule bound states are represented using the Moccia's monocentric wavefunctions. We obtain analytical expressions for the transition matrix element that enable the computation of cross sections by numerical quadratures. We compare our predictions for photon energies between 20 and 300 eV with more elaborated theoretical results and experiments. We obtain a very good agreement with experiments, in particular, at enough high energies where there is a lack of elaborated results due to their high computational cost.

I. Introduction

Ionization of water molecules is of interest in many domains such as radiochemistry and radiobiology, as living cells are composed mostly by water. In particular, ionization of water molecules was studied in several experimental and theoretical works (see, for instance, [1–9]).

The theoretical treatment of these reactions is rather involved as the nuclear degrees of freedom of the water molecule must be taken into account as well as a proper treatment of the continuum states in the final channel of the reaction. To circumvent these inconvenients, elaborated methods were developed to obtain cross sections for photoionization of water, disregarding the vibrational and/or the rotational molecular structure. For instance, Stener *et al.* [2] studied the electronic transitions within the framework of the Density Functional Theory (DFT) with two different exchange-correlation potentials: VWN and LB94. The first is the most common parametrization within the Local Density Approximation (LDA), whereas the second is a model potential with the correct asymptotic behavior at large distances, that is, it decays like $-1/r$. Moreover, they considered the ground-state (GS) configuration where all electrons are kept in the calculation and the transition-state (TS) configuration where half an electron is removed from the ionized orbital. The former configuration is computationally less CPU time consuming because the ionizations of all orbitals can be obtained with the same potential, whereas in the latter, the potential depends on the orbital which is ionized and thus continuum states must be calculated separately for each state. They obtain cross sections up to approximately 70 eV. In a recent work by Gozem *et al.* [1] in which Dyson orbitals are used for the bound states, the final continuum states are represented by Coulomb waves describing the interaction between the active electron and the residual target. This representation is usual in high energy methods (see [10,11] and references therein). The vibrational structure is considered through the Franck-

*E-mail: martini@ifir-conicet.gov.ar
†E-mail: fojon@ifir-conicet.gov.ar

1 Laboratorio de Colisiones Atómicas, Instituto de Física Rosario (CONICET-UNR), Blvd. 27 de Febrero 210 bis, Rosario 2000, Argentina.

Condon approximation. Besides, they employ partial wave expansions to obtain matrix transition elements. In addition, they use different effective charges Z_{ef} for the Coulomb waves for several molecules, such as formaldehyde, ethylene, acetylene, 1-butene, 1-butyne, methanol, and propyne. The interest was mostly the obtention of cross sections near threshold of the reactions, *i.e.*, at low incident energies. Because of the partial wave expansions, it is expected that the above mentioned models become computationally cumbersome at higher photon energies. However, cross sections at these energies are needed to contrast with experiments and to back research on matter irradiation and related fields. Consequently, other strategies must be envisaged to obtain observables of the reaction at high impact energies.

Therefore, we adopt here a different approach and develop a Coulomb Continuum (CC) model valid at intermediate and high impact energies. We avoid the use of the Lippman-Schwinger equation or any partial wave technique that may complicate the obtention of observables at high photon energies. At first, we consider that at high enough impact energies, we can simplify the problem by assuming that the reaction occurs in a lapse much smaller than the typical vibrational and rotational times. Then, we assume that the molecule is frozen at the equilibrium configuration during the collision time. In this way, cross sections for fixed-in-space molecules can be obtained. As in most experiments, water molecules are randomly oriented, one can subsequently average over all molecular orientations to contrast the theoretical predictions. Moreover, we reduce the multielectronic problem to a monoelectronic one by considering that during the collision time one electron (the active one) is ionized and the other electrons (the passive ones) remain in their initial orbitals. In this way, our model is different from previous ones where relaxation of the target is allowed through the use of Dyson orbitals [1]. Again, our approximation is justified because the collision time is smaller than the relaxation time of the molecule. All these assumptions were employed with success in previous works (see, for instance, [10, 11] and references therein), rendering the problem tractable to obtain reliable cross sections. Furthermore, we describe the bound states of the water molecule by quantum chemistry methods. In particular, Moccia's wavefunctions offer a good description of water in terms of monocentric orbitals [12]. Even if the bound states are accessible through standard methods, the continuum states of the active electron in the final channel are still a challenge to describe in a realistic way the ionization of complex molecules. The multicenter nature of the residual target introduces complications in the theoretical descriptions. However, approximations taking into account the screening provoked by the passive electrons may be employed.

We compute cross sections within the dipole approximation for several water orbitals. We employ Moccia's wavefunctions for the bound states of the target as well as Coulomb continuum wavefunctions to describe the continuum state of the photoelectron with different effective charges to include in some way the influence of the passive electrons on the reaction. Guided by physical assumptions, we consider several effective charges to better describe the experiments. The main objective of our work is to provide reliable cross sections for photoionization of water molecules at high photon impact energies (approximately greater than 70 eV) where involved methods that work properly at low photon energies may turn to be prohibitive from a computational point of view.

Unless otherwise stated, atomic units are used.

II. Theory

i. Coulomb Continuum (CC) model

In the dipole approximation, the transition matrix amplitude for photoionization by a photon of frequency ω within the velocity *gauge* reads:

$$M_{ph}^{dip} = -i\langle\psi_f|\mathbf{e}\cdot\nabla|\psi_i\rangle, \tag{1}$$

where ψ_i denotes the electron target bound state in the initial channel and ψ_f represents the electron continuum state in the final channel. Moreover, $\mathbf{e}$ is the polarization vector of the incident radiation.

Within an independent electron model in which the active electron is ionized and the passive electrons remain in their initial electronic configurations, we use the monocentric wavefunctions given by Moccia [12] to describe the different water orbitals ψ_i. This representation takes into account the electronic structure of the molecule in a basis of monocentric wavefunctions centred on the heavier atom (the oxygen atom). In addition, we work with the Coulomb Continuum (CC) model in which the final continuum state of the active electron is chosen as

$$\psi_f(\mathbf{r}) = (2\pi)^{-3/2} e^{i\mathbf{p}\cdot\mathbf{r}} N_p G(\mathbf{r}) , \tag{2}$$

where the normalization factor is given by

$$N_p = e^{\pi\nu/2}\Gamma(1 + i\nu) , \tag{3}$$

with $\nu = Z_{ef}/p$ the Sommerfeld parameter where Z_{ef} and $\mathbf{p}$ are the effective residual charge of the target and the final electron momentum respectively, and

$$G(\mathbf{r}) = {}_1F_1(-i\nu; 1; -i\,(pr + \mathbf{p}\cdot\mathbf{r})) , \tag{4}$$

is the hypergeometric confluent function.

The final wavefunction of the CC model (Eq. (2)) takes into account the Coulomb interaction of the electron with the residual target in the final channel of the reaction. It was shown that this interaction must be accounted for to describe the photoionization reaction with neutral targets in a realistic way [1, 13, 14].

By employing the Nordsieck integral, we obtain analytical expressions for the matrix transition element given by Eq. (1) (see Appendix) for each molecular orbital of the water molecule.

ii. Cross sections

Differential cross sections as a function of the photon energy and the solid angle of the ejected electron may be obtained as [15],

$$\frac{d\sigma_{ph}(\omega)}{d\Omega_e} = \frac{4\pi^2\alpha p}{\omega}|M_{ph}|^2, \tag{5}$$

where α is the fine structure constant, ω is the frequency of the incident photon, $p = |\mathbf{p}|$ is the magnitude of the asymptotic ejected electron momentum, $d\Omega_e = sin\theta\ d\theta\ d\varphi$ where Ω_e is the solid angle of the ejection direction, and M_{ph} is the matrix transition amplitude between the final and initial electronic states.

Integrating over all directions of ejection, one obtains total cross sections as a function of the photon energy,

$$\sigma_{ph}(\omega) = \int d\Omega_e \left(\frac{d\sigma_{ph}(\omega)}{d\Omega_e}\right). \tag{6}$$

By using our analytical expression for the matrix transition element, we are able to obtain reliable cross sections at intermediate and high photon energies by standard numerical quadratures with a high degree of precision and very low CPU time.

III. Results

We present our cross sections for the single photoionization of randomly oriented water molecules by photons of frequency ω linearly polarized. We compare our predictions with available theoretical and experimental results. To compare with experiments, we average our cross sections over all orientations of the water molecule.

We compute our total cross sections within the dipole approximation by using the matrix transition amplitude given by Eq. (1). We have verified that the results obtained within the length and velocity gauge are almost the same in the present energy domain. In relation to the electronic final continuum states, we have considered different effective charges Z_{ef} for the Coulomb function describing the photoelectron in the final channel. Gozem *et al.* [1] also employed effective charges for several molecules other than water.

In Fig. 1, we compare our predictions for the $1b_1$, $3a_1$ and $1b_2$ orbitals and the total contribution with previous theoretical and experimental results. We show our calculations for different effective charges Z_{ef} for the photoelectron wavefunction.

In the case of the $1b_1$ orbital, the asymptotic charge value $Z_{ef} = 1$ is expected to be the best choice. As a matter of fact, we can observe in Fig. 1(a) that cross sections computed with this value are in very good agreement with experiments. From a physical point of view, $Z_{ef} = 1$ corresponds to the asymptotic charge of the residual target with the nuclear charges of the three centers of the water molecule completely screened by the passive electrons. Although not shown here, we have checked that effective charges less than unity give a worse agreement with experiments. Then, the ejection of an electron from the most external water orbital may be viewed as if the active electron were ejected with the hydrogen atoms acting as spectators. This is due to the non-ligand character of the $1b_1$ orbital that is mostly atomic in character (in fact, no s-component from the H atoms is present

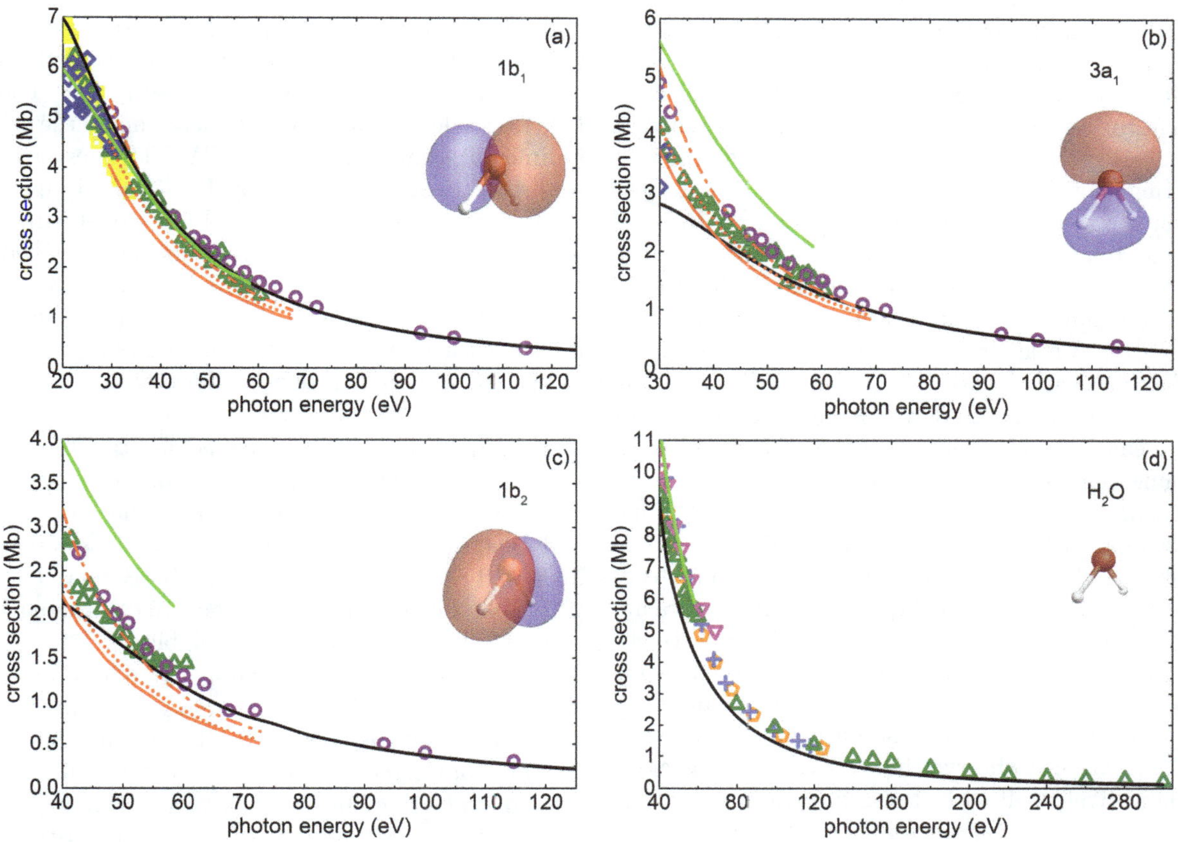

Figure 1: Total cross sections as a function of the photon energy for photoionization. Our results (▬) for (a) $1b_1$ orbital with $Z_{ef} = 1$, (b) $3a_1$ orbital with $Z_{ef} = 0.6$, (c) $1b_2$ orbital with $Z_{ef} = 0.1$ and (d) the total contribution. Calculations from [1] (▬) and from [2] for VWN GS (-·-), VWN TS (···) and LB94 GS (▬). Experiments: [3] (□), [4] (△), [5] (◊), [6] (○), [7] (⬠), [8] (+) and [9] (▽).

in the Moccia's wavefunction for this orbital). On the contrary, ejection from the $3a_1$ and $1b_2$ ligand orbitals necessarily involves the influence of the nuclear charges of the three nuclei of the molecule. If the active electron is emitted from the neighborhoods of the oxygen or hydrogen nuclei, it will experience an effective charge corresponding to the nuclear ones screened by the passive electrons. In the case of the hydrogen nuclei, this effective charge is necessarily less than unity. Then, an effective charge less than unity could take into account the influence of all the nuclear charges of the molecule in an averaged way. This may be related to the fact observed by Gozem *et al.* [1] in the description of the bound states by Dyson orbitals for several molecules, *i.e.*, the effective charge at the centroid of such orbitals is smaller than the charge of the corresponding polyatomic cation.

Therefore, in order to describe properly ionization from the $3a_1$ and $1b_2$ orbitals, effective charges less than unity should be employed. It must be mentioned that in the method employed by Gozem *et al.* [1], the effective charge for water is taken always as $Z_{ef} = 1$. This may be linked to the fact that their effective charges are related to the Mulliken population analysis in which the charges of the H atoms are summed onto heavy atoms. In the case of water with only one heavy atom (the oxygen atom), this leads to an effective charge $Z_{ef} = 1$. In this way, their cross sections for these orbitals overestimate the measurements.

Then, we compute cross sections for those inner states of water by using adequate effective charges. The optimal effective charges could be obtained formally by variational methods. Gozem *et al.* [1] develop a scheme for a one dimensional simulation with a model potential to find optimal effective charges. As a result, the optimized effective charges are very close to the ones obtained through the best fitting to the measured cross sections. It is clear that a variational procedure applied to the real three-dimensional problem may turn quite lengthy and difficult to implement due to the multicenter and multielectronic character of the molecule. However, we are not interested here in obtaining effective charges with a high numerical precision. Instead, we prefer to determine approximate effective charges and explore their physical meaning in connection with the mechanisms involved in the ionization reaction. So, we chose a simpler and practical method. After the inspection of several effective charges less than unity, we conclude that the best choices for the $3a_1$ and $1b_2$ orbitals are approximately $Z_{ef} = 0.6$ and $Z_{ef} = 0.1$, respectively.

In Fig. 1(a), we observe that our approximation for the $1b_1$ orbital with $Z_{ef} = 1$ gives an excellent agreement even at photon energies as low as 20 eV. We can observe that our calculations agree qualitatively well with the ones by Stener *et al.* [2] for photon energies greater than 30 eV where they have almost the same slope. In particular, the best quantitative agreement with our cross sections is observed for the VWN choice with the GS configuration. In addition, we can also see that our cross sections are in very good accordance with the ones obtained with Dyson orbitals [1] for photon energies between 40 and 60 eV. However, there is some discrepancies at lower energies. Our calculations agree better with the ones given by Brion *et al.* [3] whereas the ones obtained with Dyson orbitals [1] are in general accordance with the ones reported by Truesdale *et al.* [5]. Finally, we remark that our predictions agree well with the experiments at high impact energies (around 100 eV) by Banna *et al.* [6] where there is a lack of theoretical results.

We consider now the other orbitals of water, paying attention to the use of effective charges different from unity. In Fig. 1(b), we show our results for the $3a_1$ orbital together with other predictions and experiments. Our cross sections are computed with $Z_{ef} = 0.6$, obtaining a very good agreement with the experiments especially at high photon energies (greater than 70 eV). As it can be seen, the calculations by Gozem *et al.* [1] with Dyson orbitals and an effective charge $Z_{ef} = 1$ overestimate the measurements. On the contrary, the cross sections by Stener *et al.* [2] give a better description. The LB94 GS calculations tend to underestimate the experimental results whereas the VWN TS ones agree well with the measurements by Tan *et al.* [4] up to approximately 50 eV. The VWN GS model agrees well with the experiments by Banna *et al.* [6] at low energies and with the results by Tan *et al.* [4]; about 60-70 eV are very close to our predictions that give the correct trend for the results by Banna *et al.* [6] where there exist a lack of theoretical results.

A similar situation is observed for the $1b_2$ orbital in Fig. 1(c), where we show our results obtained with $Z_{ef} = 0.1$, along with other theories and experiments. Again, our cross sections reproduce well the experimental results by Banna *et al.* [6] for photon energies greater than 70 eV where no theoretical cross sections are available.

Finally, in Fig. 1(d), we present our total cross sections for water obtained summing the contribution of the previous relevant orbitals at the photon energies considered. We observe that the results by Gozem *et al.* [1] are in general agreement for experiments at low photon energies whereas our results tend to underestimate them. On the contrary, our predictions are in very good agreement with the measurements by Tan *et al.* [4] at high photon energy.

IV. Conclusions

We have computed photoionization total cross sections as a function of the photon energy within the dipole approximation for photon energies from 20 eV and up to 300 eV. To describe the final continuum state, we employ Coulomb functions with effective charges that take into account the screening of the nuclear charges by the passive electrons. We obtain very good agreement with more elaborated previous results as well as with experiments except perhaps at low incident energies where our model is not expected to be valid. In particular, our results agree very well with measurements in the region of high enough photon energy where there is a lack of theoretical predictions. In this sense, our work aims at filling a gap providing accurate cross sections in this region. Moreover, the computation of cross sections with our CC model for other molecules may be performed easily. Cross sections for molecules with a heavy central atom such as HF, CH_4, SiH_4, H_2S, HCl, NH_3, NH_4^+, PH_3, PH_4^+, for which Moccia's wavefunctions are available can be easily obtained. In addition, the versatility of our model enables the study of other molecules for which molecular orbitals may be safely described in terms of Slater basis. Finally, we think that our results may be useful in several domains. For instance, to study in detail the effects of ionizing radiations during its passage through living matter where our cross sections may be employed as the basic inputs to fed Monte Carlo codes to obtain the energy deposition in the biological medium.

Acknowledgements - Authors acknowledge financial support from the Agencia Nacional de Promoción Científica y Tecnológica (PICT No. 01912) and the Consejo Nacional de Investigaciones Científicas y Técnicas de la República Argentina (PIP No. 11220090101026).

Appendix

In this appendix, we present the calculations to obtain the matrix transition elements (Eq. (1)) corresponding to the Coulomb Continuum (CC) model presented in Section II.i.

Within the framework of the dipole approximation, the matrix element reads:

$$M_{ph}^{dip} = -i\langle\psi_f|\mathbf{e}\cdot\nabla|\psi_i\rangle. \tag{7}$$

where ψ_i is the initial electronic state, ψ_f is the final electronic state given by Eq. (2) and where $\mathbf{e}$ represents the polarization vector of the incident radiation.

We employ Moccia's wavefunctions [12] to represent the initial states ψ_i of the different water orbitals. As these functions are linear combinations of Slater orbitals, the transition matrix amplitudes for the different water orbitals are linear combinations of the transition matrix amplitudes of the Slater orbitals.

As an example, we show the calculations for the transition matrix amplitude of the simplest Slater orbital, χ_{1s}. In this case, the initial wavefunction ψ_i for the transition matrix element (7) is given by

$$\psi_{\chi_{1s}}(\mathbf{r}) = \frac{\alpha^{3/2}}{\sqrt{\pi}}e^{-\alpha r}. \tag{8}$$

Replacing Eqs. (8) and (2) in equation (7), one obtains:

$$\begin{aligned} M_{\chi_{1s}} =& - i\,(2\pi)^{-3/2}\,N_p^*\frac{\alpha^{3/2}}{\sqrt{\pi}} \\ & \int d^3r\; e^{-i\mathbf{p}\cdot\mathbf{r}}\; G^*(\mathbf{r})\,\mathbf{e}\cdot\nabla_{\mathbf{r}}\left(e^{-\alpha r}\right). \end{aligned} \tag{9}$$

Considering that $\nabla_{\mathbf{r}}\left(e^{-\alpha r}\right) = -\alpha e^{-\alpha r}\,\frac{\mathbf{r}}{r}$, one gets:

$$\begin{aligned} M_{\chi_{1s}} =& i\,(2\pi)^{-3/2}\,N_p^*\frac{\alpha^{5/2}}{\sqrt{\pi}} \\ & \int \frac{d^3r}{r}\; e^{-\alpha r-i\mathbf{p}\cdot\mathbf{r}}\; G^*(\mathbf{r})\,(\mathbf{e}\cdot\mathbf{r})\,. \end{aligned} \tag{10}$$

Now, if we consider the polarization versor $\mathbf{e}$ parallel to the z axis,

$$M_{\chi_{1s}} = i\,(2\pi)^{-3/2}\,N_p^*\frac{\alpha^{5/2}}{\sqrt{\pi}}I_1, \tag{11}$$

where we have defined

$$I_1 = \int \frac{d^3r}{r}\; e^{-\alpha r-i\mathbf{p}\cdot\mathbf{r}}\; G^*(\mathbf{r})\,r_z, \tag{12}$$

with r_z the z-component of the position vector $\mathbf{r}$.

To solve this integral, we introduce a vector $\mathbf{v}$ parallel to the z axis and we define the integral

$$\begin{aligned} I_2 =& \int \frac{d^3r}{r}\; e^{-\alpha r-i\mathbf{p}\cdot\mathbf{r}}\; e^{i\mathbf{v}\cdot\mathbf{r}}\; G^*(\mathbf{r}) \\ =& \int \frac{d^3r}{r}\; e^{-\alpha r-i\mathbf{p}\cdot\mathbf{r}}\; e^{ivr_z}\; G^*(\mathbf{r})\,, \end{aligned} \tag{13}$$

which satisfies

$$I_1 = -i\frac{\partial}{\partial v}\left(I_2\right)_{v=0}. \tag{14}$$

Therefore,

$$M_{\chi_{1s}} = (2\pi)^{-3/2}\,N_p^*\frac{\alpha^{5/2}}{\sqrt{\pi}}\frac{\partial}{\partial v}\left(I_2\right)_{v=0}. \tag{15}$$

Then, employing the Nordsieck's method [16], it is possible to obtain the integral I_2 as:

$$\begin{aligned} J(\lambda,\mathbf{q},\mathbf{v}) &= \int \frac{d^3r}{r}\, e^{-\lambda r+i(\mathbf{q}+\mathbf{v}-\mathbf{p})\cdot\mathbf{r}}\, G^*(\mathbf{r}) \\ &= 4\pi\, f(\lambda,\mathbf{q},\mathbf{v})\,, \end{aligned} \quad (16)$$

where,

$$f(\lambda,\mathbf{q},\mathbf{v}) = \frac{\left[(\mathbf{q}+\mathbf{v})^2+(\lambda-ip)^2\right]^{-i\nu}}{\left[\lambda^2+(\mathbf{q}+\mathbf{v}-\mathbf{p})^2\right]^{1-i\nu}}. \quad (17)$$

Consequently,

$$I_2 = 4\pi f(\alpha,\mathbf{0},\mathbf{v})\,, \quad (18)$$

and the transition matrix element:

$$M_{\chi_{1s}} = \frac{\sqrt{2}}{\pi} N_p^* \alpha^{5/2} \frac{\partial}{\partial v} f(\alpha,\mathbf{0},\mathbf{v})_{v=0}\,. \quad (19)$$

Finally, calculating the derivative according to equation (19), we obtain the transition matrix element corresponding to the $1s$ Slater orbital,

$$M_{\chi_{1s}} = \frac{2\sqrt{2}}{\pi} N_p^* \alpha^{5/2} \frac{(1-i\nu)\,(\mathbf{e}\cdot\mathbf{p})}{(\alpha^2+p^2)^2} e^{-2\nu\, arctg(p/\alpha)}. \quad (20)$$

The matrix elements for other Slater orbitals can be computed in an analogous way. Those corresponding to the Slater orbitals χ_{2s} and χ_{2p_0} are given by, respectively,

$$\begin{aligned} M_{\chi_{2s}} =& 2\sqrt{\frac{2}{3}}\frac{1}{\pi} N_p^* \alpha^{5/2} \frac{(1-i\nu)\,(\mathbf{e}\cdot\mathbf{p})}{(\alpha^2+p^2)^2} \\ & \left[\frac{3\alpha^2-p^2-2\nu\alpha p}{(\alpha^2+p^2)}\right] e^{-2\nu\, arctg(p/\alpha)}, \end{aligned} \quad (21)$$

and,

$$\begin{aligned} M_{\chi_{2p_0}} =& \frac{2\sqrt{2}}{\pi} i N_p^* \alpha^{5/2} \left[\frac{i\nu\alpha(\alpha+ip)^2}{(\alpha^2+p^2)^3} + \right. \\ & - \frac{2\alpha\,(2-i\nu)\,(1-i\nu)\,(\mathbf{e}\cdot\mathbf{p})^2}{(\alpha^2+p^2)^3} + \\ & \left. + \frac{\nu p-\alpha}{(\alpha^2+p^2)^2} + \frac{\alpha\,(1-i\nu)}{(\alpha^2+p^2)^2}\right] e^{-2\nu\, arctg(p/\alpha)}. \end{aligned} \quad (22)$$

---

[1] S Gozem et al, *Photoelectron wave function in photoionization: Plane wave or Coulomb wave?*, J. Phys. Chem. Lett. **6**, 4532 (2015).

[2] M Stener, S Furlan, P Decleva, *Density functional calculations of photoionization with an exchange correlation potential with the correct asymptotic behaviour*, J. Phys. B: At. Mol. Opt. Phys. **33**, 1081 (2000).

[3] C E Brion, F Carnovale, *The absolute partial photoionization cross section for the production of the X^2B_1 state of H_2O^+*, Chem. Phys. **100**, 291 (1985).

[4] K H Tan *et al.*, *Absolute oscillator strenghts (10-60 eV) for the photoabsorption photoionization and fragmentation of H_2O*, Chem. Phys. **29**, 299 (1978).

[5] C M Truesdale *et al.*, *Photoelectron angular distribution of H_2O*, J. Chem. Phys. **76**, 860 (1982).

[6] M S Banna it et al., *The photoelectron spectrum of water in the 30 to 140 eV photon energy range*, J. Chem. Phys. **84**, 4739 (1986).

[7] G N Haddad, J A R Samson, *Total absorption and photoionization cross sections of water vapor between 100 and 1000 Å*, J. Chem. Phys. **84**, 6623 (1986).

[8] L de Reilhac, N Damany, *Photoabsorption cross section measurements of some gases from 10 to 50 nm*, J. Quant. Spectrosc. Radiat. Transfer **18**, 121 (1977).

[9] E Phillips, L C Lee, D L Judge, *Absolute photoabsorption cross sections for H_2O and D_2O from λ 180 to 790 Å*, J. Quant. Spectrosc. Radiat. Transfer **18**, 309 (1977).

[10] C A Tachino *et al.*, *Ionization of water molecules by ion beams. On the relevance of dynamic screening and the influence of the description of the initial state*, J. Phys. B: At. Mol. Opt. Phys. **47**, 035203 (2014).

[11] M L de Sanctis *et al.*, *Theoretical study of the ionization of liquid water from its several initial orbitals by fast electron impact*, J. Phys. B: At. Mol. Opt. Phys. **48**, 155201 (2015).

[12] R Moccia, *One center basis set SCF MO's. III. H_2O, H_2S, HCl*, J. Chem. Phys. **40**, 2186 (1964).

[13] D I R Boll, O A Fojón, *Interferences and asymmetries in laser-assisted photoionization of diatomic molecules*, Phys. Rev. A **90**, 053414 (2014).

[14] D I R Boll, O A Fojón, *Atomic RABBITT like experiments framed as diatomic molecules*, J. Phys. B: At. Mol. Opt. Phys. **49**, 185601 (2016).

[15] G L Yudin, S Patchkovskii, A D Bandrauk, *Asymmetric interference in molecular photoprocesses*, J. Phys. B: At. Mol. Opt. Phys. **39**, 1537 (2006).

[16] A Nordsieck, *Reduction of an integral in the theory of Bremsstrahlung*, Phys. Rev. **93**, 785 (1954).

15

Application of multifractal analysis to segmentation of water bodies in optical satellite images

V. M. San Martin,[1,2]*A. Figliola[1,2]†

In this paper, we study the characteristics of multifractal spectra obtained for remote sensing images through the coarse theory and propose a method for segmentation of water bodies based on it.

In the first place, spectra of self-similar images created with Iterated Function Systems are calculated and compared with their statistical spectra. Then, optical remote sensing images are studied, emphasizing the differences between their spectra and those obtained for synthetic images. Attention is focused on the concavity of real image spectra and on its interpretation in terms of the analyzed images. This led us to the proposition of a segmentation method for water bodies.

The method is tested and the results are compared with water masks of the regions under study. Comments are made about the limitations of the proposed method.

I. Introduction

Fractal theory was initially proposed by Mandelbrot [1] and extensively developed up to these days [2–5]. Multifractal theory can be seen as an extension of fractal formalism, centered in subsets with different scaling behavior coexisting simultaneously. Since its introduction, multifractal analysis has been developed and applied in a wide range of disciplines, like medicine [6], neuroscience [7], economy [8–10], ecology [11] and theoretical physics [12–14]. Many methods have been proposed for approximating multifractal spectra. A summary of the most important ones can be found in [15].

*E-mail: victor_sanmartin1@hotmail.com

†E-mail: afigliol@ungs.edu.ar

1 National Scientific and Technical Research Council (CONICET), Argentina.

2 Institute of Human Development, National University of General Sarmiento, Juan María Gutiérrez 1150, 1613 Los Polvorines, Buenos Aires, Argentina.

Particularly, multifractal analysis has been used via different formalisms to extract information from images.

In a series of three articles, Arneodo et al. [16–18] develop a multifractal analysis method for images based on wavelet coefficients. Their approach leads to a multifractal spectra expressed in terms of Hölder exponent for functions, a solid way of studying singularities. However, the present work leads to using Hölder exponent for measures. Despite the fact that there are theoretical connections between those spectra, the passage from one to the other is not straightforward, making any comparison of numerical results quite complicated. Nevertheless, it is interesting to see how those singularity spectra are useful for characterizing captures with diverse textures. In [16], the authors propose performing a multifractal analysis of images through the so-called wavelet transform modulus maxima method and testing it in random self-affine surfaces. In [17], this group applies the proposed formalism to multifractal synthetic images simulating cloud structures, while in [18], they utilize it on Landsat images with cloud cover.

In a similar line, Wendt et al. present the use of wavelet leaders for studying regularity and obtaining the multifractal spectra of images with different textures [19].

In a classical article, Lévy Véhel and Mignot [20] characterize different types of 2D singularities via Hölder exponents for the sum measure and various capacities. Once the response of these exponents to the singularities is studied, the authors use them to obtain the edges in both, synthetic and real images.

Multifractal analysis based on measures has also been applied to medical images. Nilsson [21] studies the possibility of using this methodology to distinguish healthy tissue from a potentially tumoral one based on the differences in the width of the corresponding spectra. Vasiljevic et al. [22] use particular parameters of multifractal analysis, like some generalized dimensions or the range of Hölder exponents, for classifying the primary cancer in cases of metastasis bone disease.

Millán et al. [23] also pay attention to several selected parameters estimated from multifractal spectra and use them together with a log-Levy stable model for characterizing soil surface moisture distributions. In geophysics, Posadas et al. [24] characterize the dynamics of preferential water flow in soils and porous media with some particular parameters of multifractal spectra of MRI data. In engineering, multifractal spectra has been proposed as a method for evaluating the integrity of concrete shear walls through the study of images of fracture patterns caused by earthquakes [25]. They concluded that multifractal parameters move toward higher values as crack patterns extend and grow.

As it has been pointed out, many of the cited works oriented to image analysis do not deal with the multifractal spectrum as a whole, but rather focus only on the values of some parameters to characterize the processes under study. This is a clear difference with the present contribution, which seeks to study the shape of the spectrum for real images and how it is associated with different regions of the captures.

In recent decades, on the other hand, remote sensing has become an extremely useful tool for studying the complex dynamics of the Earth's surface. This technique gives us a huge amount of periodic and varied information about it.

Along with the rapid development of remote sensing, interest in the study of geometrical and self-similarity properties of different elements of the Earth's surface has grown. Some examples are the study of coastlines [26, 27], river basins [28], urban sprawls [29, 30] and forests [31]. Researchers have also turned their attention to the characteristics of changes introduced in the ground by various environmental disasters such as wildfire [32, 33], oil spills [34], seismic activity [35] and others. These studies, mostly based on fractal theory, show the intricate nature of the structures present in satellite images captured by different sensors.

Taking all this into account, it is reasonable to use multifractal approach on remote sensing images in order to develop segmentation tools that could allow us to differentiate various ground covers based on their textural features.

In this work, we use the so-called coarse multifractal theory [5] because it is computationally manageable and allows us to study, at the same time, regularity of each pixel of an image and the structures that points of similar regularity determine. This dual approach, local and global, becomes useful for classifying regions according to their textural properties. Such a task would be very hard to carry out by implementing a method that analyzes only globally the properties of an image.

Since this work was done in the framework of an interdisciplinary project devoted to plain rivers modeling and flood predictions, we decided to direct our attention towards segmentation of water bodies. Our motivation is to develop a reliable method for segmenting flooded areas accurately in order to use the obtained information as input in a prediction model.

This goal helped us devote our efforts in examining geometrical properties of multifractal spectra corresponding to real images and to interpret them in terms of geographical components of the scenes. In particular, concavity of spectra caught our attention because particularities on it seemed to be correlated with water covers in the regions of interest.

This article is organized as follows: Mathematical tools used in the analysis of the images are succinctly presented in section II, where central concepts of multifractal theory and Iterated Function Systems are defined. An interpretation of theoretical concepts, which will allow further considerations on results, is also presented at the end of this section.

Section III is mainly devoted to explain how theoretical concepts are adapted for image analysis and segmentation. A procedure for testing multifractal spectra obtained for synthetic images is also explained here.

At the beginning of section IV, we present the results of testing multifractal method with theoretical images. Then, we show the first spectra of remote sensing images emphasizing the differences with the ones corresponding to theoretical images. After analyzing real images, we state a segmentation criterion for flooded areas and present comparisons between its results and water masks of the images. The conclusions of this research are presented in section V.

II. Mathematical tools

In this section, we define the mathematical tools that we are going to use. This is basically done to clarify the notation and check the meaning of some theoretical concepts, which will help interpret results later. For a deeper explanation of the concepts of measure, fine and coarse multifractal theory, we refer the reader to [3, 5, 36].

i. Measures on $\mathbb{R}^n$

A measure over a set $X \subset \mathbb{R}^n$ is defined as a function which associates a value in $[0, \infty)$ to every subset of χ [5]. The support of a measure ($spt(\mu)$) is defined as the largest closed subset of the set χ for which every open neighborhood of every point has positive measure. A finite measure μ can be considered as a mass distribution over the support of the measure. By definition, a measure is additive, that is: the measure of a union of two disjoint subsets is the sum of the measures of the subsets.

The class of Borel sets is defined as the smallest collection of subsets of $X \subset \mathbb{R}^n$ which satisfy:

1. every open set and every closed set is a Borel set;
2. the union of every finite or countable collection of Borel sets is a Borel set, and the intersection of every finite or countable collection of Borel sets is a Borel set.

If the Borel sets of a set X can be measured with a measure μ, then that measure is called a Borel measure.

In this article, we are going to use "sum measure": let $g(k,l)$ be the intensity in a grey scale of a point (k,l), and let Ω be the set of all points in a neighborhood of width i centered in a point (m,n), then the measure of Ω is defined:

$$\mu_i(m,n) = \sum_{(k,l)\in\Omega} g(k,l). \tag{1}$$

ii. Hölder exponent and fine theory

Given a Borel measure μ and a point x in the support of μ, its local dimension [5] in x is

$$dim_{loc}\mu(x) = \lim_{r\to 0} \frac{\ln(\mu(B(x,r)))}{\ln r} \tag{2}$$

if the limit exists, where $B(x,r)$ is a ball of radius r centered in x. This limit is usually called the Hölder exponent (α). For $\alpha \geq 0$ we define

$$F_\alpha = \{x \in spt(\mu) : dim_{loc}\mu(x) = \alpha\}. \tag{3}$$

We define the Hausdorff fine multifractal spectrum $f_H(\alpha)$ as the function that maps α to the Hausdorff dimension dim_H of the set F_α, that is:

$$f_H(\alpha) = dim_H F_\alpha. \tag{4}$$

Although the Hausdorff dimension is more convenient for theoretical issues, in practice, it is almost impossible to estimate from experimental data. That is the reason why the so-called 'Coarse Theory' is used to obtain the multifractal spectrum.

iii. Coarse theory

Once α is calculated for all $x \in spt(\mu)$ as explained in the previous section , $spt(\mu)$ is covered with a regular lattice of n-dimensional boxes of width r (r-mesh). For $r > 0$ and $\alpha \geq 0$ we define

$$N_r(\alpha) = \#\{\text{ boxes of r-mesh such that } \mu(box) \geq r^\alpha\}. \tag{5}$$

The coarse multifractal spectrum is defined in [5] as

$$f_C(\alpha) = \lim_{\epsilon \to 0} \lim_{r \to 0} \frac{\ln(N_r(\alpha+\epsilon) - N_r(\alpha-\epsilon))}{-\ln r} \quad (6)$$

if the double limit exists.

iv. Moment sum and Legendre transformation

Another method to estimate the multifractal spectrum is based on the definition of the so-called partition function. Given a measure μ, its support is covered with an r-mesh. For $q \in \mathbb{R}$ and $r > 0$ partition function is defined as the sum of the q powers of the measures μ_i of each box in the mesh,

$$\chi_q(r) = \sum_{i=1}^{N_r} \mu_i^q, \quad (7)$$

where N_r is the total number of boxes in the r-mesh. Consistently with our previous assumptions, $\chi_q(r)$ satisfies a power law behavior with respect to its scale[5] at least for small r

$$\chi_q(r) \sim r^{-\tau(q)}, \quad (8)$$

If we discretize the range of α in intervals of width ϵ (with ϵ sufficiently small), the sum in the definition of $\chi_q(r)$ can be decomposed into a sum over regions where α is approximatively constant and of size proportional to $r^{-f(\alpha)}$. It is then easy to see that in the limit $r \to 0$, the sum is dominated by the $\alpha(q)$ that minimizes the exponent $q\alpha - f(\alpha)$. This leads to a relationship between $f(\alpha)$ and $\tau(q)$ in terms of Legendre transform, defined for $\alpha \geq 0$ by

$$f(\alpha) = \inf_{-\infty<q<\infty} (\tau(q) + q\alpha) \quad (9)$$

If $\tau(q)$ is a convex function, which ensures that the Legendre transform is uniquely determined, it is possible to calculate $f(\alpha)$ inverting the Legendre transform. The spectrum obtained with this procedure is usually known as Legendre Spectrum and denoted $f_L(\alpha)$.

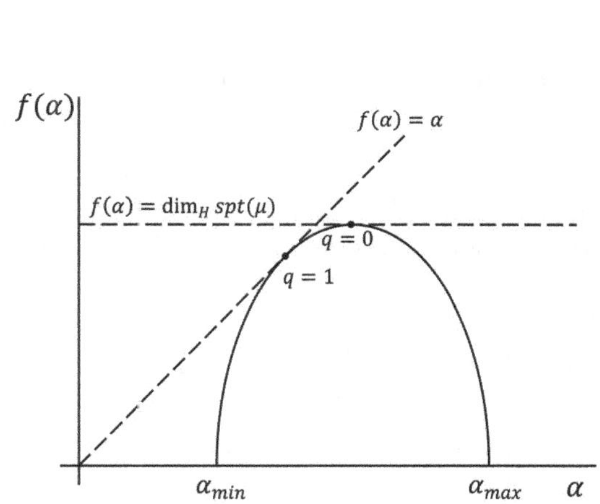

Figure 1: Typical multifractal spectrum of a set constructed by a Random Iteration Algorithm.

This method is usually more numerically manageable than the one derived from coarse theory. However, its application is limited to cases in which $\tau(q)$ is a convex function of q, otherwise the result is meaningless.

v. Iterated function systems (IFS)

Let D be a closed subset of $\mathbb{R}^n$. A mapping $S : D \to D$ is called a contraction on D if there is a number c with $0 < c < 1$ such that $|S(x) - S(y)| \leq c|x-y|\ \forall x, y \in D$. If $|S(x) - S(y)| = c|x-y|$, then S is called a contracting similarity.

An iterated function system is defined as a finite family of contractions S_m with $m \geq 2$. A non-empty compact subset $F \subseteq D$ is called attractor for the IFS if $F = \bigcup_{i=1}^{m} S_i(F)$. An IFS determines a unique attractor, which is usually a fractal.

Let us suppose we associate a probability to each contraction of an IFS. If we apply the contractions according to their probabilities on a point in the IFS attractor, we will obtain a self-similar measure. This method is known as Random Iteration Algorithm.

As shown in [5], multifractal spectrum of self-similar measures is strictly concave. In addition, proposition 11.7 of [5] states that $f_C(\alpha) = f_L(\alpha)$ for that type of measures.

vi. Interpretation of α values

A typical multifractal spectrum of a set D constructed by a Random Iteration Algorithm is sketched in Fig. 1. An interpretation of α values arises from the spectra of these IFS. In [5], it is shown that for a self-similar measure created by an IFS $\{S_i\}$, $i = 1, \ldots, m$, with contraction ratios $\{c_i\}$ and probabilities $\{p_i\}$

$$\alpha_{min} = \min_{1\le i\le m} \frac{\log p_i}{\log c_i} \tag{10}$$

$$\alpha_{max} = \max_{1\le i\le m} \frac{\log p_i}{\log c_i} \tag{11}$$

If we assume that $c_i = c\ \forall i$, α_{max} is related to the minimum p_i. Assuming also that $p_i \neq p_j$ when $i \neq j$, there is a single point in D, let us say $x_{\alpha_{max}}$, for which $\alpha = \alpha_{max}$. Every point in D has a probability greater than that point. Applying similar reasoning, the point in D for which $\alpha = \alpha_{min}$, $x_{\alpha_{min}}$ is associated with the highest probability.

From Eq. (2), balls of radius r centered in $x_{\alpha_{max}}$ have lower measures as r becomes lower, but that reduction is not proportional to r^2. Since the set is multifractal, discontinuities in the measures of the neighborhoods occur as r varies, resulting in a reduction of the measure proportional to $r^{\alpha_{max}}$, with $\alpha_{max} > 2$.

III. Description of methods

In this section, we describe the use of mathematical tools presented in section II to carry out the study of images. In the first subsection, we describe the algorithm used for synthesizing multifractal sets of well known properties. We also describe how statistical multifractal spectra of these sets are obtained through the method of moments. In subsection ii, we describe the application of coarse theory to synthetic images and how its results are compared with the statistical spectra. Finally, in subsection iii, the analysis of satellite images through the coarse theory is described.

i. Algorithm for creating synthetic multifractal images

The algorithm used to obtain multifractal sets is explained in a more detailed way in [21]. This algorithm emulates deterministically the result of a Random Iteration Algorithm over a subset of $\mathbb{R}^2$ with contractions

$$\begin{aligned} S_1(x_1, x_2) &= (x_1/2, x_2/2) && (12)\\ S_2(x_1, x_2) &= (x_1/2, x_2/2 + 1/2) && (13)\\ S_3(x_1, x_2) &= (x_1/2 + 1/2, x_2/2) && (14)\\ S_4(x_1, x_2) &= (x_1/2 + 1/2, x_2/2 + 1/2), && (15)\end{aligned}$$

and corresponding probabilities $\{p_i\}$, $i = 1, \ldots, 4$ The contraction factor is 0.5 in all cases.

The algorithm begins with a square array of $2^N \times 2^N$ pixels of value 1. The initial square is divided into four squares and each one is multiplied by the probability corresponding to the contraction that each square represents. This process is iterated over the resulting squares up to N iterations (resolution limit).

Images of 256×256 pixels were synthesized with this algorithm based on random probabilities vectors.

Because of the way in which these images are created, $\tau(q)$ must be monotonically concave in Eq. (9), allowing the use of Legendre transform for determining their multifractal spectra. Different meshes with boxes 64, 128 and 256 pixels wide were applied. Given a mesh, the sum of the values of the pixels within each box is calculated. These measures are raised to an exponent q and added, obtaining the value of the partition function for that mesh width and the exponent q. After making this calculation for each mesh, a linear fit of the logarithm of the partition functions versus the negative of the logarithms of the corresponding box widths is performed. This gives us $\tau(q)$ of Eq. (8) for each q. After applying Legendre transform to $\tau(q)$, a statistical estimate of the multifractal spectrum is obtained.

As we said in section II.v, Legendre spectra of self-similar synthetic sets must agree with coarse multifractal spectra. That is the reason for using the synthetic images as a test for the multifractal method based on coarse theory.

ii. Analysis of synthetic images through coarse theory

We choose $B(x, r)$ in Eq. (2) to be square neighborhoods of $(2k-1) \times (2k-1)$ pixels, with $k =$

$2, \ldots, 8$. Those neighborhoods are centered in the pixel whose Hölder exponent we are trying to calculate. The measure of all neighborhoods is calculated. A linear fit is then performed on the logarithms of the measures versus the logarithms of the neighborhood widths. The slope of the adjusted line is our estimate of α for that pixel. A padding of copies of the image is built for performing calculations on pixels next to the edges.

Once α values for every pixel are determined, the range of α is divided into $R = 10$ classes bounded by each pair of successive elements of the list $(\alpha_{min}, \alpha_{min} + \frac{\delta\alpha}{2} + s \cdot \delta\alpha, \alpha_{max})$, with $\delta\alpha = (\alpha_{max} - \alpha_{min})/R$ and $s = 0, \ldots, R-2$. Those classes determine subsets of pixels with similar local regularity. Then, we calculate the box counting dimension of each subset. Different meshes are applied to the image and the number of boxes necessary to completely cover the selected subset is counted. For each subset, we make a linear fit of the logarithm of the number of boxes versus minus the logarithm of the width of the boxes. The slope of the fit is our estimate of $f(\alpha_m)$, where α_m is the mean value of each class. The boxes of the meshes used in the images have $4, 8, 16, 32, 64, 128$ and 256 pixels of width.

It was checked that the multifractal spectra obtained had uniform concavity and that they fulfilled proposition 11.7 of [5]. Although proposition 11.7 states that $f_L(\alpha) = f_C(\alpha)$ for a self-similar measure, given the discretizations made in the calculation of the spectrum through the coarse theory, it is reasonable that f_C values obtained are lower than those of f_L. Particularly, it was observed that by reducing R in calculation of f_C, a better match between the spectra is obtained.

Errors in calculations of α for each pixel are obtained. Errors in α exponents will affect the errors in multifractal spectrum values. This is primarily to the possible change in the full range of α values found in the picture. In second place, if the pixels have associated different exponents, when the total range is divided into sub-ranges, this could change the set of pixels included in a class, altering the fractal dimension of the set. To account for this, multifractal spectra for three sets of α were calculated: the set of α obtained, and the sets resulting from adding and subtracting the corresponding error to the α value of each pixel. Spectra of these three configurations are obtained with their respective errors.

The largest α error between the pixels of a class is taken as the error bound of α for that class. For each class, the error bound of $f(\alpha)$ is given by the distance between that point and the farthest error bound for that class.

Concavity is analyzed by calculating an approximation of the second derivative of $f(\alpha)$. Thus, for each point, it was calculated

$$\Delta f(\alpha_i) = \frac{f(\alpha_{i+1}) - f(\alpha_i)}{\alpha_{i+1} - \alpha_i}, \quad i = 1, \ldots, 9; \tag{16}$$

$$\Delta^2 f(\alpha_i) = \frac{\Delta f(\alpha_{i+1}) - \Delta f(\alpha_i)}{\alpha_{i+1} - \alpha_i}, \quad i = 1, \ldots, 8. \tag{17}$$

If $\Delta^2 f(\alpha_i)$ is negative for all i, then the function is concave downward.

iii. Analysis of satellite images

Optical images taken by OLI (Operational Land Imager) and TIRS (Thermal Infrared Sensor) sensors of Landsat 8 satellite, provided by the U.S. Geological Survey [37] were studied. Since the work was oriented to the detection of flooded regions, the images were selected for having remarkable water bodies and a low percentage of cloud cover. Analysis was performed particularly in fragments of band 5 (wavelength of 0.851-0.879 μm) of those images.

Images were calibrated before being studied so that their pixels represent values of reflectance measured by the sensor in each position. For more information on the Landsat 8 platform we refer to [38].

Since it was found that $\tau(q)$ did not have a uniform concavity for satellite images, these images were not studied using the method of moments, and their spectra were obtained only through the coarse theory.

As for the calculations of multifractal spectra in synthetic images, we chose the $B(x, r)$ in Eq. (2) to be square neighborhoods of $(2k-1) \times (2k-1)$ pixels, with $k = 2, \ldots, k_{end}$. Calculations on the pixels near the border of the analyzed region are made using adjacent regions of the image as padding.

The range of α is also divided into R classes bounded by each pair of successive elements of the list $(\alpha_{min}, \alpha_{min} + \frac{\delta\alpha}{2} + s \cdot \delta\alpha, \alpha_{max})$, with $\delta\alpha = (\alpha_{max} - \alpha_{min})/R$ and $s = 0, \ldots, R-2$. Box counting dimension is obtained in each subset with the procedure explained early.

Selection of k_{end}, R and the size of boxes in the meshes used in box counting will be discussed in the next section.

The classification of pixels of images as flooded or not flooded is achieved through the application of thresholds on the multifractal spectra. Selection of those thresholds arose during the exploratory study of the graphs of $f(\alpha)$ obtained and will be detailed in section IV.

IV. Results and Discussion

Multifractal spectra of 600 synthetic images were obtained through coarse theory and compared with their statistical spectra. 591 spectra showed uniform concavity and fulfilled $f_C \leq f_L$. Two examples are presented in Fig. 2.

Taking into account the results of calculations on synthetic images, we decided to analyze some remote sensing images through coarse theory. Initially, we took fragments of 256 × 256 pixels and analyzed them with the same configuration as the theoretical multifractals. Four examples are presented in Fig. 4. We noticed that concavity was not uniform in the spectra of those images. Particularly, it seemed that for images where there were large water bodies such as lakes, this change in concavity occurred more markedly.

Calculation of multifractal spectra in case of images built with IFS requires taking as maximum neighborhood width the one that most closely matches the full width of the image because, otherwise, the result does not fit the statistical spectra. We suppose that this is due to the infinite nature of fractals. Therefore, to take all possible intervals within the limits of scale means to collect as much information as possible.

As evidence of this, we have performed some calculations on fractals created by IFS with 2048 × 2048 pixels, which are not showed here. When the points used in the calculus of α were plotted, we observed that, to a certain scale, the points seemed to have a slope. Then, it changed abruptly as a result of discontinuity in the measurement. This new slope was retained during some scales and then a new change occurred. However, when a larger number of scales was observed, a linear behavior was seen in the points, despite these apparent local changes in slope.

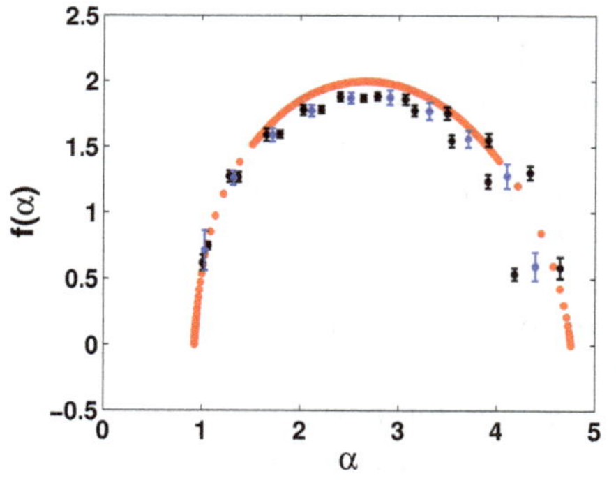

(a) $[0.526, 0.346, 0.091, 0.037]$.

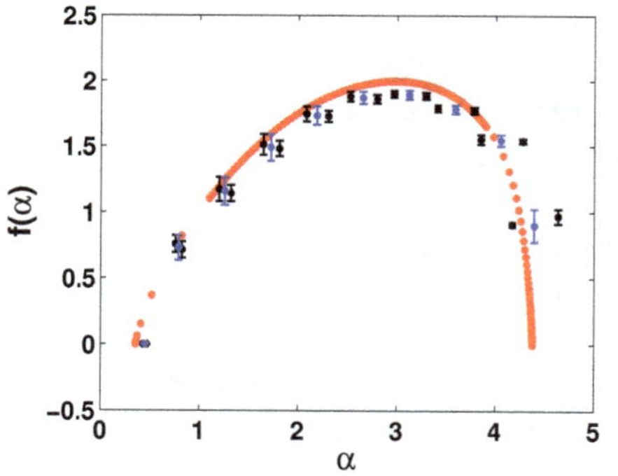

(b) $[0.781, 0.063, 0.108, 0.048]$.

Figure 2: Comparisons between multifractal spectra obtained for two synthetic images through the method of moments (red dots) and the coarse theory (blue dots). Black dots correspond to multifractal spectra resulting from α sets calculated by adding and subtracting the corresponding error to the α value of each pixel. They are used in the determination of error bonds.

The case of real images is quite different since they do not have an infinite number of possible scales. Therefore, there is no need to take all possible neighborhoods for calculating the Hölder exponent.

However, given that this work is oriented to take advantage of multifractal formalism for the detection and segmentation of water bodies, the choice of k_{end} is not trivial.

Before addressing the issue of the selection of k_{end}, we should explain the criterion for segmentation of water bodies proposed in this work. For that, note the analysis presented in Fig. 5. It was done on an image of 1024 × 1024 pixels with

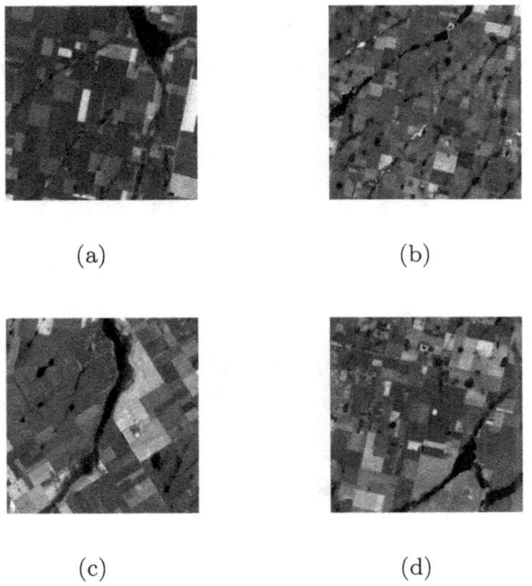

Figure 3: Fragments of 256x256 pixels of images taken by OLI - TIRS sensors of Landsat 8 satellite (band 5).

$k_{end} = 10$ and $R = 30$. In the corresponding multifractal spectrum, a central minimum is observed, let us say α_{center}. It has $f(\alpha) \approx 1$, indicating the presence of a one-dimensional structure separating two regions with different local regularity. Based on section II.vi, it can be deduced that points in an image with low values which are surrounded by regions of pixels with higher values have high α values. Then, flooded pixels in a Landsat image whose reflectance is low in band 5 should have a high Hölder exponent. Our idea is to detect α_{center} and classify pixels with $\alpha > \alpha_{center}$ as flooded.

Returning to the selection of k_{end}, we have noticed that if the largest neighborhood for each pixel in a flooded region is large enough to cover an edge of the water body, the pixels will have $\alpha \geq 2$. On the contrary, if for example the largest neighborhood centered in a pixel in the middle of a lagoon does not touch the edge of it, then that pixel will have $\alpha \approx 2$. Figure 6 presents an analysis of the same region as in Fig. 5, with $k_{end} = 8$ and $R = 30$. It is clear that the pixels from the center of the water body have an $\alpha \approx 2$, as opposed to what happens in Fig. 5. In that situation, those pixels would not be classified as water. For this reason, it is necessary to insist on the fact that k_{end} should be chosen to avoid this situation.

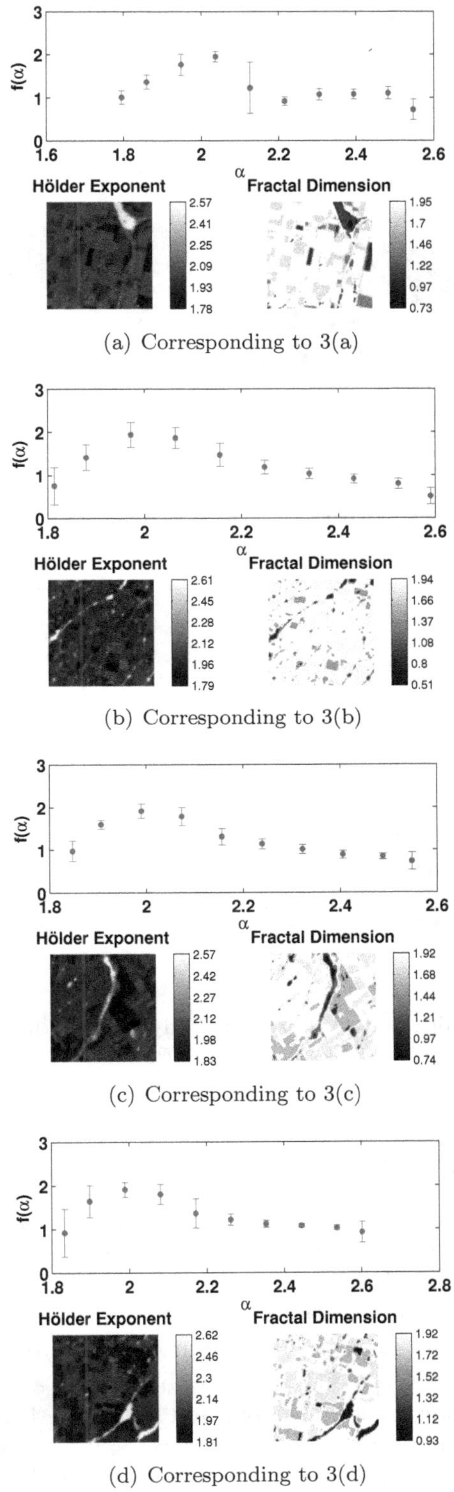

Figure 4: Analysis performed on fragments presented in Fig. 3.

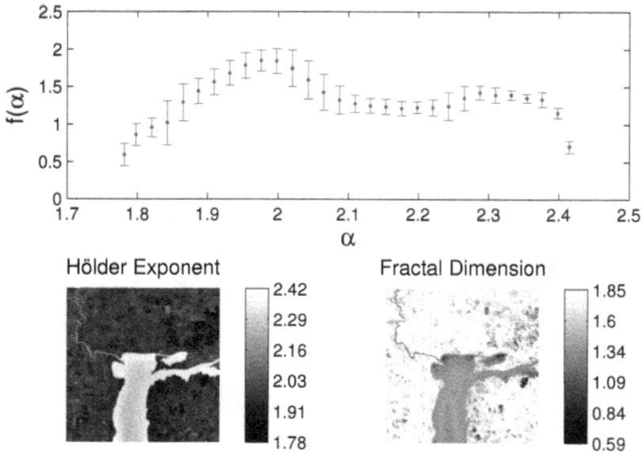

Figure 5: Analysis performed on an image of 1024× 1024 pixels with $k_{end} = 10$ and $R = 30$.

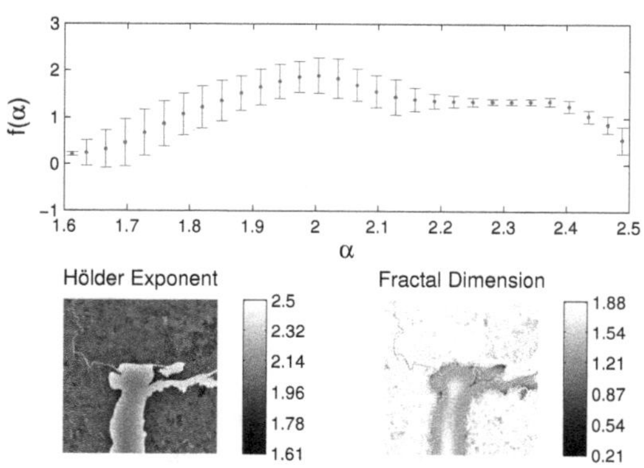

Figure 6: Analysis performed on an image of 1024× 1024 pixels with $k_{end} = 8$ and $R = 30$.

Clearly, the proposed criterion is only applicable in cases where the multifractal spectrum possesses that particular form. For example, a water body such as a river will have a high α but the fractal dimension of this region will be approximately 1. Therefore, there may be no change in concavity in that case. On the contrary, if besides a river we also have a lagoon or a water body with a fractal dimension greater than one and the greatest neighborhoods are large enough to cover any of its edges, then we will find a region with a high α and a fractal dimension greater than 1, causing a minimum in the center of the spectrum.

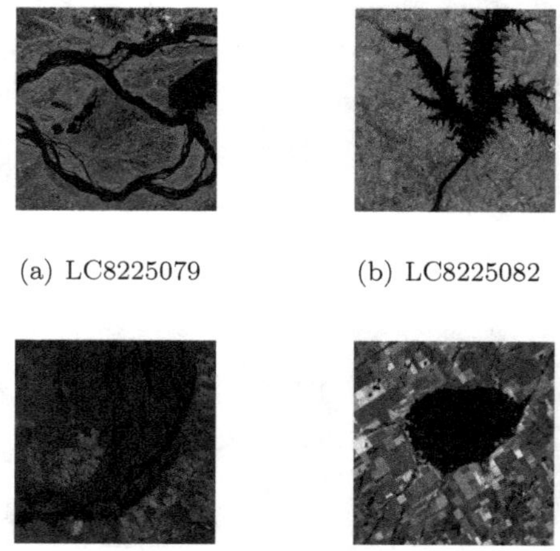

(a) LC8225079 (b) LC8225082

(c) LC8225083 (d) LC8227083

Designation:	LC8225079	LC8225082	LC8225083	LC8227083
Sensor:	OLI/TIRS			
Data Type:	L1T			
Spacial resolution:	30 meters			
Acquisition date:	19/09/2016	18/08/2016	15/07/2015	15/09/2015
WRS-2 Path:	225	225	225	227
WRS-2 Row:	079	082	083	083
Image quality:	9	9	9	9
Scene cloud cover:	.48	.03	.04	.1
Size:	1024 × 1024	1024 × 1024	1024 × 1024	512 × 512

Figure 7: Optical images studied.

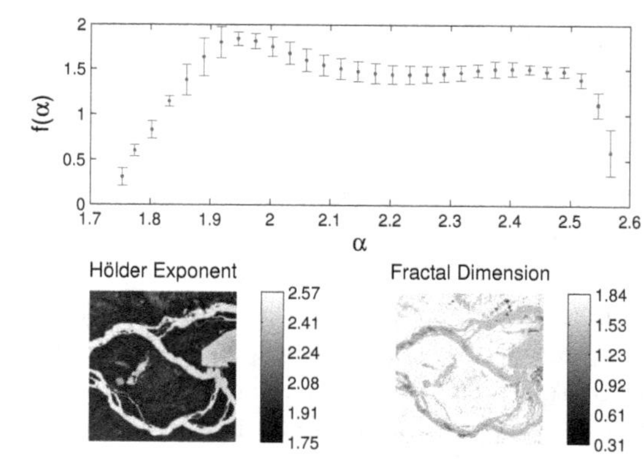

Figure 8: Analysis performed on LC8225079 with $k_{end} = 10$ and $R = 30$.

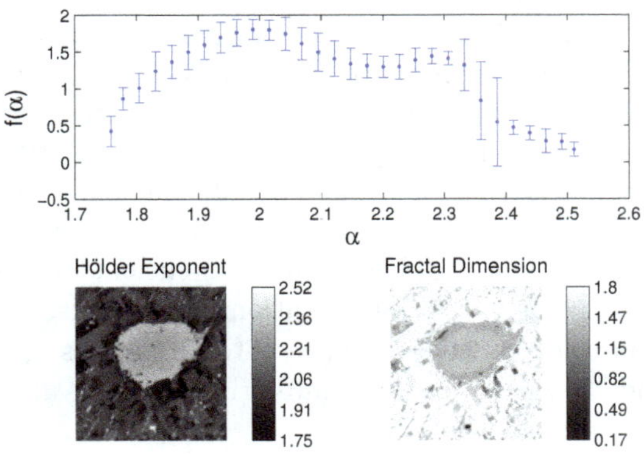

Figure 9: Analysis performed on LC8227083 with $k_{end} = 9$ and $R = 30$.

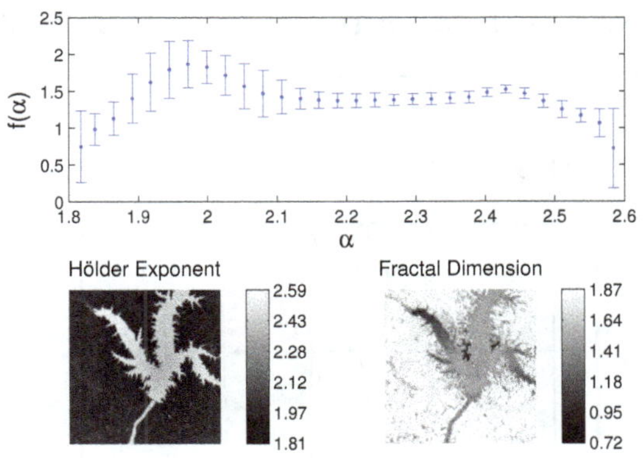

Figure 11: Analysis performed on LC8225082 with $k_{end} = 10$ and $R = 30$.

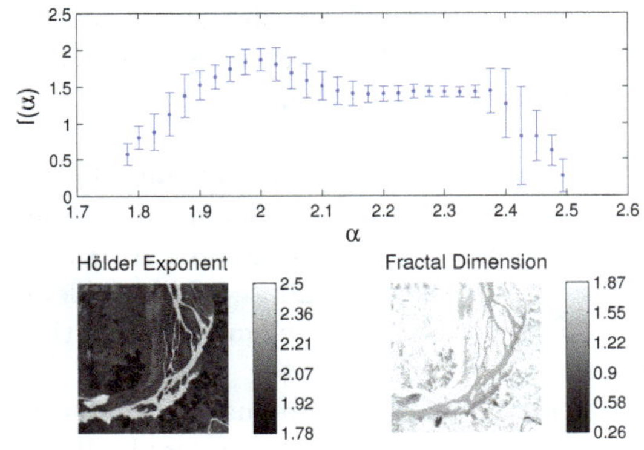

Figure 10: Analysis performed on LC8225083 with $k_{end} = 10$ and $R = 30$.

Table 1: Confusion matrix for image LC8225079.

	Total population: 1048576	Water Mask: Water	Water Mask: Not water	
Multifractal analysis	Water	236568 (22.56%)	2164 (0.21%)	PPV: 99.09%
Multifractal analysis	Not water	17080 (1.63%)	792764 (75.60%)	NPV: 97.89%
		Sensitivity: 93.27%	Specificity: 99.73%	Accuracy: 98.16%

Table 2: Confusion matrix for image LC8225082.

	Total population: 1048576	Water Mask: Water	Water Mask: Not water	
Multifractal analysis	Water	211601 (20.18%)	3425 (0.33%)	PPV: 98.41%
Multifractal analysis	Not water	13156 (1.25%)	820394 (78.24%)	NPV: 98.42%
		Sensitivity: 94.15%	Specificity: 99.58%	Accuracy: 98.42%

With all that in mind, we now present segmentations performed on some Landsat images suitably selected through visual inspection. Images are presented and described in Fig. 7.

Images LC8225079, LC8225082 and LC8225083 were analyzed with $k_{end} = 10$, $R = 30$ and meshes with boxes 4, 8, 16, 32, 64, 128, 256, 512 and 1024 pixels wide. On the other hand, for the analysis of LC8227083 we used $k_{end} = 9$, $R = 30$ and meshes with boxes 4, 8, 16, 32, 64, 128, 256 and 512 pixels wide.

Figures 8, 9, 10 and 11 show the results of the analysis. Values of α_{center} found in spectra are

Table 3: Confusion matrix for image LC8225083.

		Water Mask		
	Total population: 1048576	Water	Not water	
Multifractal analysis	Water	125959 (12.01%)	575 (0.05%)	PPV: 99.55%
	Not water	20908 (1.99%)	901134 (85.94%)	NPV: 97.73%
		Sensitivity: 85.76%	Specificity: 99.94%	Accuracy: 97.95%

Table 4: Confusion matrix for image LC8227083.

		Water Mask		
	Total population: 262144	Water	Not water	
Multifractal analysis	Water	44503 (16.98%)	278 (0.11%)	PPV: 99.38%
	Not water	4609 (1.76%)	212754 (81.16%)	NPV: 97.88%
		Sensitivity: 90.62%	Specificity: 99.87%	Accuracy: 98.14%

Table 5: Values of α_{center} found in multifractal spectra of analyzed images.

Image	α_{center}
LC8225079	2.23
LC8225082	2.18
LC8225083	2.17
LC8227083	2.20

Table 6: Cohen's kappa coefficients obtained for the correlation between segmentations and masks.

Image	Cohen's kappa coefficient
LC8225079	0.95
LC8225082	0.95
LC8225083	0.91
LC8227083	0.94

listed in table 5.

As a way of evaluating the efficiency of the proposed segmentation method, the results were compared with water masks of the regions, which were provided by our colleagues of The National Water Institute (INA): L. Giordano and J. Bianchi. Comparisons were made through confusion matrices and the calculation of Cohen's kappa coefficient for the correlation between segmentations and masks.

Confusion matrices are presented in Tables 1, 2, 3 and 4, while table 6 summarizes kappa values obtained.

V. Conclusions

In this work, we have studied properties of coarse multifractal spectra calculated for optical satellite images and we have compared them with spectra corresponding to self-similar images. We have analyzed the concavity of natural images spectra and interpreted it in terms of structures present in the images.

We have seen that, in presence of suitable water bodies, a local minimum exists in the multifractal spectra. Thanks to the dual approach of coarse multifractal analysis, we could associate Hölder exponents higher than the local minimum with regions covered by water.

Based on those results, we have introduced a new criterion for segmentation of water bodies. Of course, this criterion is far from universal. As we mentioned early, the particular shape of spectra used in segmentation depends on the presence of given structures in the studied scenes. Without those conditions, we are not able to perform an accurate segmentation.

We must remark that the whole analysis could not have been done by means of a statistical or global approach because we would have needed also a local characterization of pixels in order to detect the correlation that allowed us to state the segmentation criterion.

After discussing the applicability of the method, we tested the accuracy of the results by comparing them with water masks. The segmentations performed showed a high coincidence with the 'true' flooded regions, situation evidenced by Cohen's kappa values close to 1 presented in section IV.

It is interesting to note that the proposed method

is only applied in this work to one band of spectral information, but there is no limitation for its application to other bands or even sensors of different nature. It is reasonable then to assume that the study of the diverse remote sensing information currently available could lead to a more complete characterization of a region of interest.

Another possibility that could further extend the usefulness of this method is the implementation of different measures or even capabilities in addition to the one presented here. This could allow the detection of various structures or regions in an image, leading to more detailed segmentations.

Acknowledgements - This work was done in the framework of interdisciplinary project PIO CONICET-UNGS 14420140100011CO, on Plain Rivers Modelling and Flood Predictions. We are happy to acknowledge the assistance we received from our fellow colleagues in the research group and especially thank hydrologists L. Giordano and J. Bianchi whose inspiration was crucial during the investigation and M A. Virasoro for his contribution during the design of the study and in the analysis of the results. Finally, The authors thank CONICET and UNGS.

---

[1] B B Mandelbrot, *The Fractal Geometry of Nature*, WH Freeman, New York (1983).

[2] G A Edgar, *Classics on Fractals*, Westview Press, Boulder, Colorado (2004).

[3] H-O Peitgen, H Jürgens, D Saupe, *Chaos and Fractals: New Frontiers of Science*, Springer, New York (1992).

[4] M R Schroeder, *Fractals, Chaos, Power Laws: Minutes from an Infinite Paradise*, Dover Publications, New York (2012).

[5] K Falconer, *Techniques in Fractal Geometry*, Wiley, Chichester (1997).

[6] A N Pavlov, A R Ziganshin, O A Klimova, *Multifractal characterization of blood pressure dynamics: stress-induced phenomena*, Chaos Soliton Fract. **24**, 57 (2005).

[7] A K Maity, R Pratihar, A Mitra, S Dey, V Agrawal, S Sanyal, A Banerjee, R Sengupta, D Ghosh, *Multifractal Detrended Fluctuation Analysis of alpha and theta {EEG} rhythms with musical stimuli*, Chaos Soliton Fract. **81**, 52 (2015).

[8] L Zunino, A Figliola, B M Tabak, D G Pérez, M Garavaglia, O A Rosso, *Multifractal structure in latin-american market indices*, Chaos Soliton Fract. **41**, 2331 (2009).

[9] M Dai, J Hou, J Gao, W Su, L Xi, D Ye, *Mixed multifractal analysis of China and {US} stock index series*, Chaos Soliton Fract. **87**, 268 (2016).

[10] G Cao, W Xu, *Multifractal features of {EUA} and {CER} futures markets by using multifractal detrended fluctuation analysis based on empirical model decomposition*, Chaos Soliton Fract. **83**, 212 (2016).

[11] S Kumar, M Cuntz, Z E Musielak, *Fractal and multifractal analysis of the rise of oxygen in Earth's early atmosphere*, Chaos Soliton Fract. **77**, 296 (2015).

[12] J W Kantelhardt, S A Zschiegner, E Koscielny-Bunde, S Havlin, A Bunde, H E Stanley, *Multifractal detrended fluctuation analysis of nonstationary time series*, Physica A **316**, 87 (2002).

[13] E Serrano, A Figliola, *Wavelet Leaders: A new method to estimate the multifractal singularity spectra*, Physica A **388**, 2793 (2009).

[14] A Figliola, E Serrano, G Paccosi, M Rosenblatt, *About the effectiveness of different methods for the estimation of the multifractal spectrum of natural series*, Int. J. Bifurcat. Chaos **20**, 331 (2010).

[15] R Lopes, N Betrouni, *Fractal and multifractal analysis: A review*, Med. Image Anal. **13**, 634 (2009).

[16] A Arnéodo, N Decoster, S Roux, *A wavelet-based method for multifractal image analysis. I. Methodology and test applications on isotropic and anisotropic random rough surfaces*, Eur. Phys. J. B. **15**, 567 (2000).

[17] N Decoster, S Roux, A Arnéodo, *A wavelet-based method for multifractal image analysis. II. Applications to synthetic multifractal rough surfaces*, Eur. Phys. J. B. **15**, 739 (2000).

[18] S Roux, A Arnéodo, N Decoster, *A wavelet-based method for multifractal image analysis. III. Applications to high-resolution satellite images of cloud structure*, Eur. Phys. J. B. **15**, 765 (2000).

[19] H Wendt, S G Roux, S Jaffard, P Abry, *Wavelet leaders and bootstrap for multifractal analysis of images*, Signal Processing **89**, 1100 (2009).

[20] J Lévy Véhel, P Mignot, *Multifractal segmentation of images*, Fractals **2**, 371 (1994).

[21] E Nilsson, *Multifractal-based Image Analysis with applications in Medical Imaging*, Master's thesis, Umeå University, Department of Computing Science (2007).

[22] J Vasiljevic, B Reljin, J Sopta, V Mijucic, G Tulic, I Reljin, *Application of multifractal analysis on microscopic images in the classification of metastatic bone disease*, Biomed. Microdevices **14**, 541 (2012).

[23] H Millán, R Cumbrera, A M Tarquis, *Multifractal and Levy-stable statistics of soil surface moisture distribution derived from 2D image analysis*, Appl. Math. Model. **40**, 2384 (2016).

[24] A Posadas, R Quiroz, A Tannús, S Crestana, C M Vaz, *Characterizing water fingering phenomena in soils using magnetic resonance imaging and multifractal theory*, Nonlinear Process. Geophys. **16**, 159 (2009).

[25] A Ebrahimkhanlou, A Farhidzadeh, S Salamone, *Multifractal analysis of crack patterns in reinforced concrete shear walls*, Struct. Health Monit. **15**, 81 (2016).

[26] Z Xiaohua, C Yunlong, Y Xiuchun, *On Fractal Dimensions of China's Coastlines*, Math. Geol. **36**, 447 (2004).

[27] H Lantuit, V Rachold, W Pollard, F Steenhuisen, R Ødegård, H-W Hubberten, *Towards a calculation of organic carbon release from erosion of Arctic coasts using non-fractal coastline datasets*, Marine Geology **257**, 1 (2009).

[28] C Donadio, F Magdaleno, A Mazzarella, G Mathias Kondolf, *Fractal Dimension of the Hydrographic Pattern of Three Large Rivers in the Mediterranean Morphoclimatic System: Geomorphologic Interpretation of Russian (USA), Ebro (Spain) and Volturno (Italy) Fluvial Geometry*, Pure Appl. Geophys. **172**, 1975 (2015).

[29] I Matsuba, M Namatame, *Scaling behavior in urban development process of Tokyo City and hierarchical dynamical structure*, Chaos Soliton Fract **16**, 151 (2003).

[30] P Frankhauser, *Fractal Geometry for Measuring and Modelling Urban Patterns*, In: The Dynamics of Complex Urban Systems, Eds. S Albeverio, D Andrey, P Giordano, A Vancheri, pag. 213, Physica-Verlag HD, New York (2008).

[31] I C Andronache, H Ahammer, H F Jelinek, D Peptenatu, A-M Ciobotaru, C C Draghici, R D Pintilii, A G Simion, C Teodorescu, *Fractal analysis for studying the evolution of forests*, Chaos Soliton Fract. **91**, 310 (2016).

[32] B D Malamud, J D A Millington, G L W Perry, *Characterizing wildfire regimes in the United States*, Proc. Natl. Acad. Sci. U.S.A. **102**, 4694 (2005).

[33] M V Moreno, B D Malamud, E A Chuvieco, *Wildfire Frequency-Area Statistics in Spain*, Procedia Environ. Sci. **7**, 182 (2011).

[34] A M Tarquis, A Platonov, A Matulka, J Grau, E Sekula, M Diez, J M Redondo, *Application of multifractal analysis to the study of SAR features and oil spills on the ocean surface*, Nonlinear Process. Geophys. **21**, 439 (2014).

[35] A Carpinteri, B Chiaia, S Invernizzi, *Applications of fractal geometry and renormalization group to the Italian seismic activity*, Chaos Soliton Fract. **14**, 917 (2002).

[36] P Abry, P Gonçalves, J Lévy Véhel, *Scaling, Fractals and Wavelets*, Wiley (2009).

[37] U.S. Geological Survey, *EarthExplorer*, https://earthexplorer.usgs.gov/ (2015).

[38] U. G. S. Department of the Interior, *Landsat 8 Data User's Handbook* (2015).

16

Open-source sensors system for doing simple physics experiments

César Llamas,[1] Jesús Vegas,[1] Miguel Á. González,[2] Manuel Á. González[3*]

An open-source platform to be used in high school or university laboratories has been developed. The platform permits the performance of dynamics experiments in a simple and affordable way, combining measurements of different sensors in the platform. The sensors are controlled by an Arduino microcontroller, which can be wirelessly accessed with smartphones or tablets. The platform constitutes an economical sensing alternative to commercial configurations and can easily be extended by including new sensors that broaden the range of covered experiments.

I. Introduction

We have designed and constructed a low cost platform that can be used to do simple physics experiments combining different sensors' data easily. The purpose of this work consist in implementing a cheap system that can be used in high schools and universities regardless of their budget in diverse experiments, allowing a higher customization than equivalent commercial sets.

II. System description

The initial requirements of the system are affordability, simplicity, flexibility and extensibility. Under the affordability requirement, we have decided to employ systems that can be easily found on the Internet at low prices. Simplicity means that users with little to no training can easily use the system in physics experiments without difficulty, but also, that its construction does not require special skills. Flexibility implies that the system should be portable enough to work on different types of experiments. Finally, its extensibility will allow users to include additional sensors in the platform to enhance its capabilities and to use it in new experiments.

In order to obtain an innovative platform as well as to add sharing features to it, almost exclusively open-source hardware and software components are used. Within this approach, the physics teacher community can work in open projects that can be used by other teachers, thus reducing the work of designing experiments with the platform. Following this condition, the characteristics of the system are published under a public GNU license in the GitHub repository [1]. All the technical details of the platform hardware and software are detailed in that repository: software code, assembling diagrams and technical characteristics of the components.

We initially decided to include sensors in the platform that measure kinematic magnitudes: accelerometer, gyroscope, distance sensor and light-gate sensors. Other sensors, such as a magnetome-

*E-mail: manuelgd@termo.uva.es

[1] Departamento de Informática, Universidad de Valladolid, 47011 Valladolid, Spain.

[2] Departamento de Física de la Materia Condensada, Universidad de Valladolid, 47011 Valladolid, Spain.

[3] Departamento de Física Aplicada, Universidad de Valladolid, 47011 Valladolid, Spain.

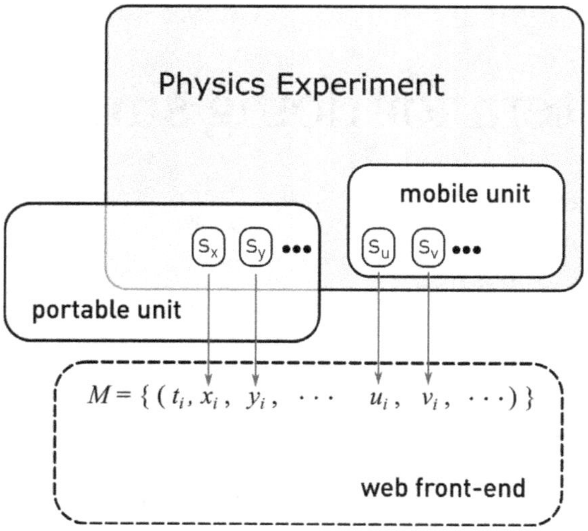

Figure 1: A conceptual diagram showing the main components of the system. It comprises a mobile unit that includes several sensors $(s_u, s_v, ...)$ controlled by an Arduino board and a portable unit with a Raspberry Pi that communicates with the Arduino and establishes a Wi-Fi network. It can also host additional sensors $(s_x, s_y, ...)$. The measurements of all the sensors can be transferred into the users' devices as series of vectors M via a web front-end implemented in the Raspberry Pi.

ter, a barometer, a thermometer, etc., can be easily added later. Figure 1 shows a diagram of the use of the platform in a physics experiment: the platform includes a mobile unit, which can be attached to a mobile body, containing sensors (s_u, s_v,... in the figure) and a portable unit responsible for the Wi-Fi connection that allows the user to access the measured data, but which can also include additional sensors (s_x, s_y,... in the figure). Measured data are collected by the portable unit and organized in vectors ($M = (t_i, x_i, ...)$) in the figure) also containing the time stamp of the measurements. Whenever it is desirable and possible, the mobile unit can be attached to a body in order to study its movement. Then, the connection between the mobile and the portable units must be based on wireless communications, allowing the studied body and the attached unit to move freely. In our case, communications between the mobile and the portable units are performed via Bluetooth.

The design of the mobile and the portable units was based on previous works of embedded systems [2] and they comprise the following elements:

- Mobile unit: An Arduino Nano v3 controls a set of sensors. The version that has been tested in laboratory experiments included an infra-red line tracker, a motion processing unit consisting of 3-axis accelerometer, gyroscope and an ultrasound distance sensor. The ATMEL serial interface of the microcontroller is connected to a HC-SR06 Bluetooth device. A LiPo battery supplies power to all the system. In its current design, the motion processing unit can be replaced by a similar one that also includes a 3-axis magnetometer together with the accelerometer and gyroscope which would permit to have a inertial measurement unit with 9 degrees of freedom. This system is programmed in Arduino 1.6 [3] and it implements a simple protocol through a serial wireless connection, which is established by using Bluetooth connectivity that allows the initialization of the microcontroller as well as starting and stopping data acquisition. The main speed limitation in data measurements of the system is due to the Bluetooth connection. Figure 2 (top) shows an implemented mobile unit used in some physics experiments described below.

- Portable unit: It consists of a Raspberry Pi2 with Raspbian Jessie [4] as an operating system. The core process of the unit is programmed in Go [5] and it governs the overall system as well as offering a web front-end. This unit supports the Wi-Fi and bluetooth connections. The data recorded in the measurements are stored in a SD card, constituting the local persistent storage unit. These data can be accessed and retrieved by the users' laptops or smartphones via Wi-Fi connection. Additionaly, the portable unit can also allocate other wired connected sensors using the GPI of the Raspberry Pi. In the current configuration of the system, four infrared emitter-receivers are connected to it. These can be used, for example, to measure instant speed at different points along the trajectory of a body. The unit can be powered by an ordinary USB battery, allowing to do physics experiments outside the laboratory. Figure 2 (bottom) shows the portable unit.

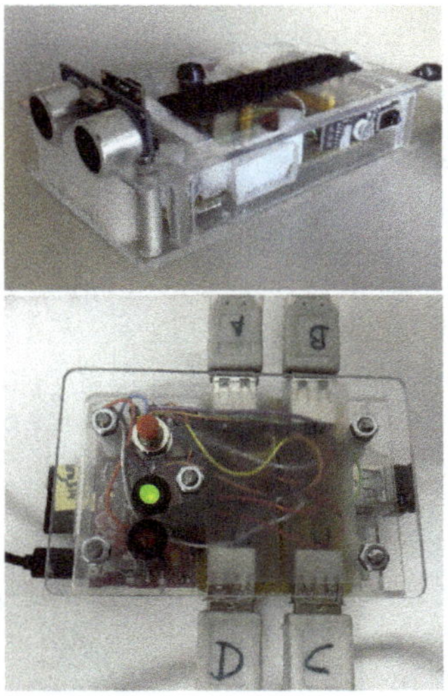

Figure 2: Mobile (top) and portable (bottom) units of the system. Portable unit shows the connections of four additional infrared sensors attached to it (connections A, B, C, and D in the figure).

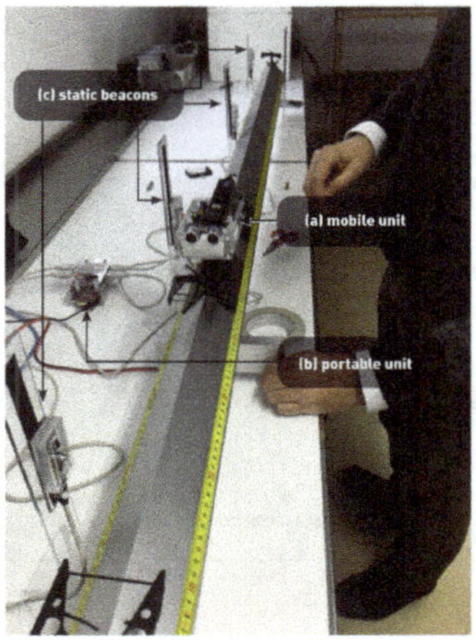

Figure 3: Photography of an air track experiment using the platform. While the mobile unit is on the moving cart, the portable unit stays on the table with four static infra-red beacons connected to it.

III. Use in the laboratory

The system described above can be used in different ways in the physics laboratory depending on the users' skill and knowledge of physics. We have tested the platform in two basic experiments: an air track in order to study an uniformly accelerated movement and a pendulum to analyze periodic motion and to obtain the gravity acceleration. Figure 3 shows the mobile and portable units being used in the air track experiment. In this experiment, the mobile unit included a 3-axis accelerometer, a 3-axis gyroscope and an ultrasonic distance sensor. In addition, four static infrared beacons were placed at fixed points of the trajectory and connected to the portable unit. The sampling frequency was about 25 samples/s ($\Delta t \approx 0.0394$ s). This frequency was the result of a compromise between the measuring capabilities of the Arduino and the data communication limitations between the Arduino board and the Raspberry, using the bluetooth connection. With the arrangement of sensors used in the experiment, users can measure the cart acceleration (accelerometer), average velocities between different points (infrared beacons), and the change in the position with time (ultrasound sensor and infrared beacons). From the data recorded by the different sensors, the user can analyze the uniformly accelerated movement and the relationships between kinematics magnitudes. On the other hand, in a pendulum experiment, users can combine the accelerometer and gyroscope measurements to study the acceleration and speed in different points and the periodic dependence of both magnitudes. Infrared beacons can also be used to measure the passage of time at different points of the pendulum trajectory and enrich the physical measurements. See [6] for more details on an air track experiment.

Recorded data are stored in the SD card as a plain CSV file with data of all the sensors used in the experiment. The CSV files are transferred to the users' computer or mobile devices using the Wi-Fi established by the portable unit. The rows of the CSV file are tagged with the acquisition time, in order to ease the analysis of the data by simply importing the file on any spreadsheet program. Figure 4 shows some results of the measurements with the air track. The measurements of the ultrasound distance sensor and the component of the accelerometer along the direction of the movement are represented as a function of time. From the data shown there, it can be observed how the experimental noise of the ultrasound sensor increases noticeably with the distance to the reflecting screen. We have checked this noise and, according to our exper-

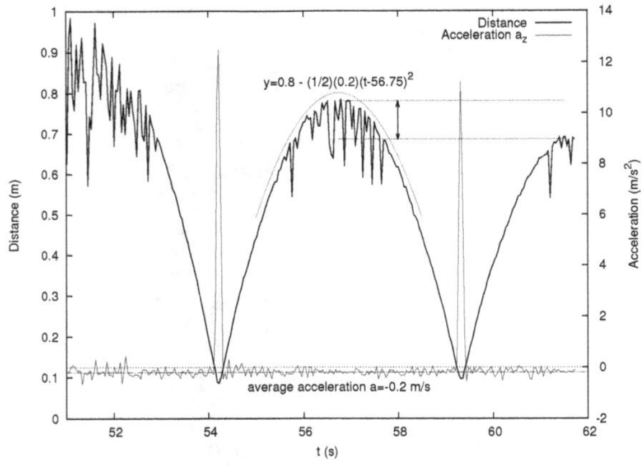

Figure 4: Example of combined data recorded with the platform in an air track experiment. The dependence of the distance and the acceleration along with the direction of movement recorded in the same experiment are shown jointly.

iments, it is due to the air exiting from the air track holes. Reducing or eliminating the airtrack pumping, that noise drastically decreases, as well as when the distance between the sensor and the reflecting screen decreases. This can be seen as a possible limitation of the used sensor, which can be solved using, for example, an optical distance sensor. The superposition of the data from the accelerometer and distance sensors allows a clear identification of the collision events between the cart and a rubber band at the end of the air track. From the diference in distances reached after consecutive rebounds (marked as example in the figure) users can obtain the restitution coefficient of the collision. Complementary, from the accelerometer data, users can obtain the change in speed during each collision and compare that result to the one obtained by analizyng the maximum distance before and after the bounce. As an illustrative result, a parabollic curve $y = 0.8 - (1/2)0.2(t - 56.75)^2$ is shown in the figure. The distance 0.8 was chosen to displace a little the theoretical curve upwards over the experimental points for clarity. The acceleration -0.2 m/s^2 corresponds to the average acceleration measured by the accelerometer sensor as shown in the figure. The agreement between different sensor data can help users to investigate relationships between different magnitudes in an experiment.

IV. Discussion

A cheap open-source system to be used in simple experiments has been developed. It comprises the hardware and software of a sensorized platform. The technical characteristics and the control software of the system can be downloaded for free from the Github repository [1]. The system permits the measurement of different magnitudes in the same experiment and the analysis of various phenomena. The electronic components used in the system can be acquired on the Internet by less than 100€, and additional cheap sensors can be also included. Unlike commercial sensor systems, this one constitutes an extensible and customizable platform that can be modified by adding other sensors and including other software characteristics. The use of the Githup repository also allows the collaboration between users that can add new characteristics to the platform.

Acknowledgements - This work has been supported by University of Valladolid teaching innovation program under grant PID2016_82.

[1] C Llamas, J Vegas, M Á González, M Á González, *https://github.com/percomp/OSHIWASP*, last accessed September 4, 2017.

[2] C Llamas, M Á González, C Hernández, J Vegas, *Open source hardware based sensor platform suitable for human gait identification*, Pervasive Mob. Comput. **38**, 154 (2017).

[3] https://www.arduino.cc/en/Main/Software, last accessed September 5, 2017.

[4] https://www.raspbian.org/, last accessed September 5, 2017.

[5] https://golang.org/, last accessed September 5, 2017.

[6] M Á González, A Gómez, M Á González, *Smartphones in the air track, examples and difficulties*, (unpublished).

17

Modeling temperature manipulations in a circular model of birdsong production

G. C. Dima,[1*] M. A. Goldin,[2] G. B. Mindlin[1]

The nature of the neural mechanisms in the birdsong motor pathway that lead to the generation of respiratory patterns are a matter of extensive debate. In a top-down control paradigm, vocal gestures emerge from a unique timescale ruled by the telencephalic nucleus HVC, which engages other brain regions downstream. Another possibility is that the generation of motor instructions is distributed throughout the neural network, flowing both upstream and downstream. In this circular architecture, the song results from the integration of more than one timescale. In order to disambiguate these views, we used local focal cooling of HVC in canaries to manipulate the timescale present there. Within the frame of the circular model, we fitted the experimental pressure patterns of different types of syllables, which form a full song. We show that at least two separate timescales must be taken into account to reproduce them, one which is manipulated by cooling while the other remains unchanged. The modifications —stretching and breaking— of the syllables were quantitatively reproduced in this frame.

I. Introduction

Birdsong is a remarkable example of how a complex behavior emerges from the interaction of peripheral motor areas and the central nervous system. The song motor pathway, which comprises the regions in the bird's brain that generate motor instructions for singing, involves several nuclei. These are located in the telencephalon, thalamus and brainstem [1–5]. The integration of the activity within them gives rise to vocal motor sequences.

*E-mail: gdima@df.uba.ar

[1] Departamento de Física, Facultad de Ciencias Exactas y Naturales, Universidad de Buenos Aires and IFIBA (CONICET), Pabellón 1, Cuidad Universitaria, 1428 Buenos Aires, Argentina.

[2] Unité de Neurosciences Information et Complexité, Centre National de la Recherche Scientifique UPR-3293, 1 Avenue de la Terrasse, 91190 Gif-sur-Yvette, France.

To generate a song, the brain motor pathway commands the activities of the respiratory system and the vocal organ called syrinx, a bipartite structure located between the bronchi and the trachea. The result is a succession of pressure gestures that are in a delicate coordination with the muscle activity in the syrinx [6–9]. These are commonly designated as syllables, which are the smallest units of a song, usually separated by silent gaps when the bird inspirates. The wall membranes in the syrinx —labia— generate sound by modulating the airflow coming from the air sacs. Therefore, since the generation of sound requires establishing airflow across the syrinx, a central aspect of motor control is its capacity to generate diverse respiratory rhythms [10]. We show in Fig. 1(A) an example of pressure in the air sacs during a song in domestic canaries (*Serinus canaria*), where different patterns can be distinguished. The song starts with a long expiratory pulse that consists of a series of small pressure

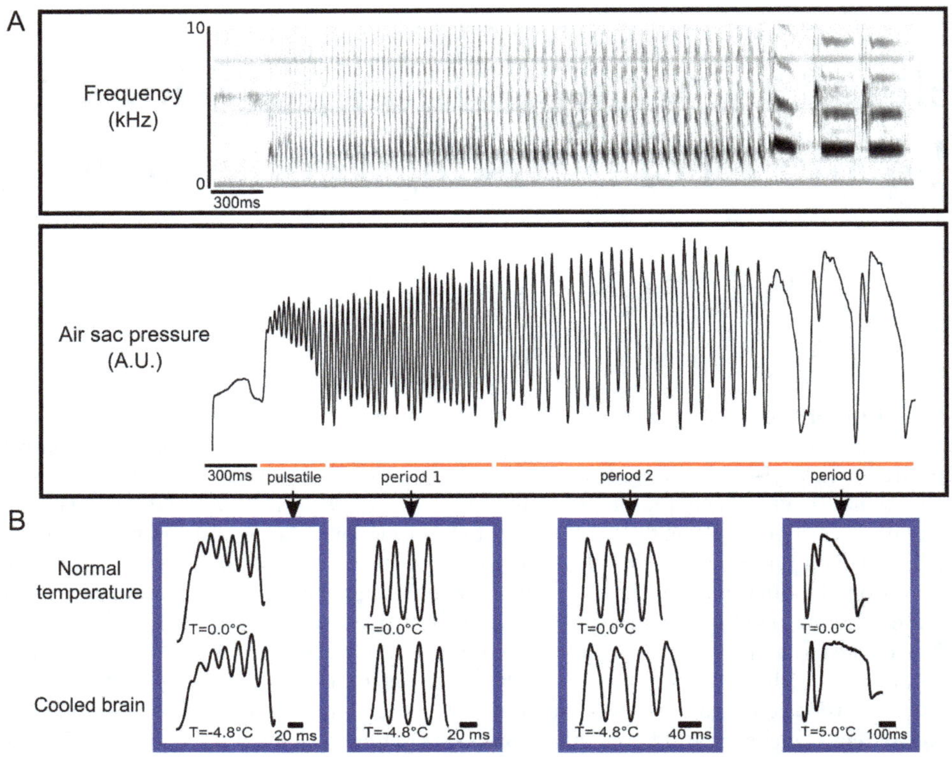

Figure 1: **Respiratory patterns of canary songs at different HVC temperatures.** Experimental measurements of pressure in the air sacs during the canary song. (A) Top: Sonogram of a complete song. Bottom: Different breathing patterns can be identified depending on their morphology and duration: Pulsatile gestures characterized by respiratory fluctuations mounted on a constant value, period 1 and period 2, displaying oscillating patterns and period 0 (repeated three times), ending with long expiratory pulse. (B) When cooling the nucleus HVC with a Peltier cooling device, a stretching in all the syllables is observed. In particular, the p0 syllables also show a change in their morphology.

pulses mounted on a DC level, each of them known as "pulsatile" pattern. It is followed by a repeating gesture of larger amplitude, with an oscillation reminiscent of harmonic pulses. These are designated as period 1 gestures (p1). Right after, we can see a p2 pattern, which also displays oscillations, but of approximately twice the duration of p1 syllables. Finally, the last syllables in the example (p0) are repeated with the lowest frequency and present a brief expiratory pulse followed by a decaying expiration. It has been shown that the vast majority of syllables can be classified into these four groups according to its frequency [11]. Fastest syllabic rates (> 27 Hz) correspond to pulsatile syllables, while syllabic repetition rates between 13 and 25 Hz correspond to p1 pressure patterns. Lower syllabic rates, ranging between 8 and 12 Hz are identified as p2 patterns whereas rates in the range of $2-5$ Hz fall in the classification of p0 patterns. Within these groups, different types of syllables can be classified according to its morphological difference such as number of pressure peaks and average time of expiration.

How these respiratory patterns emerge from the neural activity along the song motor pathway is still an open question. One hypothesis was presented as a top-down model [12–14]. In this framework, the telencephalic nucleus HVC (used as a proper name) which is synaptically far from the vocal organ (at least two synapses until reaching the brainstem), is placed as the highest element of the hierarchy. Acting like a metronome, with its neurons making a sequence of bursts of about 10 ms duration, this area is conjectured to drive the activation of the neurons in the RA (nucleus robustus of the archistriatum). RA neurons recruit in turn the brainstem motor neurons downstream. The complexity of the vocal gestures, therefore, emerges from a cascade of information ruled by a unique temporal sequence (HVC clock time) in this top-down hierarchy representation, diminishing the role of the bottom-up feedback arriving to HVC from the brainstem.

In an alternative hypothesis, the activation of the neuronal nuclei responsible for respiration may not only be due to the activity in the telencephalic area flowing downstream but also to the activation in other regions within the brainstem [15–18]. This does not mean that the timing signals in HVC do not play a crucial role, but that the system has to be analyzed in an integral fashion. A computational model was developed to represent this proposed architecture in the work of Alonso *et al.* [15]. In this framework, neural populations in the brainstem are connected upstream via the thalamus with HVC, which provides an architecture that gives the model its name: circular model. The generation of pressure patterns is therefore obtained by integrating the activity of two separated, yet connected, inputs: one from HVC and the other within the brainstem. This structure suggests the existence of more than one timescale involved, as the information coming from the brainstem is processed by HVC and then flows back into the brainstem where it combines with the original signal.

Thermal manipulation of nuclei of the motor pathway is a way that researchers in the field use to report evidence that could only be explained in one of the two presented paradigms. By locally cooling both left and right HVC nuclei in zebra finches, canaries and Bengalese finches, it was observed a significant temporal stretching across all time scales, from song motif to silent gaps [14,19–23]. From recent works exploring temperature effects at a neuronal level [24], we know that synaptic conductance propagation times coming from HVC's afferents are slowed with temperature, being then plausible to expect an expansion in the motor gesture. The stretching is then consistent with the top-down model. However, more recently, Hamaguchi *et al.* explored cooling the afferent nucleus to HVC in the thalamus, Uva thalamic nucleus uvaeformis, and showed that the reported values of stretching require an integrated architecture [22]. In addition, it was shown by Goldin *et al.* [19] that cooling HVC nuclei below 5°C results in the "breaking" of the syllables in the song of canaries into shorter motifs and in the appearance of complex respiratory patterns. In that work, the authors showed that some breaking patterns can be qualitatively explained by modeling the activity in HVC and in an unallocated brain area downstream, with a simple dynamical system driven near a bifurcation region. Although explaining some of the breaking patters within the model, not all syllable changes could be accounted for. One example of such elusive modification is shown in Fig. 1(B) (p0 "breaking"). The alteration of the timescale of a part of the entire p0 syllable as HVC is cooled pointed out the need of considering other brain areas or the exploration of different architectures for advancing the building of a model for song production.

In this work, we use the circular model to explain the different types of syllable "breakings", which conjectures that this phenomenon can be explained as a product of the interaction between two timescales. This model not only includes the one discussed in Goldin *et al.*, but it also allows to explain the structural changes of the p0 syllables. By quantitatively fitting experimental data of singing canaries at different HVC cooling temperatures, we unveil a mechanism of pulsatile, p1 and p2 stretching, as well as p0 stretching and breaking. For the last type of syllables, we show that the breaking is a natural consequence of the slowing of the activity in the telencephalon during thermal manipulation of HVC while keeping the brainstem activity unchanged. As each syllable type needs a different pattern of activity in our model, we support a recent view in which different parallel pathways are represented anatomically in HVC [25,26].

II. Materials and methods

i. Experiments

All experiments were approved by the Institutional Animal Care and Use Committee of the University of Buenos Aires (FCEN-UBA). We recorded the air sac pressure activity using a flexible cannula which was inserted through the abdominal wall into the anterior thoracic air sac under isoflurane anesthesia. A miniature piezoresistive pressure transducer

(Fujikura model FPM-02PG) mounted on the canary's back was connected to the cannula [6]. The voltage output from the transducer was amplified and recorded with a data acquisition device (National Instruments BNC2110). In addition, a cooling device that consisted in a Peltier cell was used to cool both left and right HVC nuclei [19].

Measurements consisted of recordings of song and pressure without cooling in the first place. Then, we proceeded to thermally manipulate the telencephalic nucleus by lowering the temperature by 1°C every two minutes, until the desired temperature was reached. After two minutes at the respective temperature, recording was started. Once sufficient data were collected, we continued with further cooling.

This procedure was repeated for at least four cooling temperatures in the range of 0°C to 7.4°C in three male adult canaries. A database of over 36 minutes of songs and pressure patterns was obtained. We extracted for the analysis 222 pulsatile syllables, 552 p0, 203 p1 and 1366 p2.

ii. The computational model

We present a mathematical description of the circular model which was previously introduced in Ref. [15]. The authors proposed an average description of the activity of each intervening area of the motor pathway, modeling it with a neural additive model. The equations governing the dynamics are phenomenological and connections between areas reflect a circular architecture [see Fig. 2(A)]. In what follows, we will describe in detail the variables, the connectivity and how the activity starting in the brainstem goes upstream, then downstream and integrates back with the initiating signal that traveled through another path within the brainstem.

Motor activity is identified with the dynamics of an expiratory related area ER [putatively the nucleus retroambigualis (RAm), black circle, Fig. 2(A)]. This region is part of the respiratory circuit within the brainstem and it is composed of two populations of interconnected neurons, one excitatory and another inhibitory (ER_e and ER_i, respectively). The expiratory related area receives information from an hypothetical region within the brainstem denoted as "initating area" [IA, blue circle, Fig. 2(A)]. This nucleus simultaneously sends information to both ER_e and the thalamus, represented in this model Uva [yellow circle, Fig. 2(A)]. From there, the information reaches the forebrain where HVC (green circle) is connected to RA. This latter nucleus is represented by two populations of excitatory and inhibitory neurons [RA_e and RA_i, red circle, Fig. 2(A)]. Finally RA_e projects to the respiratory circuit in the brainstem. Thus, the respiratory area processes information from IA before the contribution arrival from RA_e, resulting in a two-input-integrated cycle.

The presented architecture is based on areas anatomically connected. Several studies have shown a bottom-up connection between respiratory areas in the brainstem —such as the parambigualis nucleus (PAm) and the dorsomedial nucleus of the intercollicular complex (DM)— and the telencephalon, mediated by Uva [27,28]. Moreover, these connections are hypothesized to be necessary to coordinate the contralateral sides of HVC during the production of some syllables [29].

The activity coming from the brainstem needs at least two synapses to reach the telencephalon, since they are mediated by the thalamus (Uva), in contrast to the direct connection to ER. For this reason, it is hypothesized that the response of HVC to the activity in IA presents a time delay respect to the onset of the activity in IA [28]. The dynamics in HVC is represented as a constant level of averaged activity, which results from the nuclei intrinsic activity, which adds to the IA induced activity.

The variables in the model represent the average population activity of neurons in a set of interconnected areas [30]. The dynamics for these variables are described by a time continuous neural additive network model:

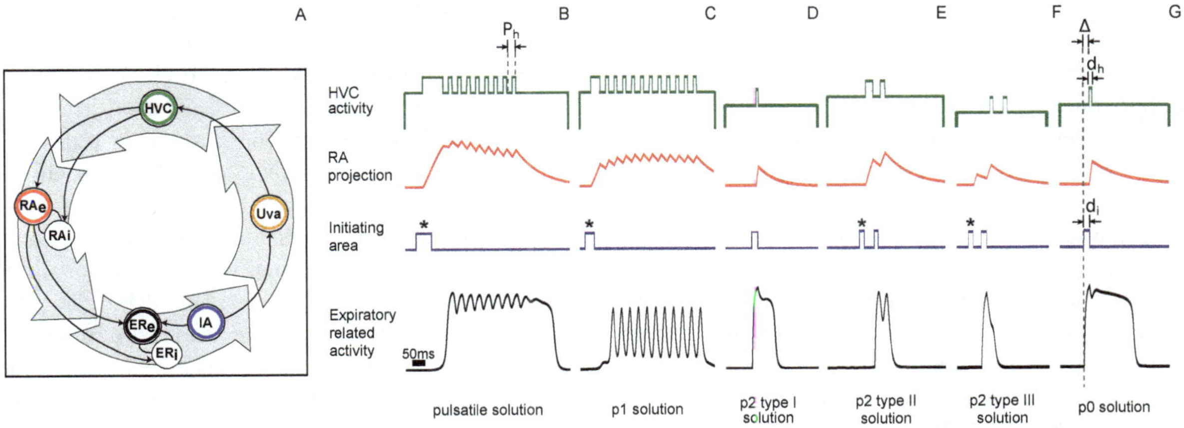

Figure 2: **Circular model elements activity that generate the singing respiratory patterns.** (A) Model architecture describing the respiratory motor pathway. The expiratory area is represented by two coupled populations of excitatory and inhibitory neurons: ER_e (black circle) ER_i. Two inputs arrive at ER_e, one from another region of the brainstem (IA, blue circle) and one from the forebrain. This last region is composed by the HVC (green circle) which has projections to the excitatory and inhibitory RA population (RA_e, red circle, and RA_i). Finally, the brainstem is connected to the telencephalon by Uva (yellow circle), which delays the signal initiated in IA. Two mechanisms are proposed to reproduce the four types of respiratory patterns of canaries, as presented in Fig. 1. (B)–(C) For syllables with a simple structure, pulsatile and p1, IA triggers an harmonic activity in HVC with a starting pulse (see asterisk). These gestures inherit the periodicity P_h of HVC dynamics. (D)–(G) For the other the syllables, the integration of activity in IA and the processed information of RA results in the composition of the gestures. The time delay between IA and HVC pulses are denoted by Δ, and their durations d_h and d_h. Variations of these parameters allowed the generation of gestures that best fit the experimental data. The constants used in the model can be found in Table 1, while the specific parameters needed to generate the gestures in (B)–(G) are: $P_h = 39.787$ ms (pulsatile); $P_h = 38.671$ ms (p1); $d_h = 11.8$ ms, $d_i = 34$ ms, $\Delta = 28$ ms (p2 type I); $d_h = 22.8$ ms, $d_i = 18$ ms, $\Delta = 31$ ms (p2 type II); $d_h = 20$ ms, $d_i = 24$ ms, $\Delta = 27$ ms (p2 type III) and $d_h = 16$ ms, $d_i = 32$ ms, $\Delta = 32$ ms (p2 type III).

$$\begin{aligned}
\frac{de_{er}}{dt} &= 249.5[-e_{er} \\
&\quad + S(\rho_{e_{er}} + 10e_{er} - 10i_{er} + \alpha_{eer,ra}e_{ra} + F_{\mathrm{IA}})], \\
\frac{di_{er}}{dt} &= 249.5[-i_{er} \\
&\quad + S(\rho_{i_{er}} + 10e_{er} + 2i_{er} + \alpha_{ier,ra}e_{ra})], \\
\frac{de_{ra}}{dt} &= 20[-e_{ra} \\
&\quad + S(\rho_{e_{ra}} + \alpha_{era,ra}e_{ra} - \beta_{era,ra}i_{ra} + F_{\mathrm{HVC},e})], \\
\frac{di_{ra}}{dt} &= 20[-i_{ra} \\
&\quad + S(\rho_{i_{ra}} + \alpha_{ira,ra}e_{ra} - \beta_{ira,ra}i_{ra} + F_{\mathrm{HVC},i})], \\
S(t) &= \frac{1}{1+e^{-t}},
\end{aligned} \tag{1}$$

where e_{er} and i_{er} represent the respiratory area (excitatory and inhibitory population) and e_{ra} and i_{ra} describe the dynamics of RA. Each area is comprised of subpopulations of neurons with different properties whose characteristics are reflected in the constants α, β and ρ (strength of the connection between excitatory and inhibitory neuronal populations, and basal activity, respectively). The dynamics of IA directly affecting the respiratory area are represented with F_{IA} while the projecting neurons of HVC to both excitatory and inhibitory regions of RA are labeled $F_{\mathrm{HVC},e}$ and $F_{\mathrm{HVC},i}$. In order to achieve long expiration patterns, we assume that RA activity is much slower than ER. This is mathematically reflected in the multiplicative constants of the right hand side of the set of equations. The values were chosen so that the ER dynamics could be directly compared with the pressure patterns. The rest of the numbers are the same as in Alonso *et al.* [15]. Different families of pressure patterns, as the ones shown in Fig. 1, can be generated within the architecture of Fig. 2(A), by different parameter values.

It has been shown that the neuronal activity in HVC during singing is composed of short bursts [12,31,32]. Within the frame of the circular model, this behavior could be produced by the activation of HVC by the brainstem. To gain in simplicity, the dynamics of IA and HVC were modeled by square pulses, since the major feature they provide is time keeping. It is important to note that the dynamics of e_{er}, which is responsible for generating the motor gestures, is externally affected by both F_{IA} signal from IA, as the contribution from RA (with weight $\alpha_{eer,ra}$. This latter nucleus receives information from HVC in form of square pulses of intensity $F_{\mathrm{HVC},e}$ and $F_{\mathrm{HVC},i}$, delayed respect to F_{IA}.

III. Results

i. Fitting the normal song pressure patterns

Respiratory gestures that generate canary songs can be classified into different classes [11]. As shown in Fig. 2, it is possible to qualitatively recreate these patterns with the model proposed. To quantitatively asses the capabilities of the model, we chose and measured a characteristic feature of each syllable type and generated patterns with the model that are as close as possible to that values. These features were chosen due to its direct impact on song, such as the oscillation frequency or duration between pressure peaks. We explored the parameters that give rise to the activity in HVC and IA in a way that allowed us to minimize the difference between the experimental and the computationally generated feature, i.e. our cost function that had to be minimized. Also, the constants that account for the properties of the subpopulations in each area were adjusted (α, β and ρ, see Table 1). In what follows we will describe the different fittings that we performed for each syllable type and the feature that we selected in each case.

a. Pulsatile

Pulsatile gestures are characterized by low amplitude oscillating pressure patterns mounted on a DC level. The oscillation frequency is almost constant throughout the syllable, making it a natural feature to characterize them. To generate these pressure patterns, we proposed that the initiating area IA begins the syllable by sending a starting pulse to both ER and HVC. It is proposed that the efferent copy reaching HVC leads to the start of a sustained harmonic activity within it. Finally, this activity excites the RA nucleus, which is responsible for forcing the respiratory area [Fig. 2(B)]. To describe the dynamics in HVC, we used a square pulse train of amplitude $(F_{\mathrm{HVC},e}, F_{\mathrm{HVC},i}) = (10, 10)$ and period P_h. In order to set each pulse duration, we computed the average period of all the dataset of pulsatile patterns. This number halved resulted to be 14 ms and was used as a constant duration for all the simulations. It was also observed that shorter pulses fail to generate periodic solutions with periodicity equal to the experimental gestures. For each syllable of this type, we adjusted the driven period P_h to generate gestures with the same average frequency observed experimentally.

b. p1 syllables

Another type of syllable with an oscillating pressure pattern is called p1. They are characterized by a greater oscillation amplitude than pulsatile gestures and by a frequency that remains almost constant for the duration of the gesture. Because of this, we used it as a characteristic feature of the syllable. Given the similarity with the pulsatile syllables, we proposed that the same mechanism is involved in the generation of this pressure gestures [Fig. 2(C)]. We modeled the activity in HVC with square pulses of 15 ms duration, period P_h and amplitude $(F_{\mathrm{HVC},e}, F_{\mathrm{HVC},i}) = (6, 0)$.

c. Oscillation onset for pulsatile and p1

In the two groups of syllables described above, the pressure gesture begins with a rapid increase before the oscillation starts. To generate this harmonic solution in our simulations, the RA area is forced periodically by HVC. However, the activity of RA has a slow initial dynamics with a pre-oscillation transient period that makes the initiation of the simulated pattern to be slow. In order to reduce this time, we began these type of syllables with a long pulse from IA that generates a prolonged ac-

Table 1: **Constants in the circular model that allowed to fit the four types of motor gestures.** The mathematical representation of the respiratory related area (ER) and RA contain parameters that allow to differentiate subpopulations of neurons within them. These constants define static properties such as the strength of the connection to excitatory and inhibitory areas (α and β respectively) and the onset level where the activity is mounted (ρ).

	Respiratory related area $(\rho_{e_{er}}, \rho_{i_{er}}, \alpha_{eer,ra}, \alpha_{ier,ra})$	RA expiratory $(\rho_{e_{ra}}, \alpha_{era,ra}, \beta_{era,ra})$	RA inhibitory $(\rho_{i_{ra}}, \alpha_{ira,ra}, \beta_{ira,ra})$
Pulsatile	$(-7.5, -11.5, 10, 3.5)$	$(-5.25, 10, 10)$	$(-12, 10, -2)$
P1	$(-6, -8, 12, 4)$	$(-5.25, 10, 10)$	$(-5, 10, -2)$
P2 *type I*	$(-7.5, -11.5, 10, 0)$	$(-3, 8, 8)$	$(-4, 3, -5)$
P2 *type II*	$(-7.5, -11.5, 10, 6)$	$(-4, 5, 5)$	$(-18, 8, -6)$
P2 *type III*	$(-7.5, -11.5, 10, 6)$	$(-2, 5, 5)$	$(-2, 5, -6)$
P0	$(-7.5, -11.5, 10, 0)$	$(-3, 6, 3)$	$(-6, 6, -6)$

tivity in HVC [see asterisk in Figs. 2(B)–(C)], that can be thought as a strong recruiting signal. We also assumed that the direct connection to the respiratory area is weak so that IA only affects the dynamics in HVC, because otherwise the morphology of the solution changes. To determine the length of these pulses, we studied the experimental p1 and pulsatile syllables with the lower average frequency of the complete dataset. For these syllables, we ran simulations varying the length of the initial burst, selecting the pulse of the shorter duration that minimizes the transitioning rise time of RA. A 80 ms recruiting pulse was used in HVC for the pulsatile solutions, while for the p1 a 40 ms duration recruiting pulse was needed, both with amplitudes $(F_{\text{HVC},e}, F_{\text{HVC},i}) = (5.5, 0)$ [Figs. 2(B)–(C), top panel]. We found that these pulse durations were enough to avoid the slow rising time in the simulated gestures of the complete dataset of p1 and pulsatile patterns.

d. *p0 and p2 syllables*

The remaining two groups of syllables (p2 and p0) present a richer internal structure than previous gestures. Basically, both syllables are composed of an initial high pitch sound generated by the tight side of the syrinx, followed by the activity of the left side of the organ [29]. In contrast to the previous syllables, we formulate an alternative process that takes into account the coordination of the two syringial sides. We first make a brief description of each type of gesture and then present the mechanisms to generate them.

e. *p2 syllables*

Period 2 (p2) pressure patterns are characterized by a length of approximately twice the period 1 gestures (between 80 and 140 ms), and three types of morphologically distinct gestures can be distinguished from the experimental patterns. The type I, with an average duration of 139±11 ms, $N = 361$, is characterized by a maximum pressure followed by a smooth decay that has a local minimum of pressure before going back down. The type II is similar to the previous one, except that after the local minimum, pressure increases again to reach a second peak before the end. Its average duration is 83 ± 6 ms, $N = 111$. Finally, type III oscillations are similar to the ones of type I, but about half their length (73 ± 5 ms, $N = 355$). The expiration is shorter, which generates a more abrupt pressure decay.

f. *p0 syllables*

Syllables lasting between 200 and 400 ms fall in the group called p0. As in the p2 type I, the pressure reaches a maximum at the beginning of the syllable, forming a characteristic peak and a local minima following. Then, after a slight increase, the expiration is practically sustained until the end of the syllable

In both groups, the pronounced peak of pressure at the beginning of the syllable followed by a slow expiration pointed to the search of a more elaborate neural mechanism for the creation of these gestures. This is also reinforced by the fact that, anatomically speaking, these features are generated by the use of only the right/left side of the syrinx (first pressure peak), followed by the left/right side (rest of the syllable) for upsweeps/downsweeps in the syllable's sound frequency content [29].

Unlike the way in which p1 and pulsatile syllables are generated with the model, where the respiratory area only integrates the processed activity of HVC, the complexity of p2 and p0 syllables is obtained by exploiting both direct connectivity of IA to ER and the contribution from the forebrain. The initiating area sends a pulse of duration d_i to both ER_e and the telencephalon. The information reaches HVC via the thalamus. It is assumed that this pulse only encodes a starting signal in HVC. Therefore, for the sake of simplicity, we modeled the dynamics in this nucleus as a pulse of duration d_h, delayed a time Δ from the initiating signal. From there, the information flows back the brainstem by passing through RA. Note that the pulse generates a slow response within this region. Finally, the respiratory area integrates both signals of RA_e and IA [Figs. 2(D)–(G)]. The first pulse arriving from the initiating area, is responsible for the formation of the first maximum of the pressure pattern, while the slow activity of RA is responsible for the sustained part of the syllable (p0 and p2 type I) and for the second maximum pressure (p2 type II). In the p2 type III, the dynamics of RA helps to smooth the pressure decrease after reaching its maximum. For these syllables, the amplitude of the pulses used were $(F_{ia}, F_{\mathrm{HVC},e}, F_{\mathrm{HVC},i}) = (10, 5, 10)$ for type I, $(F_{ia}, F_{\mathrm{HVC},e}, F_{\mathrm{HVC},i}) = (20, 14, 0)$ for type II, $(F_{ia}, F_{\mathrm{HVC},e}, F_{\mathrm{HVC},i}) = (10, 20, 0)$ for type III and $(F_{ia}, F_{\mathrm{HVC},e}, F_{\mathrm{HVC},i}) = (10, 50, 4)$ for p0. The constants for each subpopulation in the circuit are shown in Table 1.

In a similar way as the oscillating patterns, for type p2 II and type p2 III it is necessary to include a first starting pulse with the purpose of exciting the RA population. These initial pulses fail to activate through the direct connection of IA the activity of ER [see asterisk in Figs. 2(D)–(E)]. The values used were $(F_{ia}, F_{\mathrm{HVC},e}, F_{\mathrm{HVC},i}) = (2, 10, 10)$, $(d_i, d_h) = (22ms, 36ms)$ for type II and $(F_{ia}, F_{\mathrm{HVC},e}, F_{\mathrm{HVC},i}) = (2, 14, 0)$, $(d_i, d_h) = (22ms, 14ms)$ for type III.

As in the previous oscillating patterns, we found a feature which allows us to identify each of the syllables of these sets obtained experimentally. Over all the set of measured p0, we observed a significant variation in the total duration of the syllable (D_{exp}) and the position of the local minimum pressure after the first maximum (H_{exp}) [see Fig. 3(A)]. We then generated solutions that recreated these features, by only controlling the activity of HVC and IA. By varying the widths of the pulses (d_i, d_h) and the delay Δ between them, it is possible to produce morphologically similar solutions with duration D_{theo} and pressure difference H_{theo} [Fig. 3(A)]. These three parameters were used to recreate the diversity of gestures measured experimentally.

In order to quantify the goodness of our fit, D_{exp} y H_{exp} values were calculated for each experimental syllable, after normalizing the pressure in the same range as the model solution. We defined the X parameter as

$$X = \frac{|D_{exp} - D_{theo}|}{D_{exp}} \cdot \frac{|H_{exp} - H_{theo}|}{H_{exp}}. \quad (2)$$

This quantity allowed us to quantify the goodness of the theoretical fitted gesture. Therefore, by working in a region of the (α, β, ρ) space that generates solutions of the desire morphology (see Table 1), for each p0 and p2 type II syllable measured experimentally, we found the values of d_i, d_h and Δ that generated a gesture that minimized the X parameter.

For p2 type I and p2 type III syllables, given the lack of a well-defined local minimum, the relative distance between the simulated gesture duration D_{theo} and the syllable length D_{exp}, was chosen as characteristic parameter to minimize

$$X' = \frac{|D_{exp} - D_{theo}|}{D_{exp}}. \quad (3)$$

ii. Fitting of Δ

We assumed that the time of arrival and processing from IA to the telencephalon, Δ [illustrated in Fig. 2(G)], was a fixed constant for each studied bird. We determined its value by selecting the experimental recorded p0 syllable whose values D_{exp} and H_{exp} were closer to the average values $\langle D_{exp}\rangle$ and $\langle H_{exp}\rangle$ from all the dataset [$(\langle D_{exp}\rangle; \langle H_{exp}\rangle) = (317 \pm 51$ ms$, 0,07 \pm 0,03)$, $N = 103$ for bird #31, $(329 \pm 25$ ms$; 0,12 \pm 0,07)$, $N = 110$ for bird #32

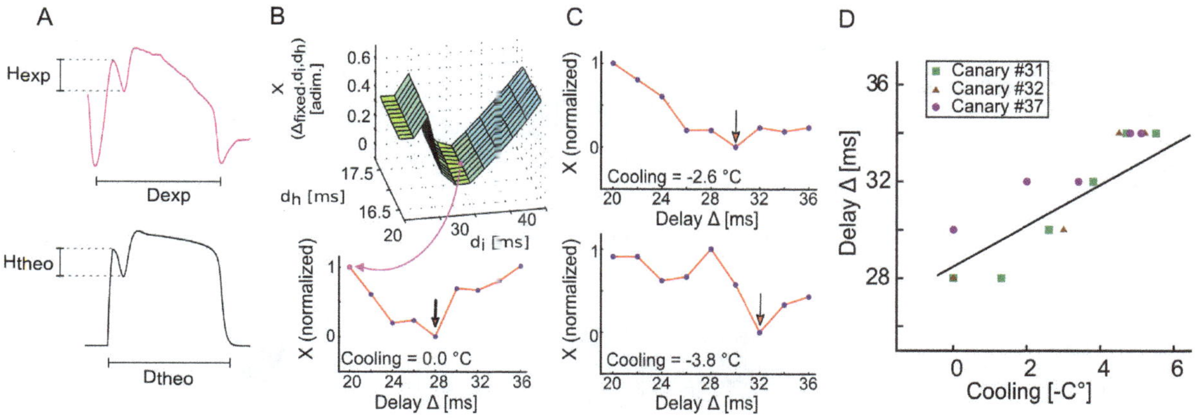

Figure 3: **Parameter space and fitting procedure for p0 syllables.** (A) Two features of p0 syllables were chosen to be fitted with the model: the total duration of gesture (D_{exp}) and the pressure difference between the first maximum and the first minimum (H_{exp}). We minimized the cost function $X = \frac{|D_{exp}-D_{theo}|}{D_{exp}} \cdot \frac{|H_{exp}-H_{theo}|}{H_{exp}}$. (B) To determine the value of Δ, the time between IA and HVC pulses, we minimized X for a representative p0 (D_{exp} and H_{exp} value closer to the population average) by fixing Δ and exploring the pulses duration space (top panel). Once the minimum was found (pink dot), the procedure was repeated by changing the time delay. The value of Δ that generated the minimum X (red arrow, lower panel) was taken as a constant for the rest of the simulations. (C) An increment of the Δ value was found as this procedure was repeated for all the cooling temperatures. (D) We summarize the results obtained for three different birds. As expected, time delays are positively correlated with cooling temperatures. Slope = (1.1 ± 0.1) ms/°C, intercept = (28.2 ± 0.5) ms, $R^2 = 0.81$.

and (470 ± 65 ms; 0,07 ± 0,01), $N = 32$ for bird #37] and we minimized X by exploring the pulse duration space (d_i, d_h) with a fixed value of Δ [Fig. 3(B), top panel]. Each pulse length was varied in the range from 2 to 65 ms (in order to recover solutions with the morphology of p0) until the absolute minimum was found [pink dot, Fig. 3(B)]. This procedure was then repeated by changing the delay Δ in the range of 20 to 36 ms, with a step of 2 ms. This is a reasonable delay range to explore for a minimum of three synapse distance connection [22,28,33]. Figure 3(B) —lower panel— shows the minimum value of X (normalized in the [0, 1] range) for each delay. We chose the time delay Δ as the minimum value of that curve, and we repeated the procedure for each bird studied. First, we searched its closer-to-the-mean p0 syllable and minimized X as a function of Δ for that gesture. The value obtained for birds #31 and #32 was 28 ms, while for the bird #37 it was 30 ms.

The procedures described in this section provided the parameters of the circular model that produced patterns that quantitatively match the experimental data. An example of the simulation of a complete song is observed in Fig. 4, along with the activity of each intervening nucleus. In the figure, the song starts with a set of pulsatile syllables followed by p1 patterns. The activity of IA in this part remains silent except for the beginning of each change of syllable type. Mounted on a constant DC level, the dynamics on HVC consists in a series of bursts that define the average frequency of the gestures. The p1 are followed by slower syllables: p2. As mentioned before, the beginning of each syllable starts in IA, followed by the activation in HVC, which has a different averaged DC with respect to the previous pattern. Finally, five p0 are generated by using different subpopulations of neurons in the presented architecture. This is reflected in the basal level of HVC as well as the slow dynamics in RA. In the end, the song ends with a set of pulsatile and p1 syllables. The continuous trace of the pressure output shows the capabilities of the model to link different simulated syllables together.

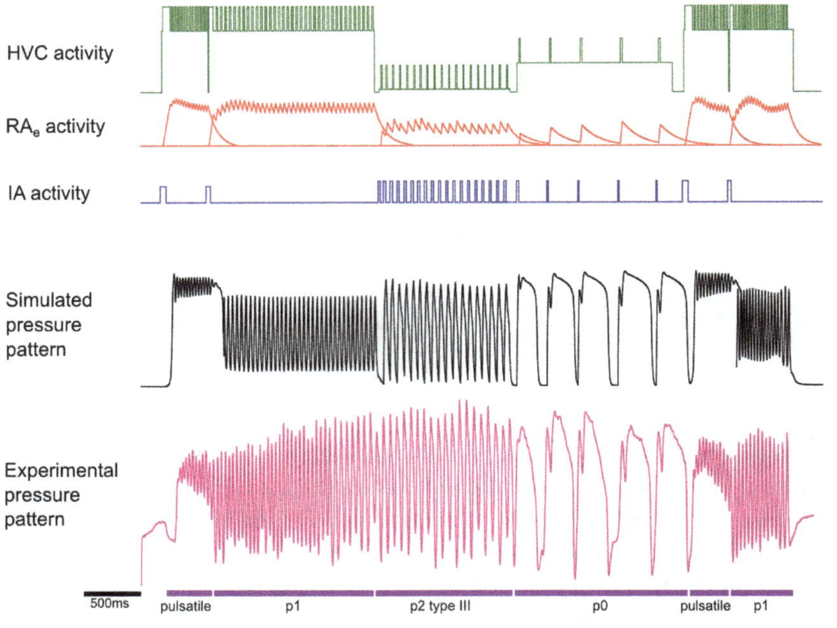

Figure 4: **A complete pressure pattern simulation based on the circular model.** In order to reproduce the respiratory pressure patterns computationally, certain features of each gesture were selected to fit the experimental data (see main text for details). For p1 and pulsatile syllables, we simulated gestures with mean oscillation frequency equal to the experimental patterns. For the p2 in this song (type III), our fitting parameter was the syllable duration. Finally, for each syllable p0, the activity in IA and HVC was varied so as to minimize X, which measure the goodness of the fit. The activity in IA is fundamentally needed to generate p0 and p2 gestures, whereas pulsatile and p1 are only produced by the activation of HVC. The baseline of the activity in HVC changes during the production of each type of syllable. The continuous trace of the pressure output shows the capabilities of the model to link different simulated syllables together. The values in Table 1 were used to generate each syllable type, while the parameters P_h, d_h, d_i were tunned for each gesture in order to reproduce its experimental feature: $P_h = 41.4112$ ms and $P_h = 33.4010$ ms (first and second p1); $P_h = 29.9720$ ms and $P_h = 33.3576$ ms (pulsatile). The values of d_i and d_h fluctuate around 20 ms and 10 ms for p2 and around 14 ms and 17 ms for p0. The time delay $\Delta = 28$ ms was fixed for all the simulations.

iii. Fitting the cooling effects

Thermal control of the telencephalon affects the respiratory patterns in two ways. For mild cooling, the song slows down, which is reflected in only a stretching of their shape. This effect is compatible with the hypothesis of a top-down representation of the song motor system, as cooling would slow down the activity in HVC. However, decreasing the temperature more has critical consequences: qualitative changes in the morphology of the pressure oscillations and even breaking of the syllables into smaller pieces. Also, the structure of the complete song is broken, presenting some complete bouts, and part of bouts or sets of syllables, sparsely distributed with extended silences between them.

Our model has the capability of translating the syllable morphological effects into changes in the parameters, since the cooling produces a slowing down of the internal dynamics of HVC, and a slowing down in the synapses from the brainstem to the telencephalic structures. Operationally, this is translated into the model by changes in the duration d_h of the driving pulses at HVC, its periodicity P_h and the delay Δ between the pulses in IA and HVC. These effects will be especially noticeable for those patterns where the two timescales interact, such as in p0. To study how these parameters

Table 2: **Linear fit results.** For each experimental syllable, a set of parameters were found to recreate its feature. We adjusted the average values of these parameters with linear regressions, for the four types of syllables. For p1 and pulsatile solutions the fit was made on the variation of the period of the activity in HVC along the temperature change. For p2 and p0, we adjusted the pulse duration of HVC and IA as a function of the time delay.

Panel	Reference	Slope	Intercept	R^2
A	Canary #31	$(8.53 \pm 1.79).10^{-4}$ ms/°C	$(3.17 \pm 0.09).10^{-2}$ ms	0.70
	Canary #32	$(1.69 \pm 0.88).10^{-4}$ ms/°C	$(1.49 \pm 0.38).10^{-2}$ ms	0.25
	Canary #37	$(6.64 \pm 0.75).10^{-4}$ ms/°C	$(1.54 \pm 0.30).10^{-2}$ ms	0.92
B	Canary #31	$(3.50 \pm 0.86).10^{-4}$ ms/°C	$(1.93 \pm 0.04).10^{-2}$ ms	0.61
	Canary #32	$(6.25 \pm 0.89).10^{-4}$ ms/°C	$(2.79 \pm 0.04).10^{-2}$ ms	0.89
	Canary #37	$(6.53 \pm 1.21).10^{-4}$ ms/°C	$(2.14 \pm 0.05).10^{-2}$ ms	0.82
C	Type I	0.20 ± 0.03	(4.99 ± 0.89) ms	0.93
	Type II	0.06 ± 0.11	(22.58 ± 3.56) ms	0.31
	Type III	0.57 ± 0.12	(-3.74 ± 3.89) ms	0.92
D	Type I	0.29 ± 0.23	(23.18 ± 7.10) ms	0.38
	Type II	-0.16 ± 0.14	(23.09 ± 4.46) ms	0.13
	Type III	-0.07 ± 0.14	(25.18 ± 4.46) ms	0.36
E	Canary #31	1.12 ± 0.25	(-9.44 ± 7.85) ms	0.82
	Canary #32	0.79 ± 0.16	(-6.68 ± 4.92) ms	0.89
	Canary #37	1.00 ± 0.03	(-8.41 ± 0.95) ms	0.97
F	Canary #31	0.08 ± 0.25	(39.72 ± 7.87) ms	0.05
	Canary #32	0.08 ± 0.14	(40.89 ± 4.23) ms	0.25
	Canary #37	0.21 ± 0.46	(28.83 ± 14.88) ms	0.17

evolve with different temperatures, we fitted experimental gestures during HVC cooling with the ones generated by the mathematical model. For cooling below temperatures of 6°C, many p0 and a small set of pulsatile gestures present multiple breakups, changing drastically the shape of the pressure patterns. These syllables have been excluded from the analysis since more complex mechanisms that escape the presented model could be involved under these extreme conditions.

For pulsatile and p1 gestures, we hypothesized that the cooling effect changes both HVC frequency and its pulse size. We set the forcing period P_h to generate solutions with the same average frequency as the one observed in the experimental patterns. Also, we explored the pulse length, increasing its size up to 30% of the pulse length of the uncooled solutions. No qualitative changes in the morphology of the patterns where found. Beyond that limit, the simulated gestures presented no periodic solutions. Figures 5(A)–(B) show a positive correlation between the temperature and the P_h parameter (averaged over all data at the same temperature). For all birds, a significant increase of the forcing period was observed while lowering the temperature of HVC.

In order to fit the rest of the syllables, we first determined how cooling affects the time between pulses IA and HVC. The relationship $\Delta(T)$ was found as previously described: we selected the value of Δ that minimizes X (of a representative p0), when both d_i and d_h are varied. This procedure was repeated for the three birds. In Fig. 3(D) a remarkable increase of the time delay is observed for all birds, while decreasing the cooling temperature. Having fixed the time delay, we minimized the X value for each p0/p2 syllable within a complete song, while varying d_i and d_h. A notable positive correlation between the size of HVC pulses and the time delay was found [Figs. 5(C)–(F)] while the activity in IA remained virtually constant for all temperatures (see Table 2 for fitting results). All

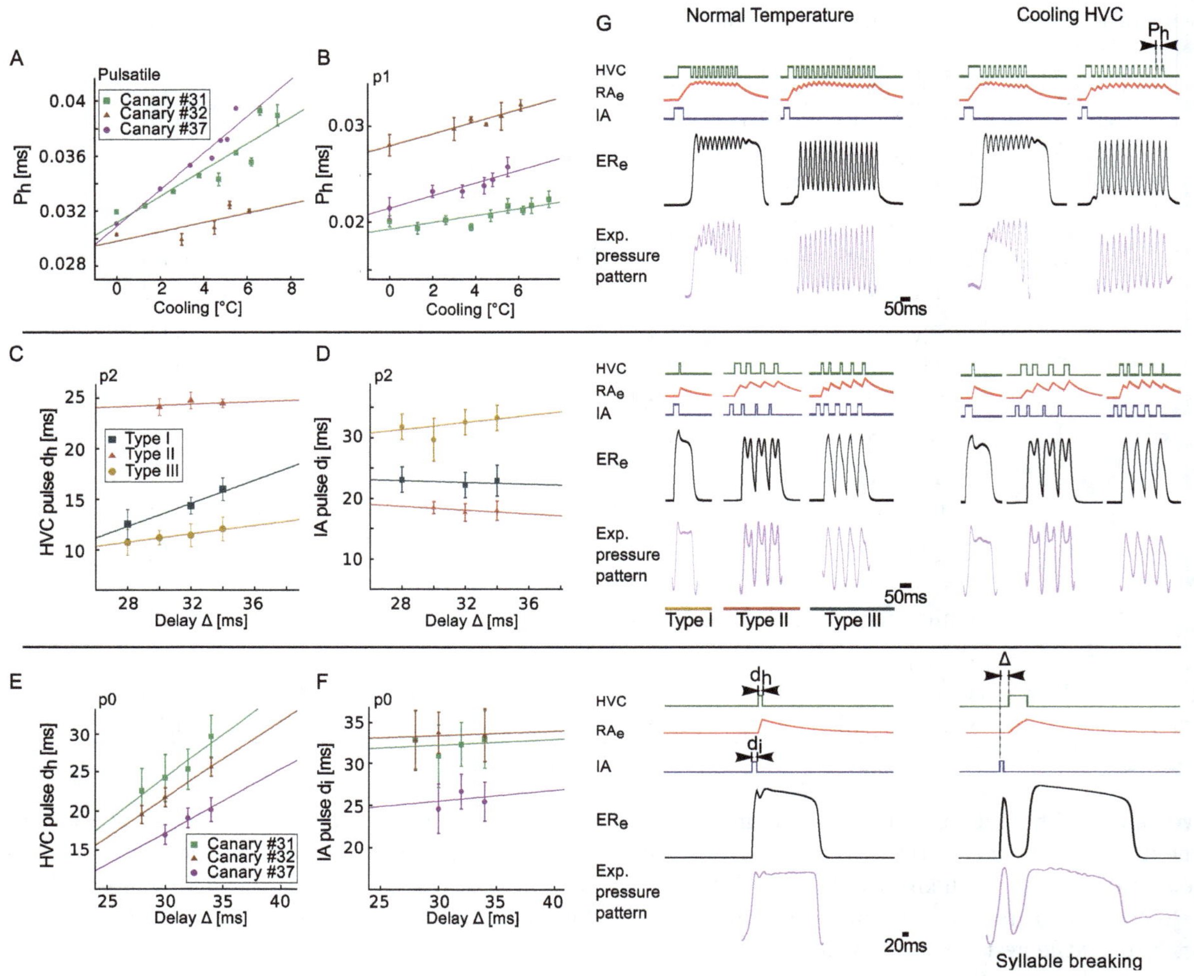

Figure 5: **Motor gesture patterns under HVC cooling.** For each syllable in the four classification groups, we set the activity in IA and HVC to recreate the selected features of the measured pressure pattern at all temperatures. For pulsatile and p1 (A)–(B), the feature varied was the forcing period P_h in HVC. For p0 and p2, it was the width of the pulses of HVC and AI, with a fixed characteristic delay at different temperatures (C)–(F). All the values are an average over all the dataset. Panel (G) shows the cooling effect for all syllables, in particular the mechanism of stretching and delay that allowed to recreate the breaking effect of a p0 syllable. Fitting values can be found in Table 2. Each gesture was generated by using the constants in Table 1 and choosing the parameters Δ, P_h, d_h, d_i that best fit the X value. For normal temperature: $P_h = 29.684$ ms (pulsatile); $P_h = 36.2818$ ms (p1); $(d_i, d_h, \Delta) = (20, 17.4, 28)$ ms for p0, $(28, 9.6, 28)$ ms (p2 type I) and values around $d_i = 12$ ms, $d_h = 24$ ms with $\Delta = 30$ ms (p2 type II) and $d_i = 24$ ms, $d_h = 13$ ms with $\Delta = 28$ ms (p2 type III). Cooling HVC: $P_h = 33.631$ ms (pulsatile); $P_h = 43,2736$ ms (p1); $(d_i, d_h, \Delta) = (14, 67.4, 34)$ ms for p0, $(34, 12, 34)$ ms (p2 type I) and values around $d_i = 12$ ms, $d_h = 27$ ms with $\Delta = 32$ ms (p2 type II) and $d_i = 24$ ms, $d_h = 14$ ms using $\Delta = 32$ ms (p2 type III).

the remarkable effects of cooling are displayed in Fig. 5(G), showing the activity in each nuclei of the model and the experimental syllable. Figure 5(G) —bottom panel— shows two syllables obtained in the simulations with the dynamics of each intervening nucleus at extreme cooling. Simulations show that only increasing the time delay the gesture fails in rising after the pulse from IA. On the other hand, stretching the activity in HVC, but leaving the delay fixed, leads to the generation of longer patterns without changes in the pressure gap.

IV. Discussion

Our starting point was a mathematical representation proposed by Alonso *et al.* [15] that qualitatively describes the mechanisms that generate the different pressure patterns observed during the song of canaries. This vision opposes to a scenario where periodic activity in HVC is enough to generate the vast syllabic repertoire. Moving a step forward in the circular model, we quantitatively fit gesture patterns by varying the interpretable parameters that account for the activity in IA and HVC. Each syllable was classified and we extracted a characteristic feature that we aimed to reproduce numerically. By using experimental data of respiratory singing activity while cooling HVC, we were able to study how the circuit dynamics change in terms of temporal coding.

Our analysis starts with two types of syllables presenting a periodic behavior: p1 and pulsatile. To simulate these patterns, we proposed a periodic activity in HVC with period P_h. Then, for each experimental syllable, we adjusted P_h so that the average frequency of the whole gesture matched the average frequency of the experimental syllable. As HVC was cooled, experimental motor gestures were stretched. Simulations showed that increasing the forcing period parameter was enough to reproduce the syllables since no qualitatively changes in the morphology of the simulated patterns were observed when varying the pulse length.

On the one hand, since the generation of these types of syllables is mainly ruled by the flow of information downstream from HVC to the brainstem, both circular and clock models are compatible. Moreover, both frameworks agree that the decrease of the frequency of HVC is an expected effect as the nucleus is cooled, resulting in a stretching of the pressure patterns. On the other hand, respiratory motor gestures with richer structure, such as p0 and p2, provided us the key to decide between the two models, since the cooling affects not only the duration of the syllable but also its internal structure. For example, at low temperatures, p0 patterns show an increase of the pressure gap after the first maximum of pressure. To explain these phenomena, we proposed that at least two timescales must be involved in the generation of the gesture. This hypothesis was tested by modeling the activity in IA and HVC with simple pulses of duration d_i and d_h respectively, separated by a fixed time Δ. If a unique timescale is involved in the generation of the motor gestures, these three parameters should change as a result of thermal manipulation of the circuit. We first associated the time delay Δ to each temperature variation by studying the morphology changes of the p0 syllables. We showed that this parameter must increase as a result of cooling HVC [Fig. 3(D)]. Letting the remaining two parameters vary independently, we searched for the ones that generated a pressure pattern having the feature of each of our experimental recorded gesture. On the one hand, Figs. 5(D)–(F) show that the dynamics of IA remains practically unalterable during cooling, in the generation of p2 and p0 syllables. On the other hand, the duration of the pulses in HVC are affected in several ways while cooling. For p2, cooling has minimal effect in type III gestures and a slight increase of duration in type II syllables [Fig. 3(C)]. However thermal manipulation in this area exerts a stronger effect in type I and p0 syllables. This result may shed light on the need of more than one timescale, as one timescale is slowed by the cooling effect, the other remains immutable and therefore only a fraction of the entire syllable is modified. Note that although for all syllables cooling HVC results in a slowdown of its activity, lower frequency syllables reflect the stretching of pulses in HVC.

Under the hypothesis of the circular model, the breaking of the p0 syllables can be easily interpreted. The first peak of pressure is the result of the arrival of the activity from IA. Pressure then rises again as the signal of RA reaches the respiratory related area. As the temperature decreases, the separation between the two signals increases, causing the pressure to drop after reaching its maximum value. A larger signal from RA is needed in order to rise the pressure back. This is accomplished by stretching the pulses within HVC. Remarkably, both effects —increasing the time delay between pulses and stretching the length of HVC pulses— are needed to recreate the breaking, and are compatible with the HVC cooling effects.

V. Conclusions

In this work, we tested the hypothesis that at least two separate timescales are responsible for the generation of the respiratory motor gestures during birdsong production. We studied quantitatively the deformations of the air sac pressure patterns of syllables under temperature changes in terms of interpretable parameters of a circular model, that account for the activity in brain areas related to motor gesture generation.

We proposed that syllables with simple internal structure (p1 and pulsatile) emerge from periodic activity in the telencephalic nucleus HVC. Decreasing the temperature in this nucleus produces a slowdown that increased the period of the pressure patterns. This is translated in the model as an increase of the time between pulses within HVC.

As HVC is cooled, gestures with richer internal structure (p2 and p0) suffer from qualitative changes in their morphology, leaving other features immutable. These phenomena may suggest the interaction of two different time scales. The second responsible timescale is proposed to emerge from a region within the brainstem (IA), which also projects to HVC. To test this hypothesis, we focused our study in the variation of the dynamics of IA and HVC needed to recreate gestures displaying the same morphology changes as observed experimentally. These dynamics were represented as sparse bursts of information of independent duration, beginning from IA and processed in HVC a time Δ later. Our results showed that the time delay between these pulses must increase in order to quantitatively fit the pressure patterns from cooled HVC. We also showed a positive correlation between the duration of the pulses from HVC and the time delay Δ, while the size of the pulses in IA remains unaltered. Therefore, reducing the temperature in HVC is represented in the model by two phenomena: a slowdown of the activity in HVC and an increase of the delay between pulses from IA. This is compatible with the hypothesis that cooling the HVC area results in delaying synaptic inputs into HVC and also delaying intrinsic synapses within HVC. Both effects are needed in order to generate quantitatively the cooled gestures. Moreover, the fact that the activity in IA results practically constant across all temperatures leads us to picture the system as a jazz band where the rhythms emerge from the interaction of many musicians instead of an orchestra with its director. This is in agreement with recent physiological experiments measuring the processing time between HVC and Uva while thermally manipulating each nucleus [22], and experimental modeling of the synaptic delays in HVC neurons [24], which supports the hypothesis of a distributed timescale among the brain song network.

Acknowledgements - We thank Marcos Trevisán for comments to this manuscript and English editing. The research reported in this work was funded by CONICET, ANCyT, UBA, and NIH through grants $R01-DC-012859$ and $R01-DC-006876$.

---

[1] H P Zeigler, P Marler, *Neuroscience of birdsong*, Cambridge University Press, Cambridge (2008).

[2] J M Wild, *Neural pathways for the control of birdsong production*, Dev. Neurobiol. **33**, 653 (1997).

[3] F Nottebohm, *The neural basis of birdsong*, PLoS Biol. **3**, e164 (2005).

[4] R C Ashmore, J M Wild, M F Schmidt, *Brainstem and forebrain contributions to the generation of learned motor behaviors for song*, J. Neurosci. **25**, 37 (2005).

[5] R H Hahnloser, A Kotowicz, *Auditory representations and memory in birdsong learning*, Curr. Opin. Neurobiol. **20**, 332 (2010).

[6] F Goller, R A Suthers, *Role of syringeal muscles in controlling the phonology of bird song*, J. Neurophysiol. **76**, 287 (1996).

[7] F Goller, R A Suthers, *Role of syringeal muscles in gating airflow and sound production in singing brown thrashers*, J. Neurophysiol. **75**, 867 (1996).

[8] G B Mindlin, T J Gardner, F Goller, R Suthers, *Experimental support for a model of birdsong production*, Phys. Rev. E **68**, 041908 (2003).

[9] M F Schmidt, F Goller, *Breathtaking songs: Coordinating the neural circuits for breathing and singing*, Physiology **31**, 442 (2016).

[10] G Angle, H Coskun, *A complete syllable dictionary for Serinus canarius*, Ecol. Inf. **20**, 67 (2014).

[11] J A Alliende, J M Mndez, F Goller, G B Mindlin, *Hormonal acceleration of song development illuminates motor control mechanism in canaries*, Dev. Neurobiol. **70**, 943 (2010).

[12] R H Hahnloser, A A Kozhevnikov, M S Fee, *An ultra-sparse code underlies the generation of neural sequences in a songbird*, Nature **419**, 6902 (2002).

[13] M S Fee, A A Kozhevnikov, R H Hahnloser, *Neural mechanisms of vocal sequence generation in the songbird*, Ann. NY. Acad. Sci. **1016**, 153 (2004).

[14] M A Long, M S Fee, *Using temperature to analyse temporal dynamics in the songbird motor pathway*, Nature **456**, 189 (2008).

[15] R G Alonso, M A Trevisan, A Amador, F Goller, G B Mindlin, *A circular model for song motor control in Serinus canaria*. Front. Comput. Neurosci. **9**, 1 (2015).

[16] A Amador, S Boari, G B Mindlin, *From perception to action in songbird production: Dynamics of a whole loop*, Curr. Opin. Syst. Biol. **3**, 30 (2017).

[17] R Bertram, A Daou, R L Hyson, F Johnson, W Wu, *Two neural streams, one voice: Pathways for theme and variation in the songbird brain*, Neuroscience **277**, 806 (2014).

[18] M F Schmidt, J M Wild, *The respiratory-vocal system of songbirds: Anatomy, physiology, and neural control*, Prog. Brain Res. **212**, 297 (2014).

[19] M A Goldin, L M Alonso, J A Alliende, F Goller, G B Mindlin, *Temperature induced syllable breaking unveils nonlinearly interacting timescales in birdsong motor pathway*, PLoS One **8**, e67814 (2013).

[20] M A Goldin, G B Mindlin, *Evidence and control of bifurcations in a respiratory system*, Chaos **23**, 043138 (2013).

[21] A S Andalman, J N Foerster, M S Fee, *Control of vocal and respiratory patterns in birdsong: Dissection of forebrain and brainstem mechanisms using temperature*, PLoS One **6**, e25461 (2011).

[22] K Hamaguchi, M Tanaka, R Mooney, *A distributed recurrent network contributes to temporally precise vocalizations*, Neuron **91**, 1 (2016).

[23] Y S Zhang, J D Wittenbach, D Z Jin, A A Kozhevnikov, *Temperature manipulation in songbird brain implicates the premotor nucleus HVC in birdsong syntax*, J. Neurosci. **37**, 2600 (2017).

[24] M A Goldin, G B Mindlin, *Temperature manipulation of neuronal dynamics in a forebrain motor control nucleus*, PLoS Comput. Biol. **13**, e1005699 (2017).

[25] D Galvis, W Wu, R L Hyson, F Johnson, R Bertram, *A distributed neural network model for the distinct roles of medial and lateral HVC in zebra finch song production*, J. Neurophysiol. **118**, 677 (2017).

[26] J Kornfeld, S E Benezra, R T Narayanan, F Svara, R Egger, M Oberlaender, W Denk, M A Long, *EM connectomics reveals axonal target variation in a sequence-generating network*, eLife **6**, e24364 (2017).

[27] M F Schmidt, R C Ashmore, E T Vu, *Bilateral control and interhemispheric coordination in the avian song motor system*, Ann. NY. Acad. Sci. **1016**, 171 (2004).

[28] R C Ashmore, J A Renk, M F Schmidt, *Bottom-up activation of the vocal motor forebrain by the respiratory brainstem*, J. Neurosci. **28**, 2613 (2008).

[29] R A Suthers, S A Zollinger, *Producing song: The vocal apparatus*, Ann. NY. Acad. Sci. **1016**, 109 (2004).

[30] F C Hoppensteadt, E M Izhikevich, *Weakly connected neural networks*, Springer, New York (1997).

[31] J S McCasland, M Konishi,*Interaction between auditory and motor activities in an avian song control nucleus*, P. Natl. Acad. Sci. USA **78**, 7815 (1981).

[32] A Amador, Y S Perl, G B Mindlin, D Margoliash, *Elemental gesture dynamics are encoded by song premotor cortical neurons*, Nature **495**, 59 (2013).

[33] R Mooney, J F Prather, *The HVC microcircuit: The synaptic basis for interactions between song motor and vocal plasticity pathways*, J. Neurosci. **25**, 1952 (2005).

18

Physical pendulum experiment re-investigated with an accelerometer sensor

C. Dauphin,[1,2*] F. Bouquet[3]

We have conducted a compound pendulum experiment using Arduino and an associated two-axis accelerometer sensor as measuring device. We have shown that the use of an accelerometer to measure both radial and orbital accelerations of the pendulum at different positions along its axis offers the possibility of performing a more complex analysis compared to the usual analysis of the pendulum experiment. In this way, we have shown that this classical experiment can lead to an interesting and low-cost experiment in mechanics.

I. Introduction

The physical pendulum experiment is the typical one to introduce the physics of oscillating systems. The usual aim of the analysis of this experiment is to determine the pendulum period and damping factor by using an angular position sensor [1, 2].

We have conducted the pendulum experiment by using a two axis accelerometer sensor. Such sensor has already been used by Fernandes et al. (2017) [3] but their study focused on the analysis of the time variation of the radial acceleration to investigate large-angle anharmonic oscillations.

Here, we have used the accelerometer to measure both radial and orbital accelerations of the pendulum at different positions along its axis, which offers the possibility of performing a more complex analysis compared to the usual single measurement of the pendulum period.

Furthermore, we use a microcontroller and an associated two axis accelerometer sensor to acquire the data. Thus, we have used this simple and low cost experiment compared to the ready to use commercial one to introduce a richer theory and data analysis tools that can lead to an interesting experiment in mechanics.

We have described the theoretical analysis of this experiment and present an example of a possible experimental setup, the analysis of the measured radial and orbital acceleration in order to acquire the moment of inertia, the center of mass, the damping factor and the period of the pendulum.

II. Example of experimental setup

We use an Arduino [4] and a two-axis accelerometer sensor as measuring devices. The Arduino is an interesting choice for an experiment, as it is an easy-to-use and low-cost microcontroller, with a large user community. Even if Arduino was not initially developed as a physicist tool, it can be used in various contexts of experimental physics activities (e.g., see references [5–10]).

*E-mail: cyril.dauphin@villebon-charpak.fr

[1] Institut Villebon-*Georges Charpak*, Université Paris-Sud (Bât. 490) - Rue Hector Berlioz – 91400 Orsay, France.

[2] Département de Physique, University Paris-Sud, Université Paris-Saclay, 91405 Orsay cedex, France.

[3] Laboratoire de Physique des Solides, CNRS, University Paris-Sud, Université Paris-Saclay, 91405 Orsay cedex, France.

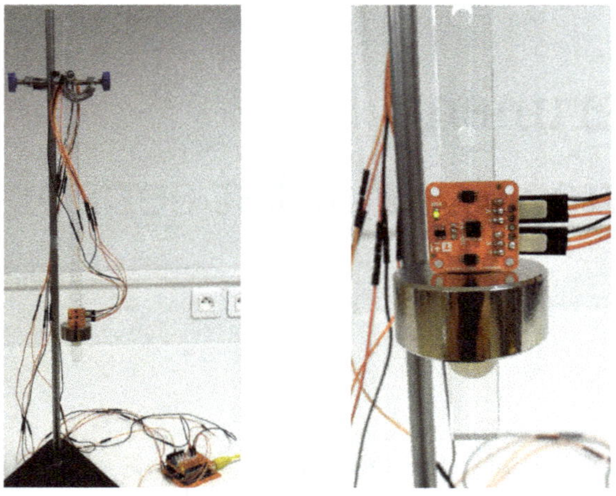

(a) *(b)*

Figure 1: (a) An example of an experimental setup. (b) An accelerometer sensor attached to the centerline of the pendulum, with one of its axes parallel to the centerline.

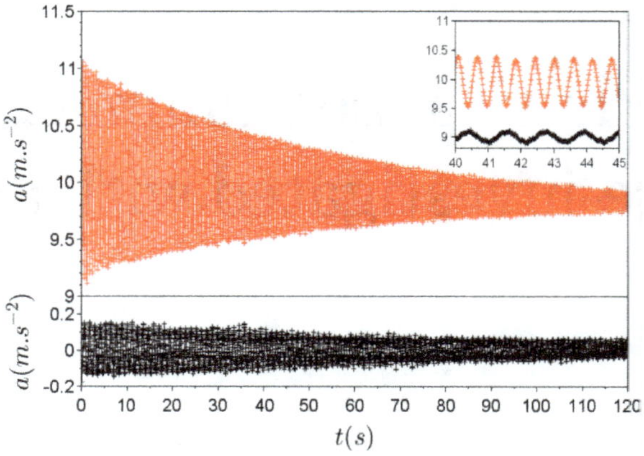

Figure 2: Radial (red) and orbital (black) accelerations measured by the accelerometer obtained with the experimental setup displayed in Fig. 1. Inset shows the temporal evolution of both accelerations between 40 s and 45 s. In order to have both curves on the same plot, orbital acceleration is shifted by a constant offset of 9 in the inset.

The experimental setup that we used here is shown in Fig. 1(a). The accelerometer sensor is a microelectromechanical inertial sensor which is precisely calibrated by using the values $+g$, $0g$ and $-g$ for each axis with $g = 9.8\,\mathrm{m\,s^{-2}}$.

The pendulum used in this experiment is composed of a bar on which masses can be attached to different positions. The accelerometer is attached to the bar and positioned in such a way that one of its measurement axes lies parallel to the bar (Fig. 1(b)). Special care should be taken so that the wires connecting the accelerometer to the board are flexible enough in order not to damp the pendulum.

Figure 2 shows an example of data acquired by the accelerometer. The main features of the graph are:

- the radial acceleration measurement decreases and goes to g as t approaches the infinity.
- when the radial acceleration reaches its maximum values, the orbital acceleration is essentially equal to zero (inset of Fig. 2).
- the radial acceleration is asymmetric about the straight line $a = g$ contrary to the orbital acceleration that is symmetric about the line $a = 0$.
- the period of the radial acceleration oscillations is twice that of the orbital acceleration oscillations (inset of Fig. 2).

In the next section, we will present the theory that explains these main features.

III. Theory

i. Expression of the acceleration components measured by the accelerometer sensor

Figure 3 shows a pendulum sketch with notations that will be used throughout this paper. O is the pivot point, G is the pendulum mass center and A the accelerometer position. L and r stands for the distance between O and G and O and A, respectively. We note I_λ the moment of inertia of the pendulum about the Oz axis.

Applying the angular momentum theorem to the pendulum and considering viscous damping leads to:

$$I_\lambda \ddot{\theta} = -MgL\sin\theta - \gamma\dot{\theta} \tag{1}$$

where θ is the angle between the pendulum axis OG and the vertical axis, M is the mass of the system and γ is the coefficient of friction. We note

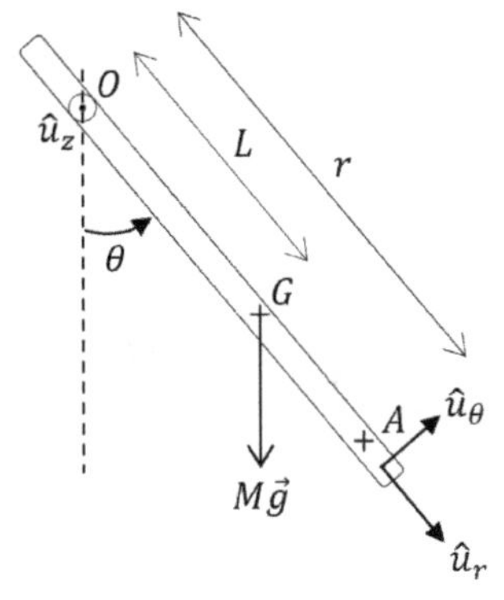

Figure 3: Sketch of the pendulum and notations used in the text. The pivot is at O, the center of mass at G and the accelerometer sensor is at A.

$I_\lambda = \alpha ML^2$ where the numerical factor α depends on the type of pendulum ($\alpha = 1$ for a simple pendulum and $\alpha \neq 1$ for a physical pendulum). We introduce the damping factor $\kappa = \frac{\gamma}{2I_\lambda}$ to obtain:

$$\ddot{\theta} = -\frac{g}{\alpha L}\sin\theta - 2\kappa\dot{\theta} \quad (2)$$

Forces acting on the proof mass (m_{sensor}) inside the accelerometer are its weight and the inertial force due to its movement. Thus, the radial and orbital components of these forces in the non-inertial reference frame of the pendulum are given by:

$$F_r = m_{sensor} r\dot{\theta}^2 + m_{sensor} g\cos\theta \quad (3)$$

$$F_\theta = -m_{sensor} r\ddot{\theta} - m_{sensor} g\sin\theta \quad (4)$$

The acceleration components, as measured by the accelerometer at a distance r from the pivot, are then given by:

$$a_r = r\dot{\theta}^2 + g\cos\theta \quad (5)$$

$$a_\theta = -r\ddot{\theta} - g\sin\theta \quad (6)$$

Including the expression of $\ddot{\theta}$ (Eq. (2)) into the expression of a_θ gives :

$$a_\theta = \frac{g}{\alpha}\left(\frac{r}{L} - \alpha\right)\sin\theta + 2\kappa r\dot{\theta} \quad (7)$$

We choose $\theta = \theta_0$ and $\dot{\theta} = 0$ as initial conditions of the pendulum movement. We only consider here small damping ($\kappa << \omega$) and the small angle approximation ($\theta_0 << \frac{\pi}{2}$), which leads from Eq. (2) to:

$$\theta = \theta_0 e^{-\kappa t}\cos(\omega t) \quad (8)$$

$$\dot{\theta} = -\omega\theta_0 e^{-\kappa t}\sin(\omega t) \quad (9)$$

$$\ddot{\theta} = -\omega^2\theta_0 e^{-\kappa t}\cos(\omega t) \quad (10)$$

with the pendulum angular frequency $\omega = \sqrt{\omega_0^2 - \kappa^2}$ and $\omega_0 = \sqrt{g/\alpha L}$. Thus, the acceleration components measured by the accelerometer for $\kappa << \omega$ and $\theta_0 << \frac{\pi}{2}$ are given by:

$$a_r = r\omega^2\theta_0^2 e^{-2\kappa t}\sin^2(\omega t) + g\cos\left(\theta_0 e^{-\kappa t}\cos(\omega t)\right) \quad (11)$$

$$a_\theta = -2\kappa r\omega\theta_0 e^{-\kappa t}\sin(\omega t) + \frac{g}{\alpha}\left(\frac{r}{L} - \alpha\right)\sin\left(\theta_0 e^{-\kappa t}\cos(\omega t)\right) \quad (12)$$

With $\theta_0 << \frac{\pi}{2}$, we can further approximate these two expressions by:

$$a_r = r\omega^2\theta_0^2 e^{-2\kappa t}\sin^2(\omega t) + g\left(1 - \theta_0^2 e^{-2\kappa t}\cos^2(\omega t)\right) \quad (13)$$

$$a_\theta = -2\kappa r\omega\theta_0 e^{-\kappa t}\sin(\omega t) + \frac{g}{\alpha}\left(\frac{r}{L} - \alpha\right)\theta_0 e^{-\kappa t}\cos(\omega t) \quad (14)$$

Some comments can be made about Eqs. (11) and (12). We first focus on the radial acceleration a_r.

- $\cos(\cos(x))$ and $\sin^2(x)$ are both π-periodic functions. So a_r is a $\frac{T}{2}$ periodic function with $T = \frac{2\pi}{\omega}$ being the pendulum period. Indeed, the pendulum reaches its maximum velocity and so its maximum radial acceleration each time θ is equal to 0.
- $g\cos\left(\theta_0 e^{-\kappa t}\cos(\omega t)\right)$ varies between between $g\cos(\theta_0)$ and g (blue curve in Fig. 4(b)) and goes to g as t approaches infinity. Physically, this function represents the projection of $\vec{g}$ onto the pendulum axis OG.
- $r\omega^2\theta_0^2 e^{-2\kappa t}\sin^2(\omega t)$ varies between $r\omega^2\theta_0^2$ and 0 and the upper envelope of this function (red

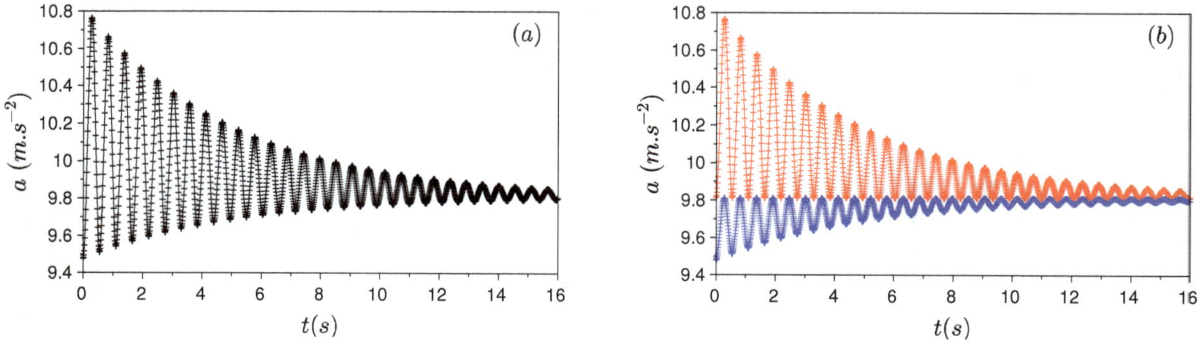

Figure 4: (a) Calculated radial acceleration versus time. (b) In red, graph of the $r\omega^2\theta_0^2e^{-2\kappa t}\sin^2(\omega t)+g$ contribution; in blue, graph of the $g\cos\left(\theta_0 e^{-\kappa t}\cos(\omega t)\right)$ contribution. $L=30\,\text{cm}$, $\theta_0=15°$, $r=45\,\text{cm}$, $\kappa=0.1\,\text{s}^{-1}$ and $\alpha=1$ for both panels.

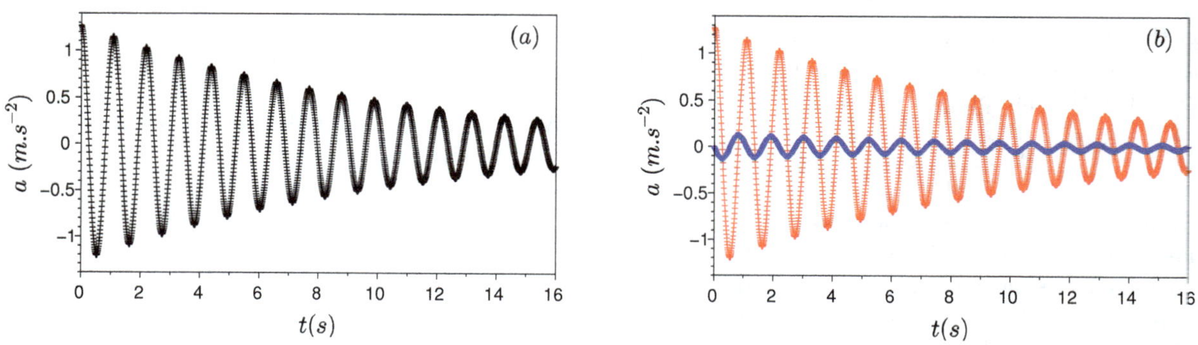

Figure 5: (a) Calculated orbital acceleration versus time. (b) In red, graph of the $\frac{g}{\alpha}\left(\frac{r}{L}-\alpha\right)\sin\left(\theta_0 e^{-\kappa t}\cos(\omega t)\right)$ contribution; in blue, graph of the part $-2\kappa r\omega\theta_0 e^{-\kappa t}\sin(\omega t)$. $L=30\,\text{cm}$, $\theta_0=15°$, $r=45\,\text{cm}$, $\kappa=0.1\,\text{s}^{-1}$ and $\alpha=1$ for both panels.

curve in Fig. 4(b), note that it has been displaced by g) decreases exponentially as $e^{-2\kappa t}$. Physically, this function represents the acceleration due to the radial centrifugal force felt by the sensor (as expected, the acceleration due to the radial centrifugal force is maximum when the pendulum is vertical). This point and the previous one explains the asymmetry of the radial acceleration about the straight line $a_r=g$, as observed in Fig. 2.

We can now study the orbital acceleration a_θ:

- $\sin(\cos(x))$ and $\sin(x)$ are both 2π-periodic functions. Therefore, a_θ is a T-periodic function with $T=\frac{2\pi}{\omega}$. Indeed, the angular acceleration $\ddot{\theta}$ reaches its highest and lowest values each time the pendulum passes through its highest points. This point explains that the period of the radial acceleration is twice that of the orbital one.
- $\frac{g}{\alpha}\left(\frac{r}{L}-\alpha\right)\sin\left(\theta_0 e^{-\kappa t}\cos(\omega t)\right)$ goes to 0 as t approaches infinity and the $\sin(\cos(x))$ function implies that the upper and lower envelope of this part of the a_θ function are symmetric about the straight line of equation $a=0$ (red curve in Fig. 5(b)).
- $-2\kappa r\omega\theta_0 e^{-\kappa t}\sin(\omega t)$ goes to zero as t ap-

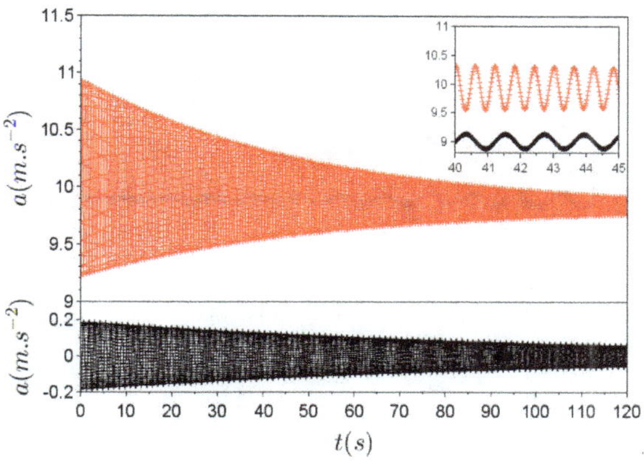

Figure 6: Radial (red) and orbital (black) accelerations obtained from Eqs. (11) and (12) with $\theta_0 = 20°$, $L = 30\,\text{cm}$, $r = 34\,\text{cm}$, $\alpha = 1.2$ and $\kappa = 0.01\,\text{s}^{-1}$. Inset shows the temporal evolution of radial (red) and orbital (black) accelerations between 40 s and 45 s. In order to have both curves on the same plot, orbital acceleration is shifted by a constant offset of 9 in the inset.

proaches infinity. The upper and lower envelopes of this part of the a_θ function are symmetric about the straight line $a_\theta = 0$ (blue curve in Fig. 5(b)) and decrease exponentially as $e^{-\kappa t}$.

Figure 6 displays the radial (red) and orbital (black) accelerations obtained from Eqs. (11) and (12). We can see that the main features inferred from the data (Fig. 2) are well reproduced.

ii. Experimentally accessible quantities

The parameters describing the pendulum and its motion can be derived from the measurements of a_r and a_θ.

- $a_r(t = 0) = a_{r,min} = g\cos\theta_0$. Thus, the measured value of a_r at $t = 0$ leads to the value of θ_0.
- A fit to the measured a_r upper envelope with an exponential function allows us to determine the damping factor κ of the pendulum from Eqs. (11) and (12). While, for small deflection angles θ, see Eqs. (13) and (14), exponential fit of any envelope of measured a_r or a_θ allows us to determine the damping factor.
- The sensor position OA can be measured with great accuracy. Thus, the position of the pendulum center of mass and moment of inertia are the two quantities which are difficult to determine experimentally. Here, we use the fit of the temporal evolution of a_r and a_θ to determine the product αL.

iii. Impact of α and r on the measured radial and orbital accelerations

Figure 7 displays the evolution of the radial and orbital accelerations with time for different values of the moment of inertia (from $\alpha = 1$ (Panel (a)) to $\alpha = 2.5$ (Panel (d))). At the difference of the radial acceleration, we can see that the orbital acceleration depends strongly on α. Indeed, a_θ expression at $t = 0$ leads to $a_\theta(t = 0) = \left(\frac{r}{\alpha L} - 1\right) g\sin\theta_0$, which is an inverse function of α. We can also note that $a_\theta(t = 0) < 0$ if $\alpha > \frac{r}{L}$ and $a_\theta(t = 0) > 0$ if $\alpha < \frac{r}{L}$. Thus, value of a_θ at $t = 0$ gives information on the α value.

Figure 8 shows the evolution of the radial and orbital accelerations with time for different values of r. The distance OA increases from Panel (a) to Panel (d). As expected, the amplitude of the radial acceleration increases with larger OA values as the centrifugal force acting on the proof mass increases and the amplitude of the orbital acceleration decreases with larger OA values as the rate of variation of $\dot{\theta}$ decreases with this distance.

In particular, acceleration components measured by the accelerometer attached to the position of the point O are given by:

$$a_r = g\cos\left(\theta_0 e^{-\kappa t}\cos(\omega t)\right) \tag{15}$$

$$a_\theta = -g\sin\left(\theta_0 e^{-\kappa t}\cos(\omega t)\right) \tag{16}$$

In this case, the only force acting on the mass inside the accelerometer is its weight and the expressions of the acceleration do not depend on α. Accelerometer sensor is used in this case as an angular position sensor.

We also note that the acceleration components measured by the accelerometer attached to the position $r = L\alpha$ are given by:

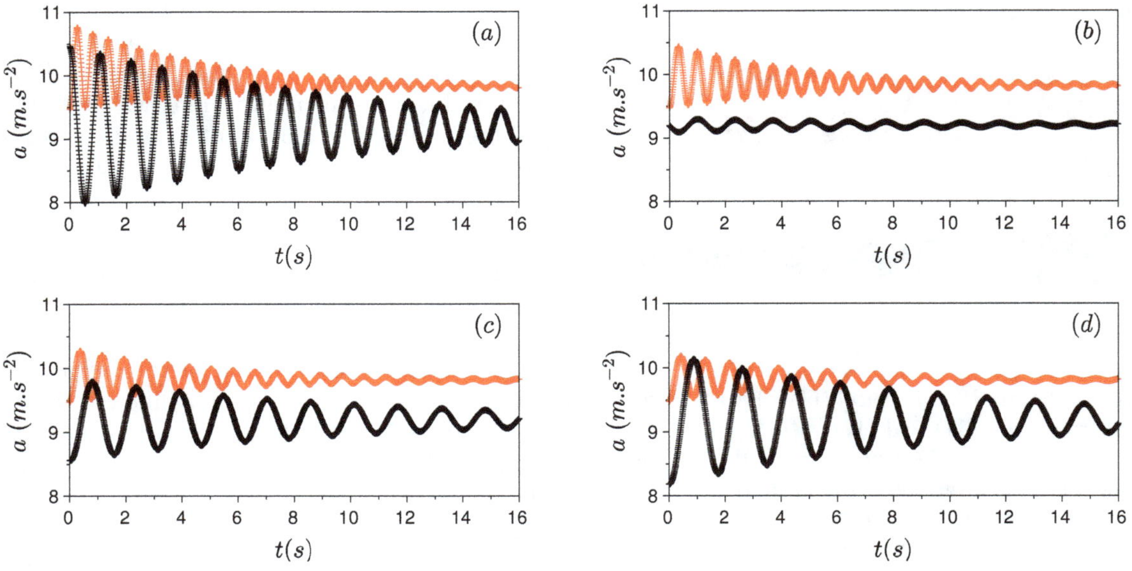

Figure 7: The calculated radial (red) and orbital (black) acceleration with time with $\alpha = 1$ (a), $\alpha = 1.5$ (b), $\alpha = 2.0$ (c) and $\alpha = 2.5$ (d). In order to have both curves on the same plot, orbital acceleration is shifted by a constant offset of 9.2. $L = 30\,\text{cm}$, $\theta_0 = 15°$, $r = 45\,\text{cm}$ and $\kappa = 0.1$ for all panels.

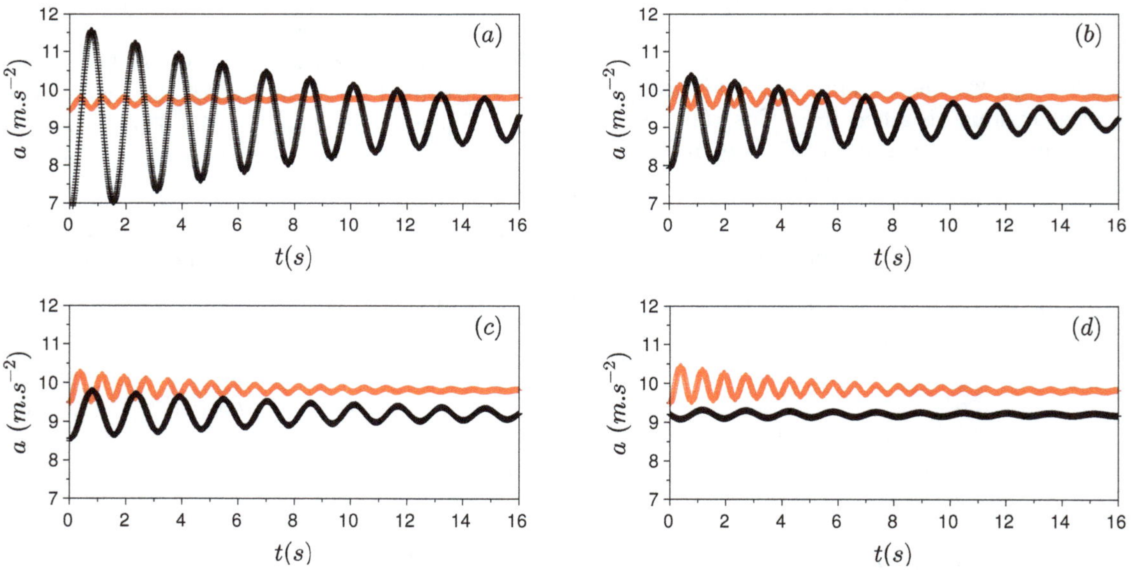

Figure 8: The calculated radial (red) and orbital (black) accelerations with time for different positions of the accelerometer relative to the center of mass G. In order to have both curves on the same plot, the orbital acceleration is shifted by a constant offset of 9.2. The accelerometer positions are (a) $r = 0\,\text{cm}$, (b) $r = 30\,\text{cm}$, (c) $r = 45\,\text{cm}$, (d) $r = 60\,\text{cm}$. $L = 30\,\text{cm}$, $\theta_0 = 15°$, $\alpha = 2$ and $\kappa = 0.1$ for all panels.

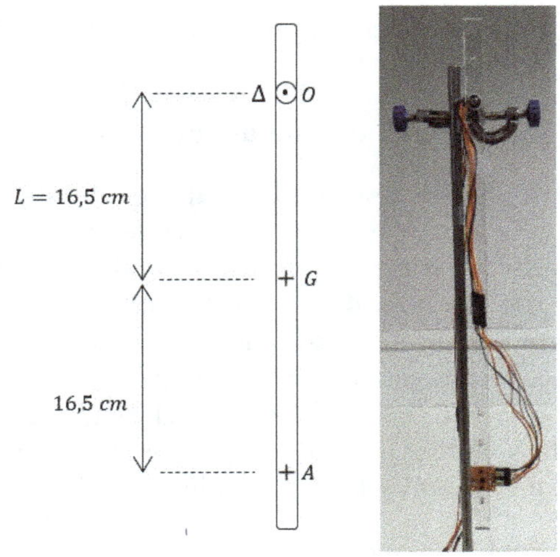

Figure 9: Experimental setup used to study a bar pendulum.

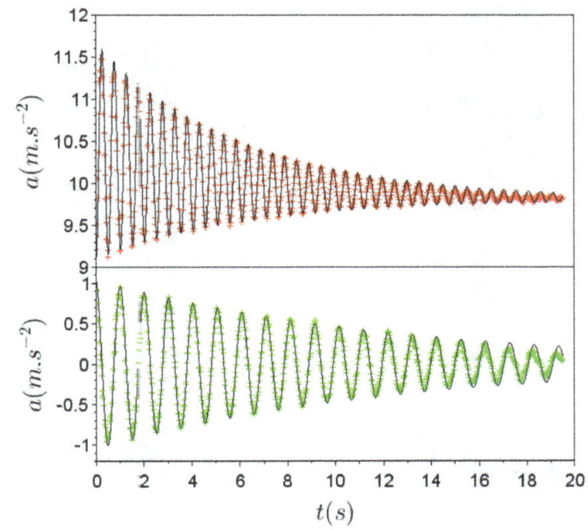

Figure 10: Radial (red) and orbital (green) accelerations measured by the accelerometer in the configuration shown in Fig. 9. Solid lines are the best fit of the data with κ and α as free parameters. The best fit is given by $1/\kappa = 12.2\,\text{s}$ and $\alpha = 1.56$.

$$\begin{aligned} a_r &= \alpha\omega^2\theta_0^2 e^{-2\kappa t}\sin^2(\omega t) \\ &\quad + g\cos\left(\theta_0 e^{-\kappa t}\cos(\omega t)\right) \end{aligned} \tag{17}$$

$$a_\theta = -2\kappa\alpha L\omega\theta_0 e^{-\kappa t}\sin(\omega t) \tag{18}$$

In this case, component of $\ddot{\theta}$ due to the gravity force is counterbalanced by the orbital component of the force of gravity acting on the accelerometer sensor. Thus, a_θ is then directly proportional to the pendulum angular velocity.

After having shown that Eqs. (11) and (12) explain the features observed experimentally, we will now use them to retrieve the pendulum parameters κ, α and L for different experimental setups.

IV. Example of data analysis

i. Example with a pendulum bar

We first focus on the pendulum shown in Fig. 9 to derive its physical parameters. We use a bar of a 45 cm length and mass 45 g with the pivot at 16.5 cm from the center of mass and the accelerometer at 33 cm.

Figure 10 shows the radial and orbital accelerations measured by the accelerometer after the pendulum has been displaced from the equilibrium position to an initial angle of 22°.

We have analyzed the data using Eqs. (11) and (12) with κ and α as free parameters. The results of the fit are shown as black curves in Fig. 10. We have derived the best fit for $1/\kappa = 12.2\,\text{s}$ and $\alpha = 1.56$.

Assuming the pendulum to be a simple homogeneous slab, we calculate that the inertia moment of the bar about the rotation axis λ is equal to $1.93 \times 10^{-3}\,\text{kg}\,\text{m}^2$, which leads to $\alpha = 1.62$. The holes in the bar, which are used for attaching the masses, together with the mass of the accelerometer, explain the difference between this and with the α value obtained from the data fit. This experiment allows us to determine with a good accuracy the moment of inertia of the pendulum.

ii. Retrieval of the mass center position

The pendulum of the previous subsection is a symmetric bar, thus, its mass center position can be determined precisely. In the general case, the mass center position can be difficult to determine and we can fit the data by using Eqs. (11) and (12) with κ, α and L as free parameters and infer the position of the center of mass.

As an example of such analysis, we use a pen-

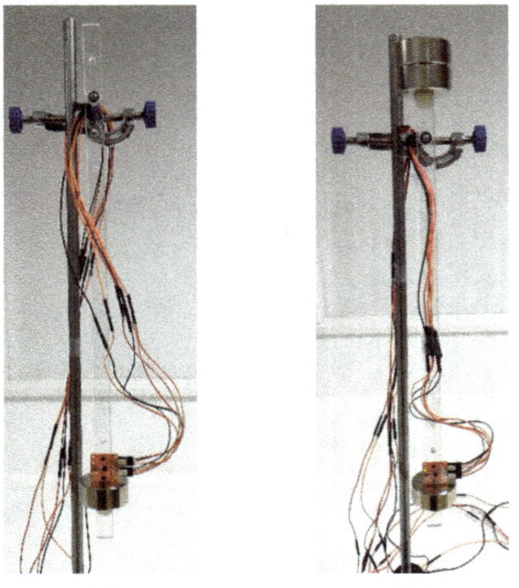

(a) (b)

Figure 11: Experimental setup used to study compound pendula, with added masses.

dulum composed of a bar on which several masses can be attached to at different positions to acquired data that corresponds to pendulums of different values of α and L (Fig. 11 (a) and (b)).

Figure 12 (a) and (b) displays the radial and orbital accelerations measured by the accelerometer after each pendulum shown in Fig. 11 has been displaced by an initial angle of 20.5°.

We fit the data by using Eqs. (11) and (12) with κ, α and L as free parameters. Results of the fit are shown as the black curves in the insets of Fig. 12. Best fit are obtained with $1/\kappa = 111\,\text{s}$, $\alpha = 1.162$ and $L = 0.293\,\text{m}$ for the pendulum of Fig. 11(a) and $1/\kappa = 101\,\text{s}$, $\alpha = 6.55$ and $L = 0.074\,\text{m}$ for the pendulum of Fig. 11(b).

For the examples displayed in Fig. 12, the evolution of the orbital acceleration from Panels (a) to (b) shows an increase of α, which is consistent with the fact that pendulum configuration goes from a configuration close to a simple pendulum ($\alpha \simeq 1$ Fig. 11(a)) to a compound pendulum ($\alpha > 1$ Fig. 11(b)).

Figure 12 shows that the angular acceleration (in green) is much more sensitive to the value of α than the radial acceleration (in red). Therefore, orbital acceleration is a good quantity to measure and to fit in order to determine the moment of inertia of a pendulum.

iii. Impact of r on the acquired data

We have shown in section III that the orbital acceleration is very sensitive to the accelerometer sensor position with respect to the pendulum rotation axis. As this position is precisely known, we can perform several measurements with different positions of the accelerometer to improve the accuracy and/or the precision of the derived pendulum parameters. As an example of such analysis, we use the experimental setups shown in Fig. 13.

Figure 14 (a) and (b) display the radial and orbital accelerations measured by the accelerometer after each pendulum shown in Fig. 13 has been displaced by an initial angle of 20.5°.

Fits of the data using Eqs. (11) and (12) with κ, α and L as free parameters are shown as black curves in Fig. 14. Best fit are obtained with $1/\kappa = 111\,\text{s}$, $\alpha = 1.162$ and $L = 29.3\,\text{cm}$ for the pendulum of Fig. 13(a) and with $1/\kappa = 100\,\text{s}$, $\alpha = 1.162$ and $L = 29.2\,\text{cm}$ for the pendulum of Fig. 13(b).

The position of the accelerometer does not affect the calculated values of the moment of inertia of the pendulum and only slightly affects the center of mass position. The new configuration of the wires connecting the accelerometer in Fig. 13(b) changes slightly the κ value. Thus, performing a second measurement with a different position of the accelerometer allow us to be more confident in the results retrieved from the first one.

We can also note that the orbital acceleration increases with lower OA values, therefore the orbital acceleration fit precision is better when the accelerometer is in the position of Fig. 13(b), while the precision of the radial acceleration fit is better when the accelerometer is in the position of Fig. 13(a).

V. Conclusions

We have shown that the pendulum experiment analyzed with an accelerometer sensor leads to a theoretical study richer than the classical one. We have derived theoretical expressions for the radial and orbital acceleration data recorded by an accelerometer and separated the contributions from the pen-

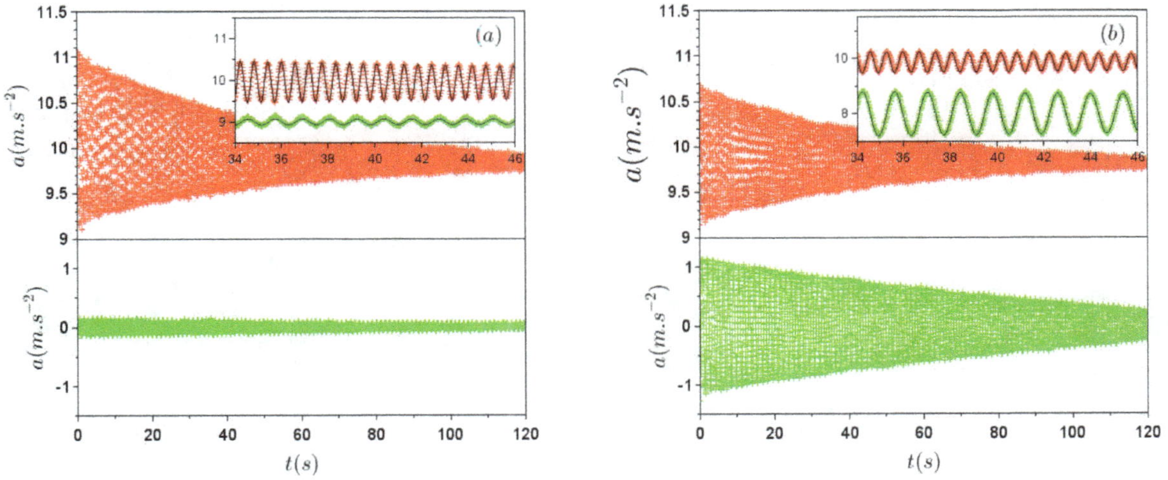

Figure 12: Radial (red) and orbital (green) accelerations measured by the accelerometer. (a) and (b) correspond to the setups displayed in Fig. 11 (a) and (b), respectively. Insets in (a) and (b) show the temporal evolution of the radial (red) and orbital (green) accelerations and the best fit (black) between 35 s and 45 s. In order to have both curves on the same plot, orbital acceleration is shifted by a constant offset of 9 in the inset of (a) and by 8 in (b).

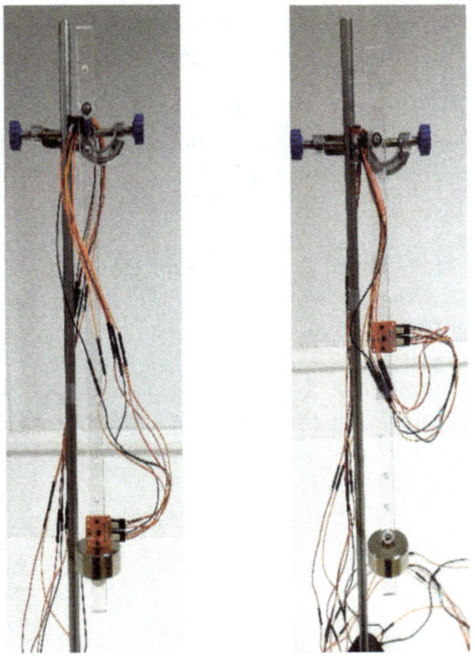

(a) (b)

Figure 13: Experimental setup used to study the impact of the acceleration sensor position on the acquired data.

dulum angular motion and the gravitational force on the proof mass.

We have also shown that the orbital acceleration is an interesting data to retrieve the moment of inertia of a pendulum.

The possibility to have different positions of the sensor allows us to perform several measurements with the same pendulum to improve the accuracy and/or the precision of the derived pendulum parameters.

In this paper, we have only focused on the classical pendulum but the device used here could also be applied to more complex systems such as chaotic pendulums.

Acknowledgements - This work benefited from a grant "Pédagogie innovante" from IDEX Paris-Saclay. The Institut Villebon - *Georges Charpak* was awarded the Initiative of Excellence in Innovative Training in March 2012 (IDEFI IVICA: 11-IDFI-002), and is supported by the Paris-Saclay Initiative of Excellence (IDEX Paris-Saclay: 11-IDEX-0003). The authors would like to thank the referee for his/her helpful comments.

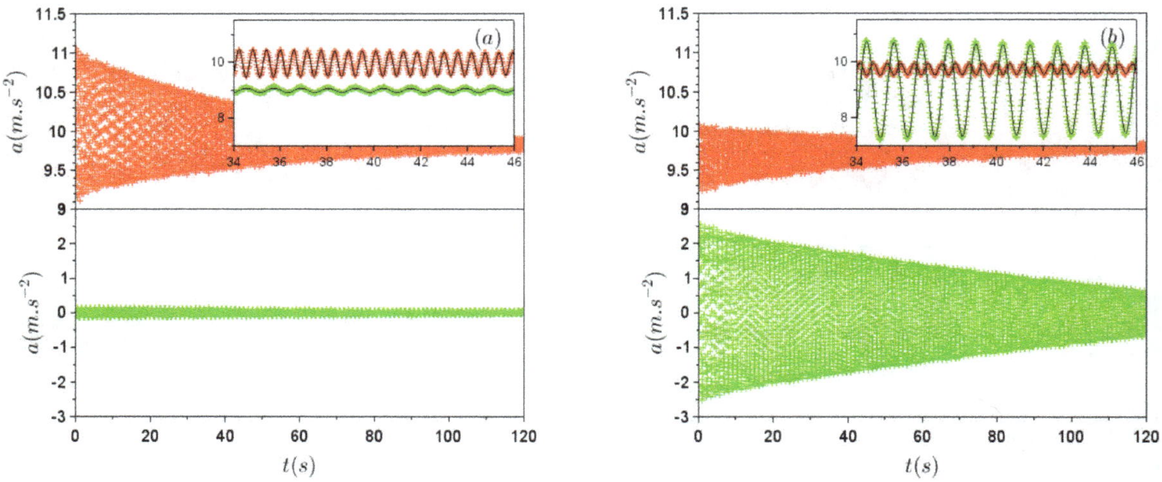

Figure 14: Radial (red) and orbital (green) accelerations measured by the accelerometer. (a) and (b) correspond to the setups displayed in Fig. 13(a) and (b), respectively. Insets in (a) and (b) show the temporal evolution of the radial (red), and orbital (green) accelerations and the best fit (black) between 35 s and 45 s. In order to have both curves on the same plot, orbital acceleration is shifted by a constant offset of 9 in the insets of (a) and (b).

[1] `http://www2.vernier.com/sample_labs/PHYS-AM-18-physical_pendulum.pdf`

[2] `https://physics.fullerton.edu/files/Labs/225/Physical_Pendulum_V4_1(1).pdf`

[3] J C Fernandes, P J Sebastião, L N Gonçalves, A Ferraz, *Study of large-angle anharmonic oscillations of a physical pendulum using an acceleration sensor*, Eur. J. Phys. **38**, 2017.

[4] `https://www.arduino.cc`

[5] Š Kubínová, J Šlégr, *Physics demonstrations with the Arduino board*, Phys. Educ. **50**, 472 (2015).

[6] C Galeriu, S Edwards, G Esper, *An Arduino investigation of simple harmonic motion*, Phys. Teach. **52**, 157 (2014).

[7] K Atkin, *Construction of a simple low-cost teslameter and its use with Arduino and MakerPlot software*, Phys. Educ. **51**, 024001 (2016).

[8] C Petry et al., *Project teaching beyond Physics: Integrating Arduino to the laboratory*, In Technologies Applied to Electronics Teaching (TAEE), pp. 1-6, IEEE (2016).

[9] R Henaff et al., *A study of kinetic friction: The Timoshenko oscillator*, Am. J. Phys. **86**, 174 (2018).

[10] F Bouquet, J Bobroff, M Fuchs-Gallezot, L Maurines, *Project-based physics labs using low-cost open-source hardware*, Am. J. Phys. **85**, 216 (2017).

19

A neuronal device for the control of multi-step computations

Ariel Zylberberg,[1–4*] Luciano Paz,[1,2†] Pieter R. Roelfsema,[4–6‡] Stanislas Dehaene,[7–10§] Mariano Sigman[2¶]

We describe the operation of a neuronal device which embodies the computational principles of the "paper-and-pencil" machine envisioned by Alan Turing. The network is based on principles of cortical organization. We develop a plausible solution to implement pointers and investigate how neuronal circuits may instantiate the basic operations involved in assigning a value to a variable (*i.e.*, $x = 5$), in determining whether two variables have the same value and in retrieving the value of a given variable to be accessible to other nodes of the network. We exemplify the collective function of the network in simplified arithmetic and problem solving (blocks-world) tasks.

I. Introduction

Consider the task of finding a route in a map. You are likely to start searching the initial and final destinations, identifying possible routes between them, and then selecting the one you think is shorter or more appropriate. This simple example highlights how almost any task we perform is organized in a sequence of processes involving operations which we identify as "atomic" (here search, memory and decision-making). In contrast with the thorough knowledge of the neurophysiology underlying these atomic operations [1–3], neuroscience is only starting to shed light on how they organize into programs [4–6]. Partly due to the difficulty of implementing compound tasks in animal models, sequential decision-making has mostly been addressed by the domains of artificial intelligence, cognitive science and psychology [7, 8].

Our goal is to go beyond the available neurophysiological data to show how the brain might sequentialize operations to conform multi-step cognitive

*E-mail: arielz@df.uba.ar
†E-mail: lpaz@df.uba.ar
‡E-mail: p.roelfsema@nin.knaw.nl
§E-mail: stanislas.dehaene@gmail.com
¶E-mail: sigman@df.uba.ar

1 These authors contributed equally to this work

2 Laboratory of Integrative Neuroscience, Physics Department, FCEyN UBA and IFIBA, CONICET; Pabellón 1, Ciudad Universitaria, 1428 Buenos Aires, Argentina.

3 Laboratory of Applied Artificial Intelligence, Computer Science Department, FCEyN UBA; Pabellón 1, Ciudad Universitaria, 1428 Buenos Aires, Argentina.

4 Department of Vision and Cognition, Netherlands Institute for Neuroscience, an institute of the Royal Netherlands Academy of Arts and Sciences, Meibergdreef 47, 1105 BA Amsterdam, The Netherlands.

5 Department of Integrative Neurophysiology, Center for Neurogenomics and Cognitive Research, VU University, Amsterdam, The Netherlands.

6 Psychiatry Department, Academic Medical Center, Amsterdam, The Netherlands.

7 Collge de France, F-75005 Paris, France.

8 INSERM, U992, Cognitive Neuroimaging Unit, F-91191 Gif/Yvette, France.

9 CEA, DSV/I2BM, NeuroSpin Center, F-91191 Gif/Yvette, France.

10 University Paris-Sud, Cognitive Neuroimaging Unit, F-91191 Gif/Yvette, France.

programs. We suppose the existence of elementary operations akin to Ullman's [9] routines, although not limited to the visual domain. Of special relevance to our report is the body of work that has grown out of the seminal work of John Anderson [8]. Building on the notion of "production systems" [10], Anderson and colleagues developed ACT-R as a framework for human cognition [8]. It consists of a general mechanism for selecting productions fueled by sensory, motor, goal and memory modules. The ACT-R framework emphasizes the chained nature of human cognition: at any moment in the execution of a task, information placed in buffers acts as data for the central production system, which feeds-back to these same modules.

Despite vast recent progress in our understanding of decision formation in simple perceptual tasks [3], it remains unresolved how the operations required by cognitive architectures may be implemented at the level of single neurons. We address some of the challenges posed by the translation of cognitive architectures to neurons: how neuronal circuits might implement a single operation, how multiple operations are arranged in a sequence, how the output of one operation is made available to the next one.

II. Fundamental assumptions and neuronal implementation

i. The basis for single operations

Insights into the machinery for simple sensory-motor associations come from studies of monkey electrophysiology. Studies of oculomotor decisions —focused primarily on area LIP [12]— have shown that neurons in this area reflect the accumulation of evidence leading to a decision [3]. In a well studied paradigm, monkeys were trained to discriminate the direction of motion of a patch of moving dots and report the direction of motion with an eye-movement response to the target located in the direction of motion [13]. Neurons in LIP which respond with high levels of activity during memory saccades to specific portions of space are recorded during the task. These neurons show ramping activity of their firing rates with a slope that depends on the difficulty of the task, controlled by the proportion of dots that move coherently in one direction [14]. In a reaction time version of the task [13], when monkeys are free to make a saccade any time after the onset of the motion stimuli, the ramping activity continues until a fixed level of activity is reached, signaling the impending saccade (Fig. 1 a). Crucially, the level of this "threshold" does not depend on the difficulty of the task or the time to respond. The emerging picture is that ramping neurons in LIP integrate sensory evidence and trigger a response when activity reaches a threshold. This finding provided strong support for accumulation or race models of decision making which have been previously postulated to explain error rates and reaction times in simple tasks, and match nicely with decision theoretical notions of optimality [15]. While experimental studies have mainly characterized the feed-forward flow of information from sensory to motor areas, evidence accumulation is also modulated by contextual and task-related information including prior knowledge about the likelihood and payoff of each alternative [16, 17]. Interestingly, a common currency —the spiking activity of neurons in motor intention areas— may underlie these seemingly unequal influences on decision formation.

ii. Sequencing of multiple operations

Brain circuits can integrate sensory evidence over hundreds of milliseconds. This illustrates how the brain decides based on unreliable evidence, averaging over time to clean up the noise. Yet the duration of single accumulation processes is constrained by its analog character, a problem pointed out earlier by von Neumann in his book *The computer and the Brain* [18]: "in the course of long calculations not only do errors add up but also those committed early in the calculation are amplified by the latter parts of it...". Modern computers avoid the problem of noise amplification employing digital instead of analog computation. We have suggested that the brain may deal with the amplification of noise by serially chaining individual integration stages, where the changes made by one ramp-to-threshold process represent the initial stage for the next one (Fig. 1 b) [11, 19].

Evidence for the ramp-to-threshold dynamics has been derived from tasks in which the decision reflects a commitment to a motor plan [3, 20]. As others [21], we posit that the ramp-to-threshold process for action-selection is not restricted to motor actions, but may also be a mechanism to select

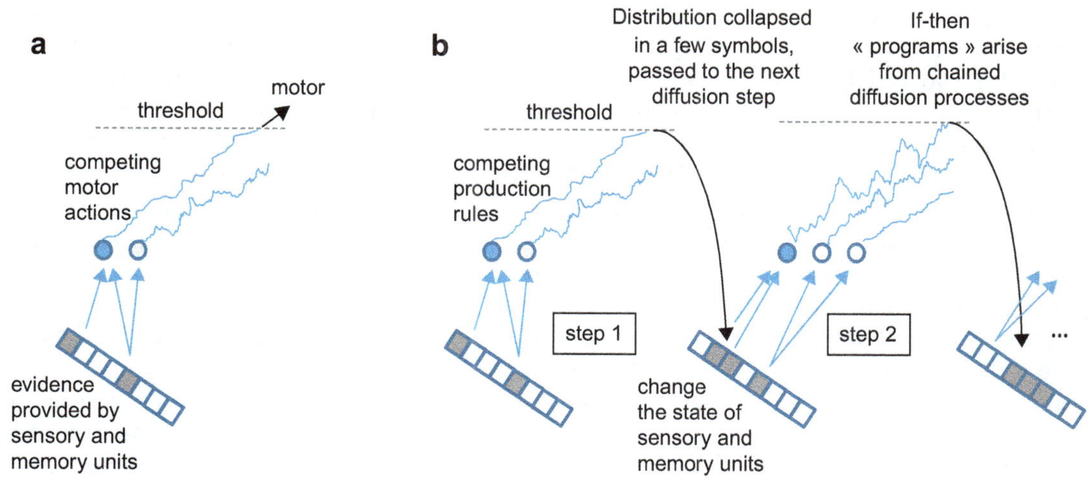

Figure 1: Diffusion processes in single and multi-step tasks. (a) In simple sensory-motor tasks, response selection is mediated by the the accumulation of sensory evidence to a response threshold. (b) In tasks involving multiple steps, there is a parallel competition between a subset of "productions rules" implemented by pools of neurons that accumulate evidence up to a threshold. The selected production can have overt effects (motor actions) as well as covert effects in modifying the state of the memory, after which a new cycle begins. Adapted with permission from Ref. [11].

internal actions like the decision of where to attend, what to store in memory, or what question to ask next. Therefore, the activation of a circuit based on sensory and mnemonic evidence is mediated by the accumulation of evidence in areas of the brain which can activate that specific circuit, and which compete with other internal and external actions. Within a single step, the computation proceeds in parallel and settles in a choice. Seriality is the consequence of the competitive selection of internal and external actions that transforms noisy and parallel steps of evidence accumulation into a sequence of discrete changes in the state of the network. These discrete steps clean up the noise and enable the logical flow of the computation.

Following the terminology of symbolic cognitive architectures [7, 8], the "ramping" neurons which select the operation to do next are referred as "production" neurons. The competition between productions is driven by inputs from sensory and memory neurons and by the spontaneous activity in the production network. As in single-step decisions [13], the race between productions concludes when neurons encoding one production reach a decision threshold. The neurons which detect the crossing of a threshold by a production also mediate its effects, which is to activate or deactivate other brain circuits. The activated circuits can be motor, but are not restricted to it, producing different effects like changes in the state of working memory (deciding what to remember), activating and broadcasting information which was in a "latent" state (like sensory traces and synaptic memories [22,23]), or activating peripheral processors capable of performing specific functions (like changing the focus of attention, segregating a figure from its background, or tracing a curve).

iii. Pointers

Versatility and flexibility are shared computational virtues of the human brain and of the Turing machine. The simple example of addition (at the root of Turing's conception) well illustrates what sort of architecture this requires. One can picture the addition of two numbers x and y as displacing y steps from the initial position x. This simple representation of addition as a walk in the number line describes the core connection between movement in space and mathematical operations. It also describes the need of operations that use variables which temporarily bind to values in order to achieve flexibility.

In this section we describe how neuronal circuits

may instantiate the basic operations involved in assigning a value to a variable (*i.e.*, $x = 5$), in determining whether two variables have the same value and in retrieving the value of a given variable. This is in a way a proof of concept, *i.e.*, a way to construct these operations with neurons. We are of course in no position to claim that this instantiation is unique (it is certainly not). However, we have tried to ground it on important principles of neurophsyiology and we believe that this construction raises important aspects and limitations which may generalize to other neuronal constructions of variable assignment, comparison and retrieval mechanisms.

Here we introduce the concept of pointers; individual or pool of neurons which can temporarily point to other circuits. When a pointer is active, it facilitates the activation of the circuit to which it is temporarily bound and which is dynamically set during the course of a computation.

A pointer "points" to a cortical territory (for instance, to V1). This cortical mantle represents a space of values that a given variable may assume. The cortex is organized in spatial maps representing features (space, orientation, color, 1-D line...) and a pointer can temporarily bind to one of these possible set of values in a way that the activation of the pointer corresponds to the activation of the value and hence functions as a variable (*i.e.*, $x = 3$). There are many proposed physiological mechanisms to temporally bind neuronal circuits [24–26]. A broad class of mechanisms relies on sustained reverberant neuronal activity [26]. A different class relies on small modifications of synaptic activity, which constitute silent memories in latent connections [23]. Here we opt for the second alternative, first because it has a great metabolic advantage allowing to share many memories at very low cost, and more importantly because it separates the processes of variable assignation and variable retrieval. As we describe below in detail, in this architecture the current state of the variable is not broadcast to other areas until it is specifically queried.

To specifically implement the binding with neuron-like elements, we follow the classic assumption that when two groups of neurons are active at the same time, a subset of the connection between them is strengthened (Fig. 2). The strengthening of the synapses is bidirectional, and it is responsible for the binding between neuronal populations. To avoid saturation of the connections and to allow for new associations to form, the strength of these connections decays exponentially within a few hundred milliseconds. Specifically, if the connection strength between a pair of populations is w_{base}, then when both populations are active the connection strength increases exponentially with a time constant τ_{rise} to a maximum connection strength of w_{max}. When one or both of the populations become inactive then the connection strength decays exponentially back to w_{base} with a time constant of τ_{decay}.

The mechanism described above generates a silent coupling between a pointer and the value to which it points. How is this value recovered? In other words, how can other elements of the program know the value of the variable X? The expression of a value stored in silent synapses is achieved by simultaneously activating the pointer circuit and forcing the domain of the variable to a winner-take-all (WTA) mode (Fig. 2). The WTA mode —set by having neurons with self-excitation and cross-inhibition [27]— assures that only one value is retrieved at a time. These neurons make stronger connections with the neurons to which they are bound than with the other neurons. When the network is set in a WTA mode, these connections bias the competition to retrieve the value previously associated with the pointer (Fig. 2). In other words, activation of the pointer by itself is not sufficient to drive synaptic activity to the neuron (or neurons) representing the value to which it points. But it can bias the activation of a specific neuron when co-activated with a tonic activation of the entire network.

This architecture is flexible and economic. Value neurons only fire when they are set or retrieved. Memory capacity is constrained by the number of connections and not by the number of neurons. But it also has a caveat. Given that only one variable can be bound to a specific domain at any one time, multiple bindings must be addressed serially. As we show later, this can be accomplished by the sequential firing of production neurons.

a. *Compare the value of two variables*

If two variables X and Y bind to instances in the same domain, it is possible to determine whether the two variables are bound to the same instance,

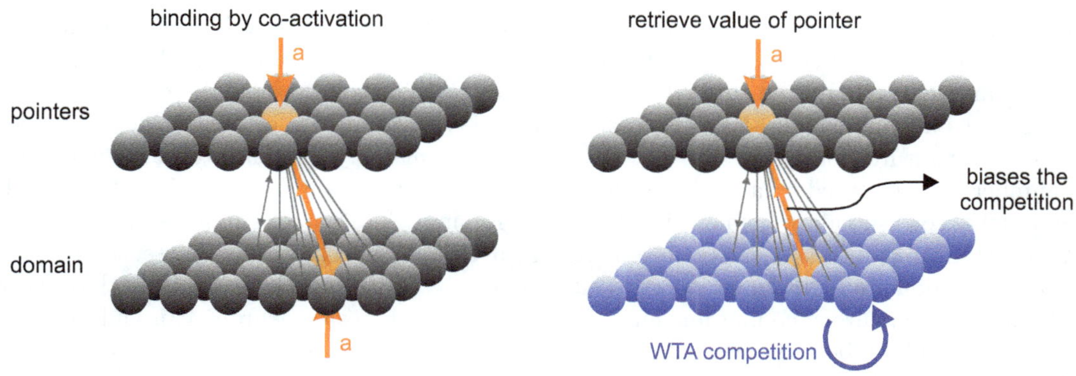

Figure 2: Instantiation of variables through plastic synapses. When two neurons become active, the connection between them is rapidly strengthened, forming a transient association which decays within a few hundreds of milliseconds. The value of a pointer can be retrieved by setting the pointer's domain in a winner-take-all mode and activating the pointer which biases the WTA competition through its strengthened connections.

i.e., whether $x = y$. The mechanics of this process is very similar to retrieving the value of a variable.

Pointer neurons[1] X and Y are co-activated. The equality in the assignment of X and Y can be identified by a coincidence detector. Specifically, this is solved by adjusting the excitability in the value domain in such a way that the simultaneous input on a single value neuron exceeds the threshold but the input of a single pointer does not. This proposed mechanism is very similar to the circuits in the brain stem which —based on coincidence detection of delay lines— encode interaural time difference [28]. This shows the concrete plausibility of generating such dynamic threshold mechanisms that act as coincidence detectors.

b. *Assign the value of one variable to another*

Similarly, to assign the value of X to the variable Y ($Y \leftarrow x$), the value of the variable that is to be copied needs to be retrieved as indicated previously, by activating the variable X and forcing a WTA competition at the variable's domain. Then, the node coding for variable Y must be activated, which will lead to a reinforcing of the connections between Y and x which will instantiate the new association.

[1]In our framework, a pointer can also be a population of neurons that functions as a single pointer.

III. Concrete implementation of neuronal programs

In the previous sections we sketched a set of principles by which brain circuits can control multistep operations and store temporary information in memory buffers to share it between different operations.

Here we demonstrate, as a proof of concept, a neuronal implementation of such circuits in two simple tasks.

The first one is a simple arithmetic counter, where the network has to count from an initial number n_{ini} to a final number n_{end}, a task that can be seen as the emblematic operation of a Turing device. The second example is a blocks-world problem, a paradigmatic example of multi-step decision making in artificial intelligence [29]. The aim of the first task is to illustrate how the different elements sketched above act in concert to implement neuronal programs. The motivation to implement blocks worlds is to link these ideas to developed notions of visual routines [9, 30, 31].

i. Arithmetic counter

We designed the network to be generic in the sense of being able to solve any instance of the problem, *i.e.*, any instantiation of n_{ini} and n_{end}. We decided to implement a counter, since it constitutes essentially a while loop and hence a basic intermediate description of most flexible computations. In the network, each node is meaningful, and all parameters were set by hand. Of course, understanding how these parameters are adjusted through a learning process is a difficult and important question, but this is left for future work.

Each number is represented by a pool of neurons selective to the corresponding numerosity value [32]. A potential area for the neurons belonging to the numerosity domain is the Intra Parietal Sulcus (IPS) [33], where neurons coding for numerosity have been found in both humans and monkeys [34]. In the model, number neurons interact through random lateral inhibitory connections and self-excitation. This allows, as described above, to collapse a broad distribution of number neurons [32] to a pool representing a single number, in a retrieval process during a step of the program. We assume that the newtwork has learned a notion of number proximity and continuity. This was implemented via a transition-network that has asymmetrical connections with the number-network. A given neuron representing the number n excites the transition neuron $n \to n+1$ population. This in turn excites the neuron that represents the number $n+1$. Again, we do not delve into how this is learned, we assume it as a consolidated mechanism.

The numbers-network can be in different modes: it can be quiescent, such that no number is active, or it can be in a winner-take-all mode with only one unit in the active state. Our network implementation of the counter makes use of two variables. The *Count* variable stores the current count and changes dynamically as the program progresses, after being initialized to n_{ini}. The *End* variable stores the number at which the counting has to stop and is initially set to n_{end}.

The network behaves basically as a while loop, increasing the value of the *Count* variable while its value differs from that of the *End* variable. To increment the count, we modeled a transition-network with units that have asymmetrical connections with the numbers-network. For example, the "$1 \to 2$" node receives excitatory input from the unit coding for number 1 and in turn excites number 2. This network stores knowledge about successor functions, and in order to become active it requires additional input from the production system. As mentioned above, here we do not address how such structure is learned in number representing neurons. Learning to count is an effortful process during childhood [35] by which children learn transition rules between numbers. We postulate that these relations are encoded in structures which resemble horizontal connections in the cortex [36–38]. In the same way that horizontal connections incorporate transition probabilities of oriented elements in a slow learning process [39, 40], resulting in a Gestalt as a psychological sense of "natural" continuity, we argue that horizontal connections between numerosity neurons can endow the same sense of transition probability and natural continuity in the space of numbers. The successor function can be as an homologous to a matrix of horizontal connections in the array of number neurons.

In a way, our description postulates that a certain number of operations are embedded in each domain cortex (orientation selective neurons in V1 for curve tracing, number selective neurons in IPS for arithmetic...). This can be seen as "compiled" routines which are instantiated by local horizontal connections capable of performing operations such as collinear continuity, or "add one". The program can control which of these operations becomes active at any given step by gating the set of horizontal connections, a process we have referred to as "addressing" the cortex [41]. Just as an example, when older children learn to automatically count every three numbers (1, 4, 7, 10, 13...) we postulate that they have instantiated a new routine (through a slow learning process) capable of establishing the transition matrix of $n \to n+3$. The repertoire of compiled functions is dynamic and can change with learning [42].

Counting requires a sequence of operations which include changes in the current count, retrieval of successor functions and numeric comparisons. The successive steps of the counting routine are governed by firing of production neurons. The order in which the productions fire is controlled by the content of the memory (Fig. 3). We emphasize that while the production selection process proceeds in parallel —as each production neuron constantly evaluates its input— the selected production strongly inhibits the other production neurons and therefore the evidence accumulated at one step is for the most part lost after a production is selected.[2] In Fig. 3 we simulate a network that has

[2]In the absence of external noise (as in the present simulations), only the production with the largest input has

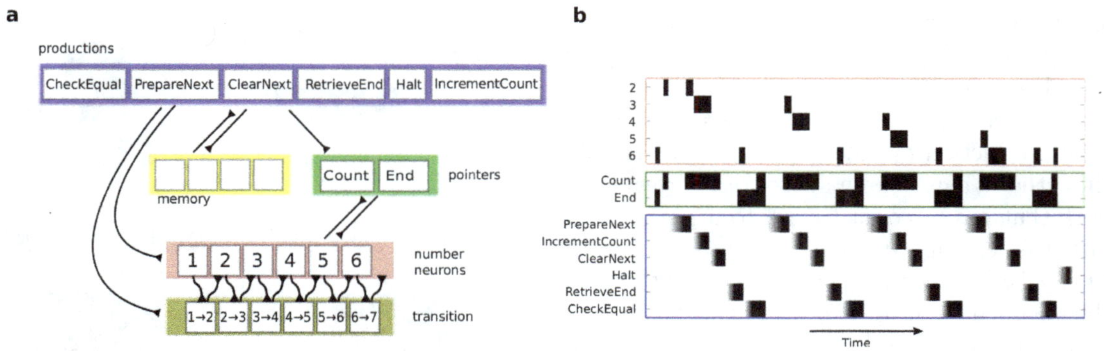

Figure 3: Sketch of the network implementing an arithmetic counter. (a) The network is divided into five sub-networks: productions, memory, pointers (or variables), numbers, and transition-networks. The order in which the productions fire is controlled by the state of the memory network, which is itself controlled by the production system. (b) Dynamics of a subset of neurons in the network. All units are binary except for the production neurons (violet) which gradually accumulate evidence to a threshold.

to count from numbers 2 to 6. Once the initial and final numbers have been bound to the *Count* and *End* variables respectively, the network cycles through six productions. The first production that is selected is the *PrepareNext* production, whose role is to retrieve the value that results from adding 1 to the current count. To this end, this production retrieves the current value of the *Count* variable, and excites the neurons of the transition-network such that the node receiving an input from the retrieved value of *Count* becomes active (*i.e.*, if 2 is active in the numbers-network, then 2 → 3 becomes active in the transition-network). To assure that the retrieved value is remembered for the next step (the actual change of the current count), neurons in the transition-network are endowed with recurrent excitation, and therefore these neurons remain active until explicitly inhibited. The same production also activates a node in the memory-network which excites the *IncrementCount* production, which is therefore selected next. The role of the next production (*IncrementCount*) is to actually update the current count, changing the binding of the *Count* variable. The *IncrementCount* production inhibits all neurons in the number network, to turn it to the quiescent state. Once the network is quiescent, lateral inhibition between number nodes is released and the asymmetrical inputs from the transition-network can activate the number to which it projects. At the same time, the *IncrementCount* production activates the *Count* variable, which is then bound to the currently active node in the numbers-network. Notice that this two-step process between the *PrepareNext* and *IncrementCount* productions basically re-assigns the value of the *Count* variable from its initial value n to a new value $n + 1$. Using a single production to replace these two (as tried in earlier versions of the simulations) required activating the number and transition neurons at the same time which lead to fast and uncontrolled transitions in the numbers-network. To increase the current count in a controlled manner, we settled for the two-productions solution. The *IncrementCount* production also activates a memory unit that biases the competition at the next stage, in favor of the *ClearNext* production. This production shuts up the activity in the transition-network, strongly inhibiting its neurons to compensate for their recurrent excitation. Shutting the activity of these neurons is required at the next step of the routine to avoid changes in the current count when the *Count* variable is retrieved to be compared with the *End* variable. After *ClearNext*, the *RetrieveEnd* production fires which retrieves the value of the *End* variable to strengthen the connections between the *End* variable and the value to which it is bound. This step is required since the strength of the plastic connections decays rapidly, and therefore the instantiation of the variables will be lost if not used or reactivated periodically. Finally, the *CheckEqual* production is selected to determine if the *Count* and *End* variables are equal. If both variables are equal, a node

higher-than-baseline activity.

in the memory network is activated which is detected by the *Halt* production to indicate that the task has been completed; otherwise, the production that is selected next is the *PrepareNext* production and the production cycle is repeated. In Fig. 3b, we show the dynamics of a subset of neurons for a network that has to count from 2 to 6.

With this example we have shown how even a seemingly very easy task such as counting (which can be encoded in up to two lines in virtually any programming language) seems to require a complex set of procedures to coordinate and stabilize all computations, when they are performed by neuronal circuits with slow building of activity and temporal decay.

ii. A world of blocks

A natural extension of the numerotopic domain used in the above example is to incorporate problems in which the actor must interact with its environment, and sensory and motor productions ought to be coordinated.

The visual system performs a great variety of computations. It can encode a large set of visual features in a parallel feed-forward manner forming its base representation [9,30,43,44], and temporally store these features in a distributed manner [45–47]. A matrix of lateral connections gated by top-down processing can further detect conjunctions of these feature for object recognition.

In analogy with motor routines, the visual system relies on serial chaining of elemental operation [9,31] to gain computational flexibility. There are many proposals as to which operations are elemental [31], but, as we have discussed above, this list may be fuzzy since the set of elementary operations may be changed by learning [41]. In this framework, atomic operations are those that are encoded in value domains.

Here, and for the purpose of implementing a neural circuit capable of solving the blocks worlds, we will focus on a simplified group of three elemental operations:

Visual search: the capacity of the system to identify the location of a given feature.

Visual cuing: the capacity of the system of highlighting the features that are present at a given location.

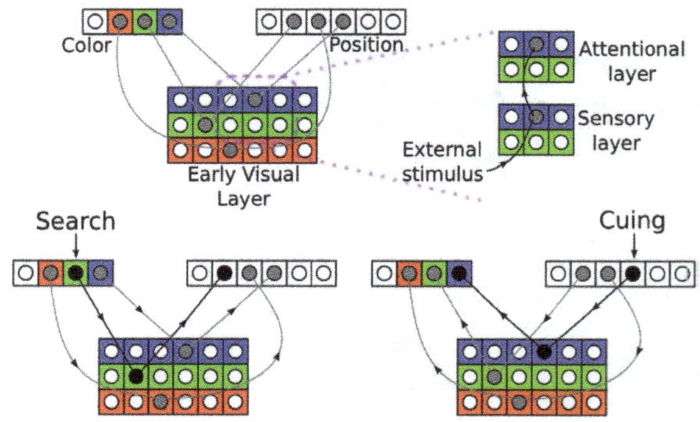

Figure 4: Simplified model of the visual system used in the blocks-world simulations. The upper portion shows the different layers and their connections. The early visual area is formed by a first sensory layer of neurons that receive stimulation from the outer world and a second attentional layer with bidirectional connections between the higher color or position tuned areas. The grayed neurons are the ones that present a higher activity. The lower panels show an example of the cuing and search operations. In the latter, a color tuned neuron is stimulated and drives an activity increase in the early visual layer. That later empowers the activity in the position layer concluding the search. The right situation shows the similar cuing operation.

Shift the processing focus: a method guided by attention to focus the processing of visual features or other computations in a given location.

Here we use a simplified representation of the visual system based on previous studies [48–50]. We assume a hierarchy of two layers of neurons. The first one is tuned to conjunctions of colors and locations in the visual field. The second one has two distinct groups of populations, one with neurons that have large receptive fields that encode color irrespective of their location and another which encodes location independently of color (Fig. 4). The model assumes that neurons in the first layer tuned to a particular color and retinotopic location are reciprocally connected to second layer units that encode the same color or location.

This architecture performs visual search of color in a way which resembles the variable assignation described above, through a conjunction mechanism between maps encoding different features. In the

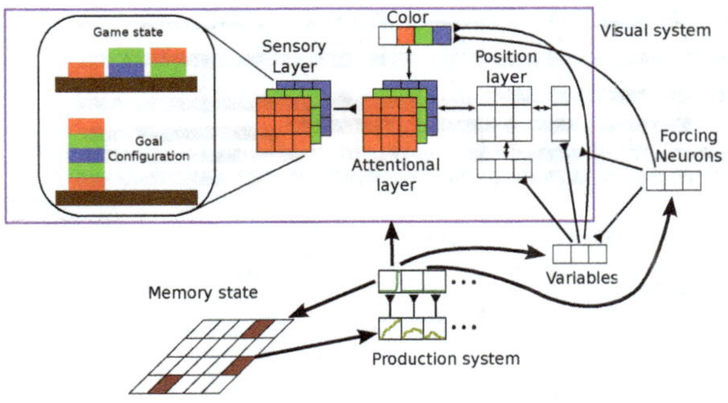

Figure 5: Sketch of the network that solves the blocks-world problem. The sensory early visual system receives input from the BW configuration and excites the first attentional layer. The latter is connected to color and position specific areas. The arrows show there is a connection between layers. The individual connections may be excitatory or inhibitory. The connections with the inverted triangle head indicate only excitatory connections exist.

model, the color cortex encodes each color redundantly for all positions forming a set of spatial vectors (one for each color). Of course, all these spatial maps selective for a given color can be intermingled in the cortex, as it is also the case with orientation columns which sample all orientations filling space in small receptive fields. If, for instance, a red square is presented in position three, the neuron selective for red (henceforth referred to as in the red map) and with a receptive field in position three will fire. This activation, in turn, propagates to spatial neurons (which are insensitive to color). Thus, if four squares of different colors are presented in positions 1 to 4, the spatial neurons in these positions will fire at the same rate. To search for the spatial position of a red block, the activity of neurons coding for red in the color map must be enhanced. The enhanced activity propagates to the early visual areas which code for conjunctions of color and space, which in turn propagates to the spatial map, highlighting the position where the searched color was present. Spatial selection is triggered by an attention layer which selects the production "attend to red", addressing the sensory cortex in a way that only locations containing red features will be routed to the spatial neurons. The color of a block at one location can be retrieved by an almost identical mechanism. In this case, the production system sets the attentional network to a given position in space and through conjunction mechanisms (because connections are reciprocal) only the color in the selected position is retrieved. This is a simple device for searching based on the propagation of attentional signals which has been used before in several models (*e.g.*, Ref. [31]) (Fig. 4).

To bridge these ideas which are well grounded in the visual literature [30, 31, 43] with notions of planning and sequential mental programs, we use this model to implement a solver for a simple set of Blocks-World problems. The blocks-world framework is a paradigmatic artificial intelligence problem that consists of a series of colored blocks piled up on top of a large surface in many columns. The goal is to arrange the blocks according to their color in a given goal configuration. The player can only move one block from the top of any column and place it at the top of another, or on the surface that supports all the blocks. We choose to construct a solver for a restricted blocks-world problem where the surface can only hold 3 columns of blocks and the goal configuration is to arrange them all into one column (that we call the target column).[3]

We implement a network with a set of memory and production neurons —analogous to the counter circuit described above— which coordinates a set of visual and motor productions (Fig. 5). The interaction between the memory layer and the production system triggers the execution of elemental visual processes, motor actions and changes in the memory configuration in order to solve any given instance of the problem.

To solve this problem, an algorithm needs to be able to find whether a block is in the correct position. For this, it requires, first, a "retrieve color" from a given location function. Normally the location that is intended to be cued is the one that is being attended to. We implement the attended location as a variable population (that we call the processing focus or PF inspired in Ullman's work [9]), so the "retrieve color" is equal to cue the color in PF's location.[4] Second, it must compare the colors in different locations. This can be done by bind-

[3] This restricted problem is equivalent to the Tower of London game [51].

[4] There are works that name PF as *Deictic Pointers* and suggest that it would be possible to store it also by keeping gaze or even a finger at the relevant location [52].

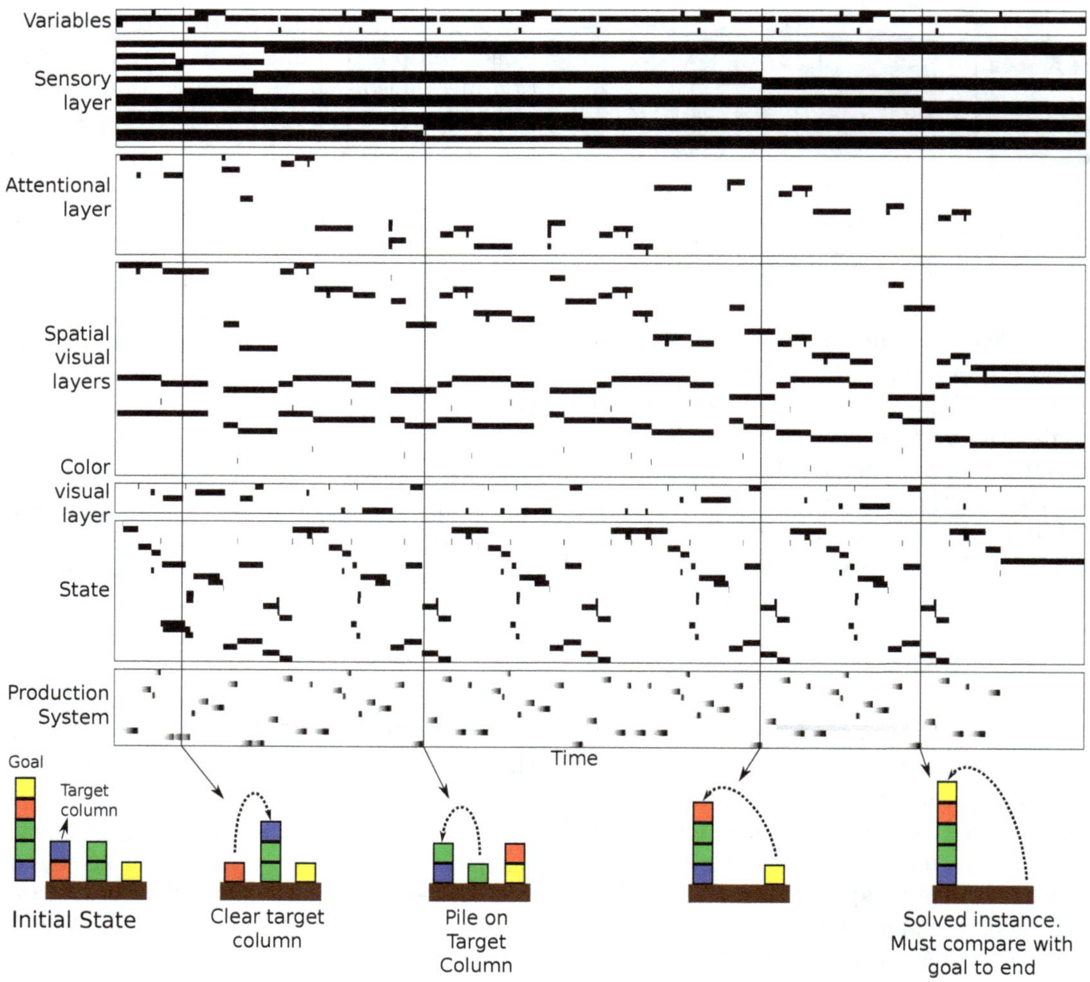

Figure 6: Mean firing rate of a subset of neuronal populations involved in the resolution of an instance of BW. The rate is normalized between 0 (white) and 1 (black). The horizontal axis represents the ellapsed time in arbitrary units. At the bottom, we show a subset of intermediate BW configurations, aligned to the execution time of the motor commands which lead to these configurations.

ing the relevant location colors to separate variables and then comparing them in the way described in section a.

As the goal is to pile all the blocks in the correct order in a given target column, a possible first step towards the goal is to compare the target column with the goal configuration from the bottom to the top. This can be done by chaining several movements of the processing focus with color retrievals and subsequent comparisons. Once the first difference is found, the target column's upper blocks must be moved away in order to replace the different colored block with the correct one. This process is carried out using several motor productions. Once the target column is free to receive the correct colored block, that color must be searched in the remaining columns. This is done as described earlier in this section. Once found, the *PF* can be moved there in order to view if there are blocks above it. If there are, motor productions must be chained in order to free the desired block and move it to the target column. After this is done, the program can loop back to comparing the target column with the goal configuration and iteratively solve the problem.

Our neuronal implementation chains the productions in a similar way as the one described above and elicits a complex activity pattern (Fig. 6). A detailed explanation of the implementation can be found in the supplementary material [53].

IV. Conclusions

Here we presented ideas aimed to bridge the gap between the neurophysiology of simple decisions and the control of sequential operations. Our framework proposes a specific set of mechanisms by which multi-step computations can be controlled by neural circuits. Action selection is determined by a parallel competition amongst competing neurons which slowly accumulate sensory and mnemonic evidence until a threshold. Actions are conceived in a broad sense, as they can result in the activation of motor circuits or other brain circuits not directly involved in a movement of the eyes or the limbs. Thresholding the action of the productions results in discrete changes to a meta-stable network. These discrete steps clean up the noise and enable a logical flow of the computation.

Comprehending the electrophysiological mechanisms of seriality is hindered by the intrinsic difficulty of training non-human primates in complex multi-step tasks. The ideas presented in this report may serve to guide the experimental search for the mechanisms required to perform tasks involving multiple steps. Neurons integrating evidence towards a threshold should be observed even in the absence of an overt response, for instance in the frontal-eye fields of awake monkeys for the control of attention. Memory neurons should show fast transitions between metastable states, on average every $\sim$ 100-250 msec, compatible with the mean time between successive productions in ACT-R [8].

As mentioned, we do not address how the productions and the order in which they are executed are learned. There is a vast literature, for instance in reinforcement learning [54–56] describing how to learn the sequence of actions required to solve a task. Deahene & Changeaux [57] showed how a neuronal network can solve a task similar to the BW that we modelled here, but where the order in which productions fire was controlled by the distance from the game state to the goal. Instead, our aim here was to investigate how the algorithm (the pseudo-code) may be implemented in neuronal circuits —once it has already been learned— from a small set of generic principles. The operation of the proposed neuronal device in a simple arithmetic task and in a neuronal network capable of solving any instance of a restricted Blocks-World domain illustrates the plausibility of our framework for the control of computations involving multiple steps.

Acknowledgements - AZ was supported by a fellowship of the Peruilh Foundation, Faculty of Engineering, Universidad de Buenos Aires. LP was supported by a fellowship of the National Research Council of Argentina (CONICET). PRR was supported by a Netherlands Organization for Scientific Research (NWO)-VICI grant, a grant of the NWO Excellence Program Brain and Cognition, and a Human Frontier Science Program grant.

---

[1] M Platt, P Glimcher, *Neural correlates of decision variables in parietal cortex*, Nature **400**, 233 (1999).

[2] J D Schall, *Neural basis of deciding, choosing and acting*, Nat. Rev. Neurosci. **2**, 33 (2001).

[3] J I Gold, M N Shadlen, *The neural basis of decision making*, Annu. Rev. Neurosci. **30**, 535 (2007).

[4] P R Roelfsema, P S Khayat, H Spekreijse, *Subtask sequencing in the primary visual cortex.*, P. Natl. Acad. Sci. USA **100**, 5467 (2003).

[5] R Romo, E Salinas, *Flutter discrimination: Neural codes, perception, memory and decision making*, Nat. Rev. Neurosci. **4**, 203 (2003).

[6] S I Moro, M Tolboom, P S Khayat, P R Roelfsema, *Neuronal activity in the visual cortex reveals the temporal order of cognitive operations*, J. Neurosci. **30**, 16293 (2010).

[7] A Newell, *Unified theories of cognition*, Harvard University Press, Cambridge, Massachusetts (1990).

[8] J R Anderson, C J Lebiere, *The atomic components of thought*, Lawrence Erlbaum, Mahwah, New Jersey (1998).

[9] S Ullman, *Visual routines*, Cognition **18**, 97 (1984).

[10] A Newell, *Productions systems: Models of control structures*, In: Visual Information Processing, Ed. W G Chase, Pag. 463, Academic Press, New York (1973).

[11] S Dehaene, M Sigman, *From a single decision to a multi-step algorithm*, Curr. Opin. Neurobio. **22**, 937 (2012).

[12] J Gottlieb, P Balan, *Attention as a decision in information space*, Trends Cogn. Sci. **14**, 240 (2010).

[13] J D Roitman, M N Shadlen, *Response of neurons in the lateral intraparietal area during a combined visual discrimination reaction time task*, J. Neurosci. **22**, 9475 (2002).

[14] M N Shadlen, W T Newsome, *Motion perception: Seeing and deciding*, P. Natl. Acad. Sci. USA **93**, 628 (1996).

[15] Y Huang, A Friesen, T Hanks, M Shadlen, R Rao, *How prior probability influences decision making: A unifying probabilistic model*, In: Advances in Neural Information Processing Systems 25, Eds. P Bartlett, F C N Pereira, C J Ca L Burges, L Bottou, K Q Weinberger, Pag. 1277, Lake Tahoe, Nevada (2012).

[16] L P Sugrue, G S Corrado, W T Newsome, *Matching behavior and the representation of value in the parietal cortex*, Science **304**, 1782 (2004).

[17] J D Wallis, K C Anderson, E K Miller, *Single neurons in prefrontal cortex encode abstract rules*, Nature **411**, 953 (2001).

[18] J Von Neumann, *The computer and the brain*, Yale University Press, New Haven, Connecticut (1958).

[19] A Zylberberg, S Dehaene, P R Roelfsema, M Sigman, *The human Turing machine: A neural framework for mental programs*, Trends Cogn. Sci. **15**, 293 (2011).

[20] G Maimon, J A Assad, *A cognitive signal for the proactive timing of action in macaque lip*, Nat. Neurosci. **9**, 948 (2006).

[21] M N Shadlen, R Kiani, T D Hanks, A K Churchland, *Neurobiology of decision making an intentional framework*, In: Better Than Conscious?, Eds. C Engel, W Singer, Pag. 71, MIT Press, Massachusetts (2008).

[22] A Zylberberg, S Dehaene, G B Mindlin, M Sigman, *Neurophysiological bases of exponential sensory decay and top-down memory retrieval: A model*, Front. Comput. Neurosci. **3**, 4 (2009).

[23] G Mongillo, O Barak, M Tsodyks, *Synaptic theory of working memory*, Science **319**, 1543 (2008).

[24] R C O'Reilly, *Biologically based computational models of high-level cognition*, Science **314**, 91 (2006).

[25] L Shastri, V Ajjanagadde, *et al.*, *From simple associations to systematic reasoning: A connectionist representation of rules, variables and dynamic bindings using temporal synchrony*, Behav. Brain Sci. **16**, 417 (1993).

[26] R Hahnloser, R J Douglas, M Mahowald, K Hepp, *Feedback interactions between neuronal pointers and maps for attentional processing*, Nat. Neurosci. **2**, 746 (1999).

[27] X j Wang, *Introduction to computational neuroscience*, Technical report Volen Center for Complex Systems, Brandeis University, Waltham, Massachusetts (2006).

[28] C Carr, M Konishi, *A circuit for detection of interaural time differences in the brain stem of the barn owl*, J. Neurosci. **10**, 3227 (1990).

[29] J Slaney, S Thiébaux, *Blocks world revisited*, Artif. Intell. **125**, 119 (2001).

[30] P R Roelfsema, V A Lamme, H Spekreijse, *The implementation of visual routines*, Vision Res. **40**, 1385 (2000).

[31] P R Roelfsema, *Elemental operations in vision*, Trends Cogn. Sci. **9**, 226 (2005).

[32] S Dehaene, J P Changeux, *Development of elementary numerical abilities: A neuronal model*, J. Cognitive Neurosci. **5**, 390 (1993).

[33] M Piazza, V Izard, P Pinel, D Le Bihan, S Dehaene, *Tuning curves for approximate numerosity in the human intraparietal sulcus*, Neuron **44**, 547 (2004).

[34] A Nieder, S Dehaene, *Representation of number in the brain*, Annu. Rev. Neurosci. **32**, 185 (2009).

[35] C Lebiere, *The dynamics of cognition: An ACT-R model of cognitive arithmetic*, Doctoral dissertation. Carnegie Mellon University, Pittsburgh, Pennsylvania (1998).

[36] D Y Ts'o, C D Gilbert, T N Wiesel, *Relationships between horizontal interactions and functional architecture in cat striate cortex as revealed by cross-correlation analysis*, J. Neurosci. **6**, 1160 (1986).

[37] B A McGuire, C D Gilbert, P K Rivlin, T N Wiesel, *Targets of horizontal connections in macaque primary visual cortex*, J. Comp. Neurol. **305**, 370 (1991).

[38] C D Gilbert, Y Daniel, T N Wiesel, *Lateral interactions in visual cortex*, In: From pigments to perception, Eds. A Valberg, B B Lee, Pag. 239, Plenun Press, New York (1991).

[39] M Sigman, G A Cecchi, C D Gilbert, M O Magnasco, *On a common circle: Natural scenes and gestalt rules*, P. Natl. Acad. Sci. USA **98**, 1935 (2001).

[40] C D Gilbert, M Sigman, R E Crist, *The neural basis of perceptual learning*, Neuron **31**, 681 (2001).

[41] C D Gilbert, M Sigman, *Brain states: Top-down influences in sensory processing*, Neuron **54**, 677 (2007).

[42] M K Kapadia, M Ito, C D Gilbert, G Westheimer, *Improvement in visual sensitivity by changes in local context: Parallel studies in human observers and in v1 of alert monkeys*, Neuron **15**, 843 (1995).

[43] V A Lamme, P R Roelfsema, *The distinct modes of vision offered by feedforward and recurrent processing.*, Trends Neurosci. **23**, 571 (2000).

[44] S Thorpe, D Fize, C Marlot, *Speed of processing in the human visual system*, Nature **381**, 520 (1996).

[45] D J Felleman, D C Van Essen, *Distributed hierarchical processing in the primate cerebral cortex*, Cereb. Cortex **1**, 1 (1991).

[46] G Sperling, *The information available in brief visual presentations*, Psychol. Monogr. Gen. A. **74**, 1 (1960).

[47] M Graziano, M Sigman, *The dynamics of sensory buffers: Geometric spatial and experience-dependent shaping of iconic memory*, J. Vision **8**, 1 (2008).

[48] F Hamker, *The role of feedback connections in task-driven visual search*, In: Connectionist models in cognitive neuroscience, Eds. D Heinke *et al.*, Pag. 252, Springer–Verlag, London (1999).

[49] F H Hamker, *A dynamic model of how feature cues guide spatial attention*, Vision Res. **44**, 501 (2004).

[50] D Heinke, G W Humphreys, *Attention, spatial representation, and visual neglect: Simulating emergent attention and spatial memory in the selective attention for identification model (saim).*, Psychol. Rev. **110**, 29 (2003).

[51] T Shallice, *Specific impairments of planning*, Phil. Trans. R. Soc. Lond. B **298**, 199 (1982).

[52] D H Ballard, M M Hayhoe, P K Pook, R P N Rao, *Deictic codes for the embodiment of cognition*, Behav. Brain Sci. **20**, 723 (1997).

[53] A Zylberberg, L Paz, P R Roelfsema, S Dehaene, M Sigman, *Supplementary Material to this paper*, available at *www.papersinphysics.org* (2013).

[54] R S Sutton, A G Barto, *Reinforcement learning: An introduction*, MIT Press, Massachusetts (1998).

[55] P R Roelfsema, A van Ooyen, T Watanabe, *Perceptual learning rules based on reinforcers and attention.*, Trends Cogn. Sci. **14**, 64 (2010).

[56] J O Rombouts, S M Bohte, P R Roelfsema, *Neurally plausible reinforcement learning of working memory tasks*, In: Advances in Neural Information Processing Systems 25, Eds. P Bartlett, F C N Pereira, C J Ca L Burges, L Bottou, K Q Weinberger, Pag. 1880, Lake Tahoe, Nevada (2012).

[57] S Dehaene, J P Changeux, *A hierarchical neuronal network for planning behavior*, P. Natl. Acad. Sci. USA **94**, 13293 (1997).

20

Invited review: Graphite and its hidden superconductivity

P. Esquinazi[1*]

We review experimental results, from transport to magnetization measurements, on different graphite samples, from bulk oriented graphite, thin graphite films to transmission electron microscope lamellae, that indicate the existence of granular superconductivity at temperatures above 100 K. The accumulated evidence speaks for a localization of the superconducting phase(s) at certain interfaces embedded in semiconducting crystalline regions with Bernal stacking order.

I. Introduction

Over the past decade, our interpretation of the magnetic and transport properties of ordered graphite bulk samples has experienced a change respect to the partially accepted general description of their intrinsic properties. The description of graphite one finds in the not-so-old literature tells us that it is a kind of (semi)metal with a finite Fermi energy and carrier (electron plus hole) densities per graphene layer at low temperatures $n_0 \sim 10^{10} \ldots 10^{12}$ cm^{-2}, see e.g., [1–3]. However, real samples are not necessarily ideal, we mean defect-free, and therefore those carrier densities are not necessarily intrinsic of ideal graphite. The exhaustive experience accumulated in gapless and narrow band semiconductors [4] already indicates us how important defects and impurities (not necessarily magnetic ones but, for example, hydrogen) are in determining some of the measured properties. Therefore, taking experimental data of real samples as intrinsic, without knowing their microstructure and/or defect concentration, was indeed a misleading assumption in the past. This assumption has drastically influenced the description of the band structure of graphite we found nowadays in several books and publications. For example, if graphite has a finite Fermi energy E_F (whatever the majority carriers are), as assumed everywhere, up to seven free parameters have to be introduced [2,5,6] to describe the apparently ideal band structure of Bernal graphite with the well-known ABAB stacking order of the graphene layers.

The impact of well defined two-dimensional interfaces inside graphite samples [7,8] had not been realized until recent studies of the transport properties as a function of thickness of the graphite sample provided a link to the microstructure of the samples obtained by transmission electron microscope (TEM) studies. We also have to add the sensitivity of the graphite transport properties to very small amount of defects [9]. Those results [7,9] do not only indicate us that at least a relevant part of the carrier densities measured in graphite is not intrinsic but also that the metallic-like behavior of the electrical resistance does not reflect ideal, defect-free graphite [10]. An anomalous vanishing of the amplitude of the Shubnikov-de Haas (SdH) oscillations decreasing the thickness of the graphite samples was published, more than 10 years

*E-mail: esquin@physik.uni-leipzig.de

[1] Division of Superconductivity and Magnetism, Institute for Experimental Physics II, Fakultät für Physik und Geowissenschaften, Universität Leipzig, Linnéstrasse 5, D-04103 Leipzig, Germany.

ago, [11] without attracting the necessary attention, although those results already suggested that the SdH oscillations are probably not intrinsic of the graphite structure. These results are supported by the absence of SdH oscillations, i.e., no evidence for the existence of a Fermi surface, found recently in bulk oriented samples of high grade and high purity but without internal interfaces [12]. All these results indicate that the internal microstructure of the graphite samples play an important role, a microstructure that was neither characterized nor considered in the discussion of the measured properties of different graphite samples, from highly oriented pyrolytic graphite (HOPG) to Kish or natural graphite, even in nowadays literature [6,13,14].

What does this have to do with superconductivity? If we start searching for superconductivity in graphite by measuring the behavior of the electrical resistance (R) with temperature (T) and magnetic field (H), for example, it should be clear that the knowledge of the intrinsic, normal state dependence is needed. Otherwise, we may misleadingly interpret an anomalous behavior due to, for example, the influence of non-percolative, granular superconducting regions embedded in a (normal state) graphite matrix, as intrinsic of the material, clearly missing an interesting aspect of the sample. A reader with expertise in superconductivity might not be convinced that such a mistake could be ever made. However, the ballistic transport characteristics of the graphene layers in ideal graphite with their huge mobility and mean free path [15–18] provide a high conductivity path in parallel; such that it is not at all straightforward by simple experiments to realize and prove the existence of superconductivity at certain regions in some, not all, graphite samples. One needs indeed to do systematic experiments decreasing the size of the graphite samples (but not too much) to obtain clear evidence for the embedded or "hidden" superconductivity.

A note on samples: The internal ordering or mosaicity of the graphite crystalline regions inside commercial HOPG samples is given usually by the grade. For example, the highest ordered pyrolytic graphite samples have is a grade "A", which means a rocking curve width $\Delta \sim 0.4° \pm 0.2°$ ("B", $\Delta \sim 0.8°$, etc.). Interestingly, and due to the contribution of two dimensional highly conducting internal interfaces between crystalline regions [7,10], the highest grade, i.e., smaller rocking curve width, does not always mean that the used sample provides the intrinsic transport properties of ideal graphite. The characterization of the internal structure of usual HOPG samples, as well as the thickness dependence of $R(T)$ to understand the transport and the magnetic properties of graphite, indicate that these two dimensional interfaces are of importance. The existence of rhombohedral inclusions [19,20] (stacking order ABCABC instead of ABAB of the usual Bernal graphite structure) in HOPG as well as in Kish graphite samples can also have a relationship with the hidden superconductivity in graphite, following the theoretical work in Ref. [21]. According to literature (see e.g., Fig. 2-2 in Ref. [8]), the density of interfaces parallel to the graphene layers in Kish graphite, in regions of several microns length, is notable. Therefore, quantifying the perfection of any graphite sample through the resistivity ratio between 300 K and 4.2 K [8] is not necessarily the best criterion to be used if we are interested on the intrinsic properties of the graphene layers in graphite, because of the high conductivity of the interfaces in parallel to the graphene layers of the sample [10]. Two examples of the interfaces we are referring to can be seen in Fig. 1. On the other hand, commercial HOPG bulk samples are of high purity with average total impurity concentrations below 20 ppm. Especially the existence of magnetic impurities are of importance if the Defect-Induced Magnetism (DIM) is the main research issue. Their concentration remains below a few ppm for high grade HOPG samples [22].

The graphite flakes discussed in this work were obtained by exfoliation of HOPG samples of different batches, by careful mechanical press and rubbing the initial material on a previously cleaned substrate. As substrate, we used p-doped Si with a 150 nm SiN layer on top. We selected the flakes using microscopic and micro-Raman techniques to check their quality. More details on the preparation can be taken from Ref. [7] and other publications cited below.

This review is organized as follows. In the next section we discuss the experimental data for $R(T,H)$ from different graphite samples published in the last 12 years and argue that the first hints on unusual superconducting contribution can be already found in those measurements. In section III, we discuss the anomalous hysteresis in the magnetoresistance, a first clear indication for embedded granular superconductivity. Section IV, deals with the Josephson behavior measured in TEM lamellae whereas section V deals with the granular superconducting behavior found in the magnetization of water-treated graphite powder as well as in bulk HOPG samples for fields normal to the interfaces found inside those samples. In the last section, section VI, before the conclusion, we discuss possible origins for the superconducting signals on the basis of earlier and recent experimental and theoretical work.

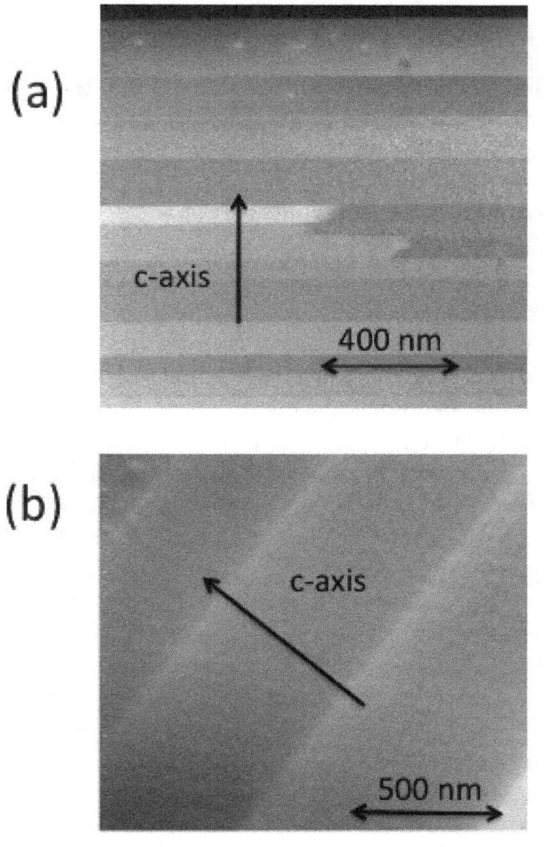

Figure 1: Transmission electron microscope pictures of two different kinds of interfaces and their distribution in HOPG samples. The TEM pictures were taken from two different lamellae, each about 300 nm thick and with the electron beam nearly parallel to the graphene planes of the samples. (a) The interfaces are recognized at the borders of crystalline regions of different gray colors. Taken from [7]. (b) Interfaces found in a HOPG sample used for magnetization measurements (see section V)that reveals hysteretic behavior in field and temperature. Taken from [29].

II. The behavior of the resistance vs. temperature at different applied magnetic fields

In this section, we discuss the behavior of the resistance $R(T, H)$ of different HOPG samples including Kish graphite. The data we present here were taken from [7, 23, 24] and a quick search in literature demonstrates that these data are reproducible and can be found in different publications, see, e.g., [2, 25–27]. Figure 2(a) shows the $R(T)$ for different bulk graphite samples of different grades (rocking curve width) and for one sample (HOPG-1) at zero and under a magnetic field applied normal to the main area, i.e., normal to the graphene planes of the sample. This figure reveals a general behavior, namely that the lower the resistivity ρ of the HOPG sample, the more metallic-like its temperature dependence. It is appealing to assume that these characteristics, low ρ, low Δ and the metallic behavior are clear signs for more ideal graphite. Therefore, from the measured $R(T)$, we may conclude that sample HOPG-3 is more ideal than sample HOPG-1 and the latter being more ideal than sample HOPG-2, see Fig. 2(a). This is indeed the usual interpretation found in several reviews in the literature, see e.g., [8, 28].

From a quick look at all the curves in Fig. 2, however, one recognizes a striking similarity between them, although we are comparing different samples with different thickness and some of the curves were measured under a magnetic field applied normal to the graphene layers of the samples; also normal to the interfaces commonly found in some ordered samples [7, 8].

Let us start discussing the metallic-like behavior of $R(T)$ of sample HOPG-3 in Fig. 2(a). This sample behaves as the HOPG-UC sample shown in (c) at zero magnetic field, having also a maximum at $T \sim 150$ K. A "better" metallic character shows the HOPG sample in (b) or the Kish graphite sample in (d), without any maximum in the shown temperature range. Is this metallic-like behavior really intrinsic of ideal graphite?

The following experimental evidence does not support such interpretation:
First, for samples from the same batch, the metallic character of $R(T)$ vanishes in the whole temperature range when the sample thickness is below ~ 50 nm [7, 11, 27], see Fig. 2(b).
Second, the metallic-like behavior vanishes in the whole temperature range after applying a magnetic field of the order of 1 to 2 kOe, an interesting behavior known as the Metal-Insulator Transition (MIT) [23, 25, 26], see Figs. 2(a), (c) and (d). Note that such magnetic field strength influences mainly the metallic-like region, see e.g., the change of sample HOPG-1 in Fig. 2(a) at zero and at 1 kOe field, an interesting behavior noted first in [30] and interpreted as due to superconducting instabilities.

At those field strengths, i.e., $H \sim 1$ kOe, the obtained $R(T)$ curves, for samples showing at zero field a metallic-like behavior, resemble the semiconducting-like curves obtained for sample HOPG-2 (Fig. 2(a)) or for samples with small thickness (Fig. 2(b)). At fields higher than a few kOe, the rather large magnetoresistance of graphite starts to play the main role and the $R(T)$ curve increases in the whole temperature range.

Finally, all these results added to the existence of well defined interfaces in the metallic-like HOPG samples as well as in Kish graphite, with distances in the $c-$axis direction usually larger than ~ 30 nm, indicate that the metallic-like behavior is due to the contribution of these interfaces and it is not intrinsic of the graphene layers of ideal graphite [7, 10]. Therefore, explanations of the MIT based on ideal graphite band models with a large number of free parameters [25, 26] are certainly not the appropriate ones.

All the different $R(T)$ curves for different samples shown in Fig. 2 and at zero field can be very well understood assuming the parallel contribution of semiconducting graphite paths in parallel to the one from the highly conducting interfaces [10]. The saturation of the resistance at $T \to 0$ K is interpreted as due to the finite resistance of the sample surfaces (the free one and the one on the substrate) short circuiting the intrinsic behavior of the bulk graphene layers at low enough temperatures.

The question is now whether parts of these interfaces hide superconducting regions. It is certainly appealing to suggest that the huge MIT at fields normal to the interfaces (and below ~ 2 kOe) is related to Josephson-coupled superconducting regions embedded in some of the interfaces. Note that the huge anisotropy of the MIT (parallel fields to the interfaces main plain do not affect the electrical transport) already implies that the regions responsible for the MIT must be laying parallel to the graphene layers [31]. Without the knowledge on the existence of these interfaces, an interpretation of the low field MIT based on the influence of superconductivity has been discussed in detail in the reviews [24, 32]. In those reviews, one can recognize the remarkable similarity between the scaling approaches used to characterize the magnetic-field-induced superconductor-insulator quantum phase transition [33] or of the field-driven MIT in 2D electron (hole) systems [34] and the one obtained for the MIT observed in graphite. As we will see in the next sections, the experimental evidence obtained in the last recent years indicates that granular superconductivity exists within some of those interfaces, indeed.

If superconducting patches exist embedded in parts of the interfaces or in other two dimensional regions of the bulk ordered samples, one expects to measure some signs of granular superconductivity, as for example nonlinear $I-V$ curves or hysteresis in the magnetoresistance. However, this is not really observed in bulk large samples. There are at least two reasons for the apparent absence of these expected phenomena. One is the distribution of the input current between the ballistic channels given by the graphene layers [17, 18], the metallic, normal conducting parts of the interfaces and the regions where the superconducting patches exist. In other words, the usual maximum currents used in transport experiments reported in bulk samples may have been small enough so that the current through the superconducting regions remained below the critical Josephson one. The other reason is the experimental voltage sensitivity to measure the possible irreversibility in the magnetoresistance due to the existence of pinned vortices or fluxons. We will see in the next section that part of these problems can be overcome decreasing the sample size; in this way, one obtains the voltage signals from the regions of interest with enough sensitivity.

Apart from the large magnetic field sensitivity of the metallic-like resistance measure in bulk graphite samples with interfaces, is there any further hint for the existence of granular supercon-

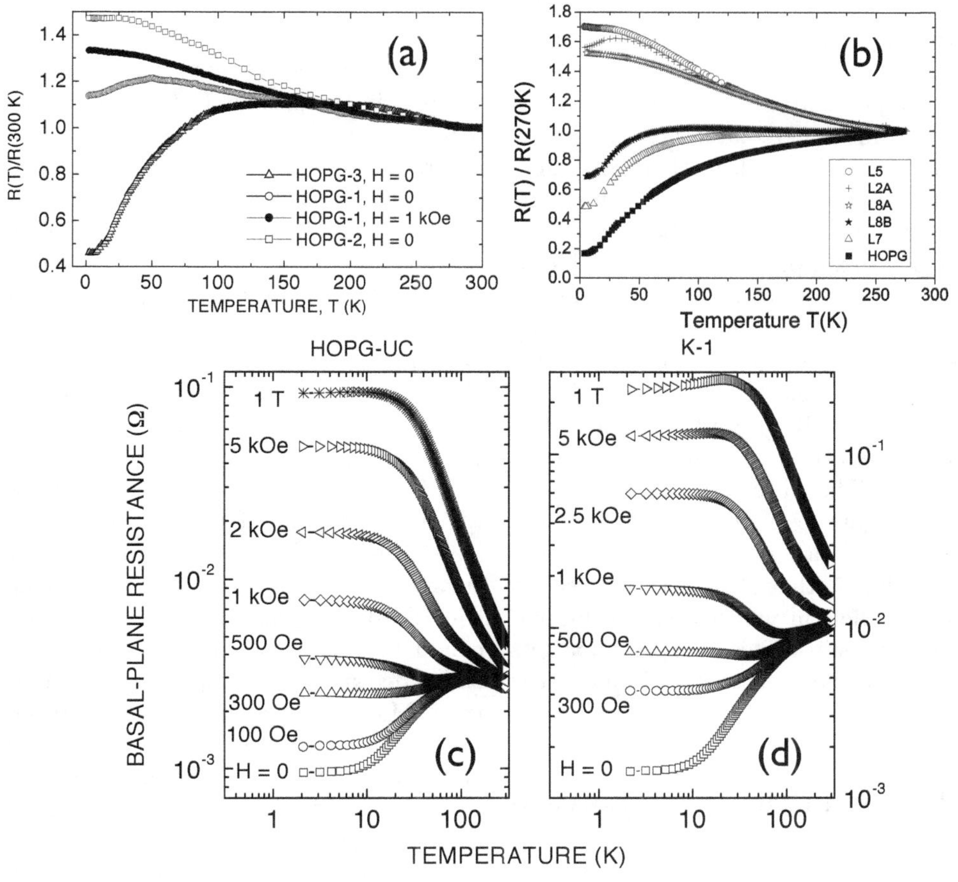

Figure 2: (a) Normalized resistance vs. temperature for three different HOPG bulk samples. The bottom, metallic-like curve corresponds to the sample HOPG-3, the curves above correspond to HOPG-1 ($H = 0$), HOPG-1 ($H = 1$ kOe), and HOPG-2. The grade and resistivity values are HOPG-1 ($\Delta = 1.4°$, resistivity at 300 K ρ(300K) = 45 $\mu\Omega$cm), HOPG-2 ($\Delta = 1.2°$, ρ(300K) = 135 $\mu\Omega$cm) and HOPG-3 ($\Delta = 0.5°$, ρ(300K) = 5 $\mu\Omega$cm). Taken from [23]. (b) Similar to (a) but for HOPG samples from the same batch but with different size, namely (thickness × length × width) L5: 12 ± 3 nm, 27 μm, 14 μm, L2A: 20 ± 5 nm, 5 μm, 10 μm, L8A: 13 ± 2 nm, 14 μm, 10 μm, L8B: 45 ± 5 nm, 3 μm, 3 μm, L7: 75 ± 5 nm, 17 μm, 17 μm, HOPG: 17 ± 2 μm, 4.4 mm, 1.1 mm, taken from [7]. (c) and (d) Resistance of bulk graphite samples vs. temperature at different applied fields normal to the graphene layers. The sample in (c) is a HOPG bulk sample from Union Carbide of grade A and the sample in (d) is Kish graphite. Taken from [24].

ductivity in those $R(T)$ curves? Yes, this hint is related to the thermally activated function ($\propto \exp(-E_a/k_BT)$ with E_a, a sample dependent effective thermal barrier ~ 30 K) one needs in order to fit the metallic-like contribution below $T \sim 200$ K [10]. This function is relevant in spite of only a factor five increase of the resistance between low and high temperatures, see Fig. 2. Skeptical readers can convince themselves about its relevance taking a similar example, as the exponential function used to fit the increase, by a similar factor, of the ultrasonic attenuation with temperature below T_c in conventional superconductors. We note that this exponential function has already been used to describe the increasing resistance of bulk graphite samples with temperature and it was speculated to be related to some superconducting-like behavior in graphite [32]. It is clear that this function is not the usual one, one expected for metals or semimetals and that cannot be understood within the usual electron-phonon interaction mechanisms, nor in two dimensions. A similar dependence has been observed in granular AlGe [35], which shows for a particular Al concentration a superconductor-semiconductor transition similar to that reported in Ref. [10] or, after an appropriate scaling in temperature, to some of the curves shown in Fig. 2. The observed thermally activated behavior might be understood on the basis of the Langer-Ambegaokar-McCumber-Halperin (LAMH) model [36, 37] that applies to narrow superconducting channels in which thermal fluctuations can cause phase slips. This interpretation gets further support from the evidence we discuss in the following sections.

III. Hysteresis in the magnetoresistance

In order to reveal by transport measurements the existence of granular superconductivity in some regions of the graphite samples, we need to increase the sensitivity of the measured voltage to those regions. To achieve this, we decrease the size of the sample enhancing in this way the probability to get some measurable influence of this phenomenon in the voltage. The work in Ref. [38] reported the first observations of an anomalous irreversible behavior in the magnetoresistance (MR) in a few tens of nm thick and several micrometer large multigraphene samples. Hysteresis in the magnetoresistance is a key evidence on the existence of either magnetic order (domains with their walls, for example) or vortices/fluxons and therefore on the existence of superconductivity. Because defects as well as hydrogen can trigger magnetic order in graphite, a first attempt would be to relate the measured hysteresis in the MR with the existence of magnetic order and magnetic domains, for example. However, the data exhibited anomalous hysteresis loops in the MR [38], similar to those observed in granular superconductors with Josephson-coupled grains [39–41]. The anomalous hysteresis was observed only for magnetic fields perpendicular to the planes, whereas in the parallel to the planes direction, the MR remains negligible. This fact already points out to a remarkable large anisotropic response of the superconducting phase(s) in agreement with the hypothesis that these superconducting regions might be embedded in some of the interfaces found inside some bulk graphite samples [7, 8]. The amplitude of the hysteresis in the MR reported in Ref. [38] vanishes at temperatures $T \sim 10$ K, clearly below the temperature at which the resistance shows a maximum, as it is the case for samples HOPG-1 in Fig. 2(a) or sample L2A in Fig. 2(b).

It is clear that thermal fluctuations can prevent the establishment of a coherent superconducting state in parts of the sample and therefore zero resistance state is not so simple to be achieved if the superconducting distribution is a mixture of superconducting patches at the interfaces and these are embedded in a multigraphene semiconducting matrix. Moreover, we should take also into account that the voltage electrodes are usually connected at the top surface of the graphite samples picking the voltage difference coming from a non-negligible normal conducting path.

One possibility to increase the sensitivity of the measured voltage to the field hysteresis these regions produce is to make a constriction in the middle of the two voltage electrodes, see inset in Fig. 3(a). In this case, we expect a locally narrower distribution of superconducting and normal regions at the constriction such that averaging effects should be less important. Simultaneously, through the constriction the main part of the voltage drop depends mostly on the region at the constriction, see Fig. 2(c) in Ref. [16]. Then, a higher sensitivity to the superconducting paths can be achieved in case they remain at or near the constriction. This idea has been successfully realized in [42] and its main results will be reviewed in this section.

Let us take two slightly different samples, 1 and 2, with $R(T)-$curves as shown in Fig. 3(a). The aim of the experiment is to study the hysteresis in the MR those samples might show below the temperature at which a maximum in the resistance

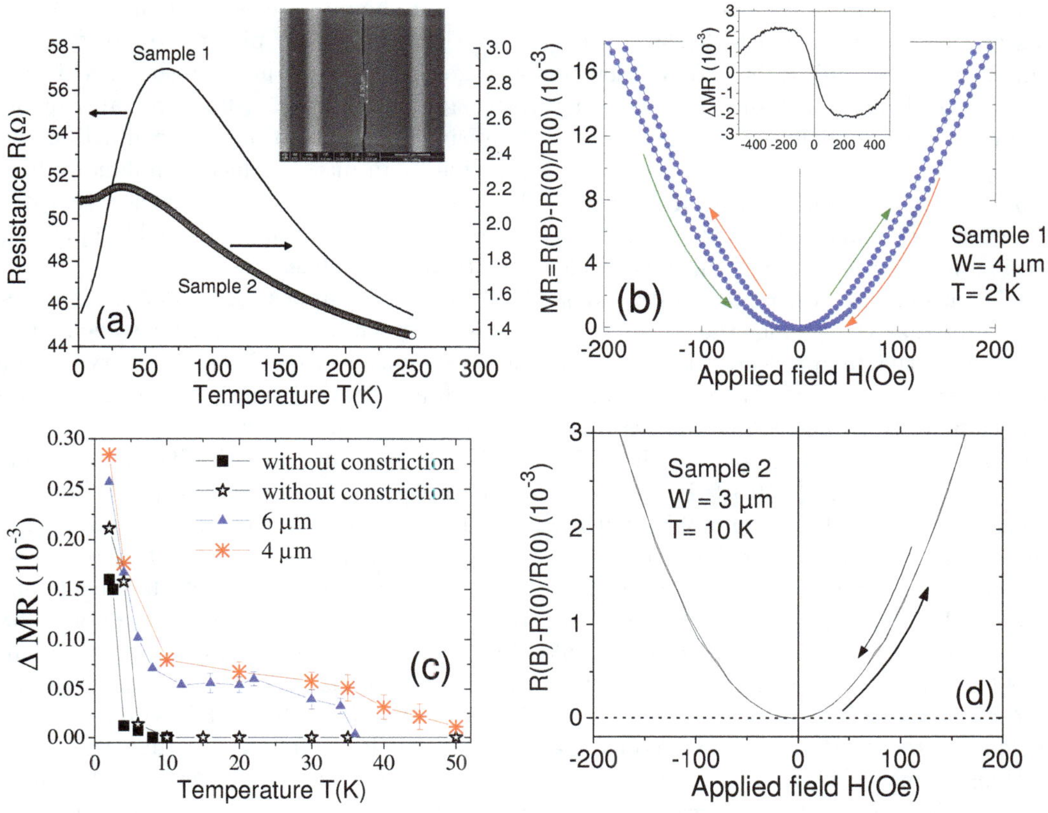

Figure 3: (a) Resistance vs. temperature, without constrictions and at zero applied field, for two graphite flakes of size (distance between voltage electrodes × width × thickness) for sample 1 (2): $13 \times 16 \times 0.015$ ($2.6 \times 6 \times 0.040$) μm^3. The observed temperature dependence remains for all constrictions widths. The inset shows a scanning electron microscope picture of sample 1 with a constriction width of 4.3 μm between the two voltage electrodes. The scale bar is 5 μm. (b) Magnetoresistance (MR) vs. applied magnetic field for sample 1 with a 4 μm constriction width and at 2 K. The input current was 1 μA. Note the clear hysteresis in the MR when the field is swept from $|H_{\max}| = 1000$ Oe. The inset shows the difference ΔMR between the curve obtained starting from $H_{\max} = +1000$ Oe and the return curve measured from $H_{\min} = -1000$ Oe. (c) The absolute difference between the two MR curves of the hysteresis loop obtained at a fixed magnetic field of 16.6 Oe for sample 1 without constriction ($\star$) and for two different constriction widths. The figure also shows the corresponding data for another graphite sample without constrictions (■) from [38]. (d) Magnetoresistance measured from a starting maximum field of 1.4 kOe at 10 K for sample 2 with a constriction width of 3 μm. Some of the figures and the data were taken from [42].

is measured, in case that maximum is related to the Josephson coupling between superconducting regions. Figure 3(b) shows one example of the anomalous hysteresis loop in the MR. The going down curve (from high, positive to low, negative fields, red arrow), for example, runs below the going up curve (green arrow) in the same quadrant as the field sweep was started, showing a minimum at positive fields of the order of 20 Oe, see also similar curves in Ref. [38]. To present the anomalous behavior clearly, the inset in Fig. 3(b) shows the difference between the two curves. This difference is in clear contrast to the usual hysteresis in superconductors as well as ferromagnets [38, 39], where the minimum (or maximum) in the MR is observed in the opposite field quadrant, and the increasing field resistance curve is usually below the decreasing field one.

Figure 3(c) shows the temperature dependence of the difference in the MR between the decreasing and increasing field curves at a fixed magnetic field for sample 1, without and with two constrictions. The results show that the smaller the constriction width, the higher the temperature at which the anomalous hysteresis is observed, decreasing below the sensitivity limit at $T > 50$ K for a constriction width of 4 μm, whereas the maximum in the $R(T)$ curve is at $\sim$ 70 K, see Fig. 3(a). The absence of any hysteresis in the MR for sample 2 with a constriction width of 3 μm and at $T = 10$ K indicates that the hysteresis does not come from some artifact due to the used focused ion beam method [42, 43] or due to an artifact in the measurement of the real field applied to the sample. As expected from the $R(T)$ curve, see Fig. 3(a), sample 2 shows the anomalous hysteresis in the MR at lower temperatures than sample 1 [42].

Summarizing this section, the observation of the anomalous hysteresis in the MR – together with the MIT and the relatively large MR at temperatures below the maximum in $R(T)$ – provides already striking hints that granular superconductivity is at work in some regions of these samples. The increase in the temperature region where the hysteresis is observed, decreasing the constriction width, demonstrates the problem of current averaging and voltage sensitivity limits usual experiments with large samples have.

IV. Direct evidence for Josephson behavior in the transport properties of graphite: Measurements in TEM lamellae

If the embedded interfaces (or some other quasi two dimensional regions) inside the measured graphite samples have superconducting properties, the best way to check them would be contacting electrodes as near as possible to those interfaces or interface regions and study the behavior as a function of any useful parameter one can take to influence their response. It should be clear that one cannot simply open the graphite sample at the interface and put voltage electrodes at the open surfaces of the interface, simply because it will not remain anymore. A tentative approach to put the contacts as near as possible to an interface has been done in Ref. [46]. Indeed, the observed behavior at low temperatures and as a function of magnetic field appeared to be superconducting-like.

A better and appealing evidence for the superconducting behavior embedded in some graphite ordered samples can be obtained by trying to locate the voltage electrodes directly at the inner edges of the interfaces. The work in Ref. [45] prepared TEM lamellae from bulk HOPG samples and using lithography and focused ion beam techniques, current and voltages electrodes at different positions of the samples were prepared. In this way, one tries to contact several of those interface edges simultaneously, as shown in Fig. 4(a). We note, however, that a thin surface layer of disordered graphite exists due to the Ga^+ ion irradiation used to cut the lamella from the bulk HOPG sample. This layer has a much larger resistance than the one of the graphene layers or of the interfaces [43] and therefore the input current goes through the lowest resistance path as well as the voltage electrodes pick up the response of the graphite sample with its interfaces. One can see this comparing first the $R(T)$ curves obtained at large enough currents in the lamellae (Fig. 4(c)) with those of graphite samples with top electrodes (Fig. 2). The fact that a zero resistance state (minimum voltage noise $\pm$ 5 nV upon sample) is obtained at low currents with $I-V$ characteristic curves that resemble the one expects for Josephson coupled grains leaves little doubt about the origin of the obtained signals. In the TEM picture of Fig. 4(b), one can see the graphite single crystalline regions (different gray colors) oriented differently between them about the common $c-$axis and having well defined two dimensional interfaces, as high resolution TEM studies revealed [47].

Figure 4(c) shows the voltage vs. temperature measured in a TEM lamella of oriented graphite [45] at different input DC currents, from 100 nA to 10 μA. The clear sharp transition, observed at $\sim$ 150 K at the lowest current, shifts to lower temperatures increasing the input DC current. For the largest input currents, the temperature dependence of the resistance of the contacted lamella shows a

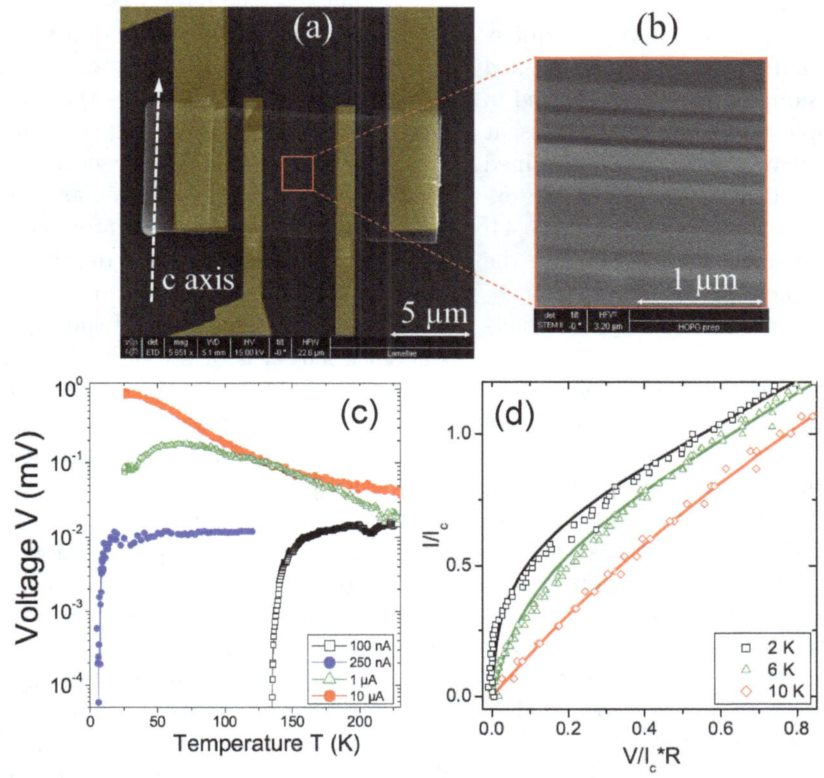

Figure 4: (a) Scanning Electron Microscopy (SEM) image of a lamella of 300 nm thickness on a Si/SiN substrate where the yellowish colored areas are the electrodes. A four-point configuration has been prepared with the outer electrodes used to apply current and the inner ones to measure the voltage drop. The $c-$axis runs parallel to the substrate surface and normal to the current direction. (b) Transmission Electron Microscopy (TEM) image of a HOPG lamella. The different brightness corresponds to a different orientation within the $a-b$ plane of the crystalline regions with thickness $>$ 30 nm. (c) Voltage vs. temperature at different input currents for a lamella of $\sim$ 800 nm thickness and with Van der Pauw contact configuration. (d) Current-Voltage characteristics at different temperatures for a lamella of $\sim$ 300 nm thickness in reduced coordinates, where R is the normal state resistance, I the input current, and I_c the critical Josephson current. The continuous curves are fitted to the model proposed in Ref. [44] with $I_c(T)$ as the only free parameter. Figures taken from [45].

maximum or follows the intrinsic semiconducting behavior of the graphene layers. This behavior already suggests the existence of high temperature granular superconductivity at some parts of the sample. The study reported in Ref. [45] shows that the transition temperature depends on the prepared sample. This indicates a sample dependent distribution of the superconducting regions and/or some influence of the preparation process or sample size on the superconductivity [47]. We also note that the observed sharp decrease in the measured voltage does not necessarily indicate the critical temperature of the superconducting regions but the temperature below which a percolative granular system shows negligible resistance due to the Josephson coupling at the used input current.

Current-voltage characteristic curves at different temperatures and in different lamellae obtained from different HOPG samples have been studied in Ref. [45]. An example of this $I-V$ curves at three temperatures is shown in Fig. 4(d) obtained for a different lamella. The curves follow the expected dependence for a Josephson junction [44] with a temperature dependent critical current, the only free parameter in the fit.

Further evidence that speaks for a superconducting origin of the $I-V$ curves is given by the expected detrimental effect of a magnetic field on the superconducting state. This effect can be due to an orbital depairing effect or due to the alignment of the electron spins at much higher fields, in case of singlet coupling. The effect of a magnetic field applied normal and parallel to the interfaces has been studied in detail for thick and thin lamellae in Ref. [45]. Upon sample size (thickness, i.e., width of the graphene planes inside the lamella) the observed effects are from the usual vanishing of the zero resistance state or no effect at all for thin lamellae. A magnetic field of a few kOe applied normal to the interfaces is enough to destroy the Josephson coupling at low temperatures, an effect compatible with the MIT observed in several graphite samples, see section II. Whereas a field applied parallel to them does not influence the $I-V$ curves at all, a fact that speaks for the two dimensionality of the superconducting regions.

Nevertheless, the influence of a magnetic field in HOPG samples with interfaces is not as "simple" as in conventional superconductors. For high fields applied normal to the interfaces, the $I-V$ curves show a recovery to the zero resistance state. The observed reentrance appears to be related to the magnetic-field driven reentrance observed at low temperatures in the longitudinal resistance at high enough magnetic fields [48]. This interesting behavior as well as the insensitivity of the $I-V$ curves to magnetic fields in very thin lamellae [45] deserve further studies.

We would like to note here that the possible effects of a magnetic field on the superconducting state of quasi two-dimensional superconductors, or in case the coupling does not correspond to a singlet state, are not that clear as in conventional superconductors. For example, results in two different two-dimensional superconductors, including one produced at the interfaces between non superconducting regions [49], show that superconductivity can even be enhanced by a parallel magnetic field. In case the pairing is $p-$type [50], the influence of a magnetic field is expected to be qualitatively different from the conventional, singlet coupling behavior [51,52] with even an enhancement of the superconducting state at intermediate fields. In case the London penetration depth is much larger than the size of the superconducting regions at the interfaces of our lamellae or if the superconducting coherence length is of the order or larger than the thickness of the lamella, the influence of a magnetic field should be less detrimental.

Through these studies, and taking into account that in samples without these interfaces no signature of superconducting or metallic-like behavior has been observed (see also section V) it is appealing to suggest that superconductivity is somewhere hidden at some of those interfaces or interface regions. It should be also clear that not all those interfaces have superconducting regions with similar critical parameters. Those interfaces are formed during the preparation of the HOPG samples based on treatments at very high temperatures ($T > 3400°$C) and high pressures ($P \sim 10$ kg/cm^3) and in a non-systematic way [8]. Actually, they are not at all an aim of the production but actually the opposite, they should be avoided in order to enhance the crystal perfection of the bulk HOPG material. It is even possible that, upon the procedure used to control the structure and texture of the graphite sample, the near surface region, for example, can have a different degree of graphitization as inside the bulk HOPG sample [8]. This means that one may obtain different results from different parts of the same HOPG sample. Therefore, disconcerting situations and an apparent lack of reproducibility are preprogrammed in case the research studies are done without taking care of the internal microstructure of the studied samples.

V. Magnetization measurements

In this section, we present and discuss magnetization measurements done in bulk HOPG samples with and without embedded interfaces and in water treated graphite powders. One of the main problems in interpreting magnetization data for fields applied parallel to the c−axis of the graphite structure, i.e., normal to the graphene layers and interfaces, is the need of subtraction of a large diamagnetic background. Due to the small amplitude of the superconducting-like signals in the studied samples, the subtraction of this linear in field background is not so simple, because it is not known with enough certainty to obtain the true field hysteresis after its subtraction. That means that we always have a certain arbitrariness in the shape of the obtained field hysteresis, a situation that will improve with the increase of the amount of material responsible for those superconducting-like signals. The small SQUID signals of interest imply that one should take additional efforts to minimize or rule out possible SQUID artifacts [53–55]. Therefore, systematic studies of samples of different or equal geometry and magnetic background, with and without interfaces, are necessary.

Taking into account: the overall shape of the hysteresis, the slope of the virgin curve at low fields where the subtraction does not affect too much, and the overall experience with ferromagnetic graphite [22, 56], one can rely to a certain extent on the obtained hysteresis shape. Certainly, not only the field hysteresis but also other evidence one gets from magnetization measurements as, e.g., the remanence at zero field as a function of the maximal field applied (see for example measurements for $YBa_2Cu_3O_7$ in Ref. [57]) and the hysteresis in temperature dependent measurements helps to convince oneself about the existence of some kind of granular superconductivity. The hysteresis between the field cooled (FC) and zero-field cooled (ZFC) curves can help to discern between a superconducting or ferromagnetic-like behavior. The most obvious evidence that speaks against a simple ferromagnetic order of the hysteresis observed as a function of temperature and field is the two dimensionality of the obtained hysteretic signals [29], i.e., the superconducting-like signals are mainly measured for fields normal to the interfaces. This fact is not compatible with any kind of magnetic order including shape or magneto crystalline anisotropy, whatever large they might be. We note that the ferromagnetic response of graphite due to DIM is mostly measured for fields parallel to the graphene layers, parallel to the main area of the samples [22].

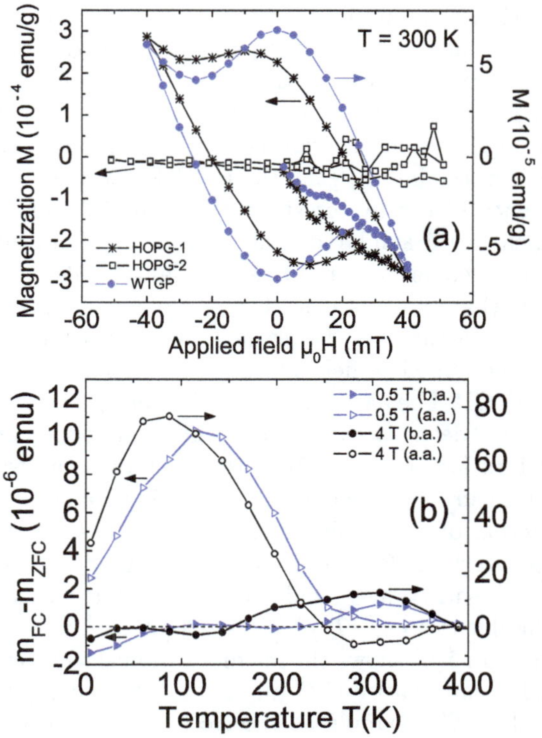

Figure 5: (a) Magnetization of two HOPG bulk samples (HOPG-1 and HOPG-2) after subtraction of a diamagnetic background and of water treated graphite powder (WTGP, right y−axis) at 300 K. The HOPG-2 sample shows no hysteresis in contrast to the other two samples. (b) Temperature dependence of the difference between FC and ZFC magnetic moments of the HOPG-1 sample before (b.a.) and after (a.a.) warming the sample up to $\simeq$ 600 K, at two constant applied fields, 0.5 T (left y−axis) and 4 T (right y-axis). The field was applied always normal to the interfaces or graphene planes of the samples. Data taken from [29].

i. Bulk graphite samples

Figure 5(a) shows the field hysteresis, after subtraction of the corresponding diamagnetic linear background, at 300 K of two bulk HOPG samples, HOPG-1 and HOPG-2 and a water treated graphite powder (WTGP) (right y−axis). A TEM characterization of the internal microstructure of the HOPG-1 sample shows clear evidence for well defined interfaces running parallel to the graphene layers, in contrast to the HOPG-2 sample [29], see Fig. 1(b). These results clearly indicate that the origin of the hysteresis is related to the existence of the interfaces in the HOPG-1 sample. The absence of the hysteresis in the HOPG-2 sample, which has a similar diamagnetic background and overall geometry as the HOPG-1 sample, also indicates that the hysteresis is not due to an obvious SQUID artifact or an artifact in the background subtraction. The field hysteresis is similar to that of WTGP. The narrowing of the hysteresis observed at high fields is expected for granular superconductors [58, 60–62]. From the hysteresis, as well as measuring the remanent magnetic moment as a function of the applied field [29], one obtains the characteristic Josephson critical fields $h^J_{c1}(T)$ and $h^J_{c2}(T)$ with values similar to the WTGP [58] and a similar ratio $h^J_{c2}(T)/h^J_{c1}(T) \sim 3$ [29].

Figure 5(b) shows the magnetic moment hysteresis in temperature (FC minus the ZFC curve) for the HOPG-1 sample as received (b.a.) and after sweeping the temperature up to 500 K (a.a.) [29], at two applied fields. We would like to stress the following features: The hysteresis for the as-received sample starts from the turning point (390 K) and it is positive. The hysteresis in temperature at both applied fields are qualitatively similar, showing a crossing to negative values at low temperatures. Larger ZFC values (smaller in absolute value) than FC ones in the magnetic moment are usually not observed, neither in superconductors nor in ferromagnets and it does appear to be a SQUID artifact [29]. This negative hysteresis in temperature would suggest that the superconducting properties can be enhanced to some extent under a magnetic field, an effect that might be related to the reentrance we have shortly mentioned in section IV. A slight annealing of the HOPG-1 sample of less than one hour at $\sim$ 500 K changes drastically the observed hysteresis for both fields (open symbols in Fig. 5(b)). The hysteresis appears to be shifted to lower temperatures but with negative values at high temperatures and high fields. We note that annealing at similar temperatures for several hours produced a decrease in the overall hysteresis observed in WTGP (see supporting information of Ref. [58]). At the state of this research, it is unclear whether pinning properties of vortices and/or of fluxons or

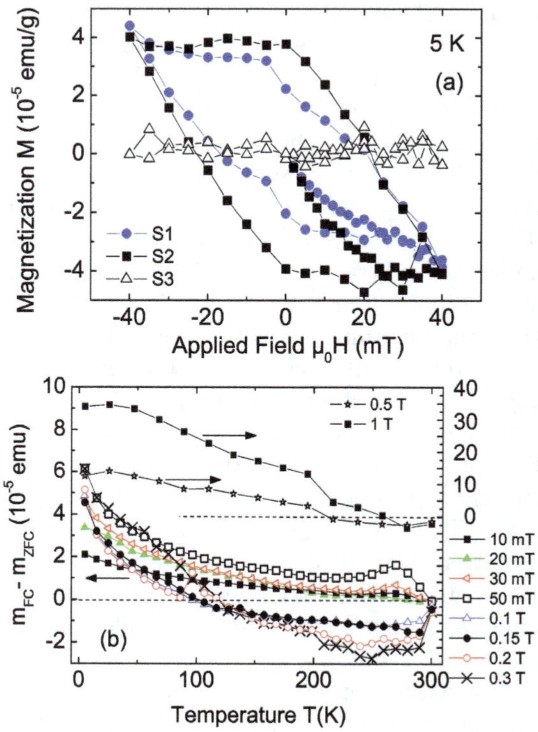

Figure 6: (a) Field hysteresis at 5 K for a maximum applied field of 40 mT for the water treated graphite powder (S1), the same powder but after pressing it in a pellet with a pressure of 18 ± 5 MPa (S2) and after pressing it again with a pressure of 60 ± 20 MPa (S3). The corresponding diamagnetic linear backgrounds were subtracted from the measured data. (b) Difference between the FC and ZFC curve at different applied fields for a water treated graphite powder. Data taken from Ref. [58].

the existence of different superconducting phases play a main role in the hysteresis that is observed for fields applied normal to the interfaces.

ii. Water treated graphite powder

The work of Ref. [58] reports on the magnetic response of WTGPs. The main message of that work is that the WTGP shows a hysteretic behavior in field and temperature compatible with granular superconductivity. As an example, we show in Fig. 6(a) the field hysteresis at 5 K of a WTGP (S1, lose powder without applying significant pressure) and the same WTGP but after pressing it into a pellet with two different pressures (S2,S3). After the diamagnetic background subtraction, the field hysteresis is similar to that obtained for bulk HOPG sample with interfaces, see Fig. 5(a) for similar data but at 300 K. The fact that the hysteresis vanishes after applying pressure to the powder rules out simple SQUID artifacts (the diamagnet background does not diminish after making a pellet from the graphite powder, but the contrary) and also it rules out that the hysteresis is due to a ferromagnetic response due to impurities.

Figure 6(b) shows the difference in the magnetic moment between the ZFC and FC curves, as in Fig. 5(b). The behavior of this difference as a function of the applied field appears to be compatible with the one expected for granular superconductors [58]. Note the following features: The hysteresis increases at all T for fields $\mu_0 H \lesssim 50$ mT, showing a maximum near the turning point of 300 K, similar to the HOPG-1 sample in the as-received state, see Fig. 5(b). At fields 0.1 T $\lesssim \mu_0 H \lesssim$ 0.2 T the difference decreases at all T and remains rather field independent. At higher fields, however, it increases showing a shift of the crossing point (from negative to positive values) to higher T . This behavior is at odds to the one expected for ferromagnets, even for ferromagnetic nanoparticles [63] as well as for superconductors with a pinning force that decreases with applied field in the shown field range. From the results in [58], and using basic concepts of vortex pinning, we would then conclude that if an upper critical field exists, then it should be clearly larger than 7 T in the temperature range of the figure.

In spite of some interesting differences between the behavior obtained for bulk HOPG and WTGP, the similarities already suggest that the water treatment helps to produce a certain amount of interfaces between graphite grains, being the origin for the whole hysteresis. Thermal annealing as well as pressing the WTGP are detrimental indicating that defects and/or hydrogen or oxygen at the interfaces could play an important role in the observed phenomena.

VI. Discussion

Superconductivity in carbon-based systems is a rather old, well recognized fact. This phenomenon was probably first observed in the potassium intercalated graphite C_8K [64] back in 1965. Since then, a considerable amount of studies reported this phenomenon in carbon-based systems, reaching critical temperatures $T_c \sim 10$ K in intercalated graphite [65, 66] and above 30 K - though not percolative - in some HOPG samples [59] as well as in doped graphite and amorphous carbon systems [67–70]. Traces of superconductivity at $T_c = 65$ K have been recently reported in amorphous carbon powder that contained a small amount of sulfur [71]. Superconductivity was found also in carbon nanotubes with $T_c = 0.55$ K [72] and 12 K [73] or possibly even higher critical temperatures [74, 75]. Superconductivity with $T_c \sim 4$ K in boron-doped diamond [76] and in diamond films with $T_c \sim 7$ K [77] belong also to the recently published list of carbon-based superconductors. We should note, however, that superconductivity at room temperature in a disordered graphite powder has been already reported in 1974 [78], see also [79], a work that did not attract the necessary attention in the community.

Whether quasi two dimensional interfaces play a role in the above mentioned carbon-based superconductors, one can probably rule out only for the intercalated graphite and doped diamond compounds, where the three dimensional superconductivity is characterized by a relatively low critical temperature. We may speculate that the traces of superconductivity found in doped amorphous carbon, disordered or ordered graphite powders may be related to some interfaces between well ordered graphite regions. The experience of the high temperature superconducting oxides already suggests that two dimensionality is advantageous to achieve higher critical temperatures.

Apart from the usual transport and magnetization measurements used to characterize the superconducting state, there are scanning tunneling spectroscopy (STS) results obtained on certain disordered regions of a HOPG surface at $T = 4.2$ K that revealed an apparent energy gap ~ 100 meV [80]. Although the overall curves resemble a superconducting-like density of states, the authors suggested that the gap originates from charging effects. See further STS results and the discussion in [70].

Theoretical works that deal with superconductivity in graphite as well as in graphene have been published in recent years. For example, p-type superconductivity has been predicted to occur in inhomogeneous regions of the graphite structure [50] or $d-$wave high-T_c superconductivity [81] based also on resonance valence bonds [82], or at the graphite surface region with rhombohedral stacking due to a topologically protected flat band [83].

For the graphite structure, the experimental evidence obtained in the last years suggests that high temperature superconductivity exists at certain interfaces or interface regions within the usual Bernal structure although the structure of the superconducting regions remains unknown. One can further speculate that due to the high carrier concentration that can be localized at those interfaces, they should be predestined to play a role in triggering superconductivity. Following a BCS approach in two dimensions (with anisotropy), for example, a critical temperature $T_c \sim 60$ K has been estimated if the density of conduction electrons per graphene plane increases to $n \sim 10^{14}$ cm^{-2}, a density that might be induced by defects and/or hydrogen adatoms [84] at the interfaces, or by Li deposition [85]. Further predictions for superconductivity in graphene support the premise that $n > 10^{13}$ cm^{-2} in order to reach $T_c > 1$ K [86, 87]. On the other hand, the possibility to have high temperature superconductivity at the surface of or in the rhombohedral graphite phase [21, 83] – a phase that sometimes is found in graphite samples [19, 20] – stimulates further careful studies of these hidden interfaces. In the last years, superconductivity has been found at the interfaces between oxide insulators [88] as well as between metallic and insulating copper oxides with $T_c \gtrsim 50$ K[89]. Also, interfaces in different Bi bicrystals show superconductivity up to 21 K, although Bi bulk is not a superconductor [90, 91].

Finally, we think that some of the interfaces are also the origin for the metallic-like behavior of graphite samples as well as for the quantum Hall effect (QHE) found in some HOPG samples [48,92]. Because the existence, density as well as the intrinsic properties of these interfaces depend on sample, we can now understand why the reproducibility of the QHE in bulk HOPG samples is rather poor.

VII. Conclusion

In this review, we have discussed the following experimental evidence:
Firstly, the temperature and magnetic field dependence of the electrical resistance of bulk and thin films of graphite samples and its relation with the existence of two dimensional interfaces.
Secondly, the Josephson behavior of the current-voltage curves with an apparent zero resistance state at high temperatures in especially made TEM lamellae.
Thirdly, the anomalous hysteresis in the magnetoresistance observed in graphite thin samples as well as its enhancement restricting the current path within the sample.
Finally, the overall magnetization of bulk graphite samples, with and without interfaces, as well as water treated graphite powders.

All this experimental evidence as a whole indicates the existence of superconductivity located at certain interfaces inside graphite samples. Although we cannot rule out other interpretations for some of the observations discussed in this work, the whole evidence suggests that superconductivity should be the origin for all the phenomena discussed here. Clearly, the situation is still highly unsatisfactory because several open questions remain, namely, the characteristics of the superconducting phase(s), from the structure to the main superconducting parameters, as "simple" as the critical temperature and critical fields, the coherence and penetration lengths, etc. It is clear that further studies are necessary in the future but the overall work done until now shows us the way to go.

Acknowledgements - The author acknowledges the support provided by the Deutsche Forschungsgemeinschaft under contract DFG ES 86/16-1 and the ESF-Nano under the Graduate School of Natural Sciences "BuildMona". The results presented in this review were part of the Ph.D. thesis of Heiko Kempa (section II), Srujana Dusari (section III) and Ana Ballestar (section IV) as well as the master thesis of Thomas Scheike (section V) done in the Division of Superconductivity and Magnetism of the Institute for Experimental Physics II of the University of Leipzig. The author thanks Dipl.

Kris. Annette Setzer, Dr. José Barzola-Quiquia and Dr. Winfried Böhlmann for their experimental assistance and support. The permanent support as well as the discussions with Nicolás García are gratefully acknowledged. Special thanks go to Yakov Kopelevich with whom we started in the year 1999 and in a rather naive way the research of a new and unexpected world behind graphite.

Note added in proof

Since the submission of this manuscript, some new works related to the subject of this review were published. Tight-binding simulations done in Ref. [93] support the work done in Ref. [21] and found that surface superconductivity is robust for ABC stacked multilayer graphene, even at very low pairing potentials. Through the observation of persistent currents in a graphite filled ring-shaped container immersed in alkanes, the author in Ref. [94] claimed possible room temperature superconductivity. For completeness, we include them in the reference list.

---

[1] J W McClure, *Energy band structure of graphite*, IBM J. Res. Dev. **8**, 255 (1964).

[2] B T Kelly, *Physics of graphite*, Applied Science Publishers, London (1981).

[3] A Grüneis, C Attaccalite, T Pichler, V Zabolotnyy, H Shiozawa, S L Molodtsov, D Inosov, A Koitzsch, M Knupfer, J Schiessling, R Follath, R Weber, P Rudolf, R Wirtz, A Rubio, *Electron-electron correlation in graphite: A combined angle-resolved photoemission and first-principles study*, Phys. Rev. Lett. **100**, 037601 (2008).

[4] I M Tsidilkovski, *Electron spectrum of gapless semiconductors*. Springer Series in Solid-State Sciences Vol. 116, Springer Verlag (1997).

[5] R O Dillon, I L Spain, J W McClure, *Electronic energy band parameters of graphite and their dependence on pressure, temperature and acceptor concentration*, J. Phys. Chem. Solids **38**, 635 (1977).

[6] J M Schneider, M Orlita, M Potemski, D K Maude. *Consistent interpretation of the low-temperature magnetotransport in graphite using the Slonczewski-Weiss-McClure 3D band-structure calculations*, Phys. Rev. Lett. **102**, 166403 (2009).

[7] J Barzola-Quiquia, J L Yao, P Rödiger, K Schindler, P Esquinazi, *Sample size effects on the transport properties of mesoscopic graphite samples*, Physica Status Solidi A **205**, 2924 (2008).

[8] M Inagaki, *New carbons: Control of structure and functions*, Elsevier (2000).

[9] A Arndt, D Spoddig, P Esquinazi, J Barzola-Quiquia, S Dusari, T Butz, *Electric carrier concentration in graphite: Dependence of electrical resistivity and magnetoresistance on defect concentration*, Phys. Rev. B **80**, 195402 (2009).

[10] N García, P Esquinazi, J Barzola-Quiquia, S Dusari, *Evidence for semiconducting behavior with a narrow band gap of Bernal graphite*, New J. Phys. **14**, 053015 (2012).

[11] Y Ohashi, K Yamamoto, T Kubo, *Shubnikov - de Haas effect of very thin graphite crystals*, In: Carbon'01, An International Conference on Carbon, Pag. 568, The American Carbon Society, Lexington, KY, United States, (2001).

[12] B C Camargo, Y Kopelevich, S B Hubbard, A Usher, W Böhlmann, P Esquinazi, *Effect of structural disorder on the quantum oscillations in graphite* (unpublished). In this work the authors show that in certain HOPG samples (SPI) of high grade, the density of interfaces is much lower than in, for example, Advanced Ceramics HOPG ZYA samples. In this new HOPG samples basically no SdH oscillations are found and the temperature dependence of the resistance shows a semiconducting behavior with saturation a low temperatures (2013).

[13] M Orlita, C Faugeras, G Martinez, D K Maude, M L Sadowski, J M Schneider, M Potemski, *Magneto-transmission as a probe of Dirac fermions in bulk graphite*, J. Phys: Cond. Mat. **20**, 454223 (2008).

[14] N A Goncharuk, L Nádvorník, C Faugeras, M Orlita, L Smrčka, *Infrared magnetospectroscopy of graphite in tilted fields*, Phys. Rev. B **86**, 155409 (2012).

[15] J C González, M Muñoz, N García, J Barzola-Quiquia, D Spoddig, K Schindler, P Esquinazi, *Sample-size effects in the magnetoresistance of graphite*, Phys. Rev. Lett. **99**, 216601 (2007).

[16] N García, P Esquinazi, J Barzola-Quiquia, B Ming, D Spoddig, *Transition from ohmic to ballistic transport in oriented graphite: Measurements and numerical simulations*, Phys. Rev. B **78**, 035413 (2008).

[17] S Dusari, J Barzola-Quiquia, P Esquinazi, N García, *Ballistic transport at room temperature in micrometer-size graphite flakes*, Phys. Rev. B **83**, 125402 (2011).

[18] P Esquinazi, J Barzola-Quiquia, S Dusari, N García, *Length dependence of the resistance in graphite: Influence of ballistic transport*, J. Appl. Phys. **111**, 033709 (2012).

[19] Q Lin, T Li, Z Liu, Y Song, L He, Z Hu, Q Guo, H Ye, *High-resolution TEM observations of isolated rhombohedral crystallites in graphite blocks*, Carbon **50**, 2369 (2012).

[20] C H Lui, Z Li, Z Chen, P V Klimov, L E Brus, T F Heinz, *Imaging stacking order in few-layer graphene*, Nano Lett. **11**, 164 (2011).

[21] N B Kopnin, M Ijäs, A Harju, T T Heikkilä, *High-temperature surface superconductivity in rhombohedral graphite*, Phys. Rev. B **87**, 140503 (2013).

[22] P Esquinazi, J Barzola-Quiquia, D Spemann, M Rothermel, H Ohldag, N García, A Setzer, T Butz, *Magnetic order in graphite: Experimental evidence, intrinsic and extrinsic difficulties*, J. Magn. Magn. Mat. **322**, 1156 (2010).

[23] H Kempa, Y Kopelevich, F Mrowka, A Setzer, J H S Torres, R Höhne, P Esquinazi, *Magnetic field driven superconductor-insulator-type transition in graphite*, Solid State Commun. **115**, 539 (2000).

[24] Y Kopelevich, P Esquinazi, J H S Torres, R R da Silva, H Kempa, *Graphite as a highly correlated electron liquid*, In: Advances in Solid State Physics, Vol. 43, Ed. B Kramer, Pag. 207, Springer-Verlag, Berlin (2003).

[25] T Tokumoto, E Jobiliong, E Choi, Y Oshima, J Brooks, *Electric and thermoelectric transport probes of metal-insulator and two-band magnetotransport behavior in graphite*, Solid State Commun. **129**, 599 (2004).

[26] X Du, S W Tsai, D L Maslov, A F Hebard, *Metal-insulator-like behavior in semimetallic bismuth and graphite*, Phys. Rev. Lett. **94**, 166601 (2005).

[27] Y Zhang, J P Small, W V Pontius, P Kim, *Fabrication and electric-field-dependent transport measurements of mesoscopic graphite devices*, Appl. Phys. Lett. **86**, 073104 (2005).

[28] See several reviews in: Graphite and precursors, World of carbon series, Vol. 1, Ed. P Delhaes, Gordon and Breach Science Publishers (2001).

[29] T Scheike, P Esquinazi, A Setzer, W Böhlmann, *Granular superconductivity at room temperature in bulk highly oriented pyrolytic graphite samples*, Carbon **59**, 140 (2013).

[30] Y Kopelevich, V Lemanov, S Moehlecke, J Torres, *Landau level quantization and possible superconducting instabilities in highly oriented pyrolitic graphite*, Phys. Solid State **41**, 1959 (1999).

[31] H Kempa, H C Semmelhack, P Esquinazi, Y Kopelevich, *Absence of metal-insulator transition and coherent interlayer transport in oriented graphite in parallel magnetic fields*, Solid State Commun. **125**, 1 (2003).

[32] Y Kopelevich, P Esquinazi, J Torres, R da Silva, H Kempa, F Mrowka, R Ocaña, *Metal-insulator-metal transitions, superconductivity and magnetism in graphite*, In: Studies of High Temperature Superconductors, Vol. 45, Chap. 3, Pag. 59, Nova Science Publishers Inc. (2003).

[33] M P A Fisher, *Quantum phase transitions in disordered two-dimensional superconductors*, Phys. Rev. Lett. **65**, 923 (2000).

[34] E Abrahams, S V Kravchenko, M P Sarachik, *Metallic behavior and related phenomena in two dimensions*, Rev. Mod. Phys. **73**, 251 (2001).

[35] Y Shapira, G Deutscher, *Semiconductor-superconductor transition in granular* Al-Ge, Phys. Rev. B **27**, 4463 (1983).

[36] J S Langer, V Ambegaokar, *Intrinsic resistive transition in narrow superconducting channels*, Phys. Rev. **164**, 498 (1967).

[37] D E McCumber, B I Halperin, *Time scale of intrinsic resistive fluctuations in thin superconducting wires*, Phys. Rev. B **1**, 1054 (1970).

[38] P Esquinazi, N García, J Barzola-Quiquia, P Rödiger, K Schindler, J L Yao, M Ziese, *Indications for intrinsic superconductivity in highly oriented pyrolytic graphite*, Phys. Rev. B **78**, 134516 (2008).

[39] L Ji, M S Rzchowski, N Anand, M Thinkam, *Magnetic-field-dependent surface resistance and two-level critical-state model for granular superconductors*, Phys. Rev. B **47**, 470 (1993).

[40] Y Kopelevich, C dos Santos, S Moehlecke, A Machado, *Current-induced superconductor-insulator transition in granular high-T_c superconductors*, arXiv:0108311 (2001).

[41] I Felner, E Galstyan, B Lorenz, D Cao, Y S Wang, Y Y Xue, C W Chu, *Magnetoresistance hysteresis and critical current density in granular* $RuSr_2Gd_{2-x}Ce_xCu_2O_{10-\delta}$, Phys. Rev. B **67**, 134506 (2003).

[42] S Dusari, J Barzola-Quiquia, P Esquinazi, *Superconducting behavior of interfaces in graphite: Transport measurements of micro-constrictions*, J. Supercond. Nov. Magn. **24**, 401 (2011).

[43] J Barzola-Quiquia, S Dusari, G Bridoux, F Bern, A Molle, P Esquinazi, *The influence of Ga^+ irradiation on the transport properties of mesoscopic conducting thin films*, Nanotech. **21**, 145306 (2010).

[44] V Ambegaokar, B I Halperin, *Voltage due to thermal noise in the DC Josephson effect*, Phys. Rev. Lett. **22**, 1364 (1969).

[45] A Ballestar, J Barzola-Quiquia, T Scheike, P Esquinazi, *Evidence of Josephson-coupled superconducting regions at the interfaces of highly oriented pyrolytic graphite*, New J. Phys. **15**, 023024 (2013).

[46] J Barzola-Quiquia, P Esquinazi, *Ferromagnetic- and superconducting-like behavior of the electrical resistance of an inhomogeneous graphite flake*, J. Supercond. Nov. Magn. **23**, 451 (2010).

[47] A Ballestar, P Esquinazi, *Highly oriented pyrolytic graphite TEM lamellae preparation to study transport properties of the internal interfaces*, J. Visual. Exp. (in press).

[48] Y Kopelevich, J H S Torres, R R da Silva, F Mrowka, H Kempa, P Esquinazi, *Reentrant metallic behavior of graphite in the quantum limit*, Phys. Rev. Lett. **90**, 156402 (2003).

[49] H J Gardner, A Kumar, L Yu, P Xiong, M P Warusawithana, L Wang, O Vafek, D G Schlom, *Enhancement of superconductivity by a parallel magnetic field in two-dimensional superconductors*, Nat. Phys. **7**, 895 (2011).

[50] J González, F Guinea, M A H Vozmediano, *Electron-electron interactions in graphene sheets*, Phys. Rev. B **63**, 134421 (2001).

[51] K Scharnberg, R A Klemm, *p-wave superconductors in magnetic fields*, Phys. Rev. B **22**, 5233 (1980).

[52] A Knigavko, B Rosenstein, *Spontaneous vortex state and ferromagnetic behavior of type-II p-wave superconductors*, Phys. Rev. B **58**, 9354 (1998).

[53] N Casan-Pastor, P Gomez-Romero, L C Baker, *Magnetic measurements with a squid magnetometer: Possible artifacts induced by sample holder off centering*, J. Appl. Phys. **69**, 5088 (1991).

[54] A Ney, T Kammermeier, V Ney, K Ollefs, S Ye, *Limitations of measuring small magnetic*

signals of samples deposited on a diamagnetic substrate, J. Magn. Magn. Mater. **320**, 3341 (2008).

[55] M Sawicki, W Stefanowicz, A Ney, *Sensitive SQUID magnetometry for studying nanomagnetism*, Semicond. Sci. Tech. **26**, 064006 (2011).

[56] J Barzola-Quiquia, W Böhlmann, P Esquinazi, A Schadewitz, A Ballestar, S Dusari, L Schultze-Nobre, B Kersting, *Enhancement of the ferromagnetic order of graphite after sulphuric acid treatment*, Appl. Phys. Lett. **98**, 192511 (2011).

[57] M W McElfresh, Y Yeshurun, A P Malozemoff, F Holtzberg, *Remanent magnetization, lower critical fields and surface barriers in an $YBa_2Cu_3O_7$ crystal*, Physica A **168**, 308 (1990).

[58] T Scheike, W Böhlmann, P Esquinazi, J Barzola-Quiquia, A Ballestar, A Setzer, *Can doping graphite trigger room temperature superconductivity? Evidence for granular high-temperature superconductivity in water-treated graphite powder*, Adv. Mater. **24**, 5826 (2012).

[59] Y Kopelevich, P Esquinazi, J Torres, S Moehlecke, *Ferromagnetic- and superconducting-like behavior of graphite*, J. Low Temp. Phys. **119**, 691 (2000).

[60] S Senoussi, C Aguillon, S Hadjoudj, *The contribution of the intergrain currents to the low field hysteresis cycle of granular superconductors and the connection with the micro- and macrostructures*, Physica C **175**, 215 (1991).

[61] M Borik, M Chernikov, V Veselago, V Stepankin, *Anomalies of the magnetic properties of granular oxide superconductor $BaPb_{l-x}Bi_xO_3$*, J. Low Temp. Phys. **85**, 283 (1991).

[62] B Andrzejewski, E Guilmeau, C Simon, *Modelling of the magnetic behaviour of random granular superconductors by the single junction model*, Supercond. Sci. Tech. **14**, 904 (2001).

[63] R Prozorov, Y Yeshurun, T Prozorov, A Gedanken, *Magnetic irreversibility and relaxation in assembly of ferromagnetic nanoparticles*, Phys. Rev. B **59**, 6956 (1999).

[64] N B Hannay, T H Geballe, B T Matthias, K Andres, P Schmidt, D MacNair, *Superconductivity in graphitic compounds*, Phys. Rev. Lett. **14**, 225 (1965).

[65] T E Weller, M Ellerby, S S Siddharth, R P Smith, T Skippe, *Superconductivity in the intercaled graphite compounds C_6Yb and C_6Ca*, Nat. Phys. **1**, 39 (2005).

[66] N Emery, C Hérold, M D'Astuto, V Garcia, C Bellin, J F Marêché, P Lagrange, G Loupias, *Superconductivity of bulk CaC_6*, Phys. Rev. Lett. **95**, 035413 (2005).

[67] R R da Silva, J H S Torres, Y Kopelevich, *Indication of superconductivity at 35 K in graphite-sulfur composites*, Phys. Rev. Lett. **87**, 147001 (2001).

[68] Y Kopelevich, R R da Silva, J H S Torres, S Moehlecke, M B Maple, *High-temperature local superconductivity in graphite-sulfur composites*, Physica C **408**, 77 (2004).

[69] I Felner, Y Kopelevich, *Magnetization measurement of a possible high-temperature superconducting state in amorphous carbon doped with sulfur*, Phys. Rev. B **79**, 233409 (2009).

[70] Y Kopelevich, P Esquinazi, *Ferromagnetism and superconductivity in carbon-based systems*, J. Low Temp. Phys. **146**, 629 (2007).

[71] I Felner, O Wolf, O Millo, *High-temperature superconductivity in sulfur-doped amorphous carbon systems*, J. Supercond. Nov. Magn. **25**, 7 (2012).

[72] M Kociak, A Y Kasumov, S Guéron, B Reulet, I I Khodos, Y B Gorbatov, V T Volkov, L Vaccarini, H Bouchiat, *Superconductivity in ropes of single-walled carbon nanotubes*, Phys. Rev. Lett. **86**, 2416 (2001).

[73] I Takesue, J Haruyama, N Kobayashi, S Chiashi, S Maruyama, T Sugai, H Shinohara, *Superconductivity in entirely end-bonded multiwalled carbon nanotubes*, Phys. Rev. Lett. **96**, 057001 (2006).

[74] Z K Tang, L Zhang, N Wang, X X Zhang, G H Wen, G D Li, J N Wang, C T Chan, P Sheng, *Superconductivity in 4 angstrom single-walled carbon nanotubes*, Science **292**, 2462 (2001).

[75] G M Zhao, Y S Wang, *Possible superconductivity above 400 k in carbon-based multiwall nanotubes*, arXiv:0111268 (2001).

[76] E A Ekimov, V A Sidorov, E D Bauer, N N Mel'nik, N J Curro, J D Thompson, S M Stishov, *Superconductivity in diamond*, Nature **428**, 542 (2004).

[77] Y Takano, M Nagao, I Sakaguchi, M Tachiki, T Hatano, K Kobayashi, H Umezawa, H Kawarada, *Superconductivity in diamond thin films well above liquid helium temperature*, Appl. Phys. Lett. **85**, 2851 (2004).

[78] K Antonowicz, *Possible superconductivity at room temperature*, Nature **247**, 358 (1974).

[79] K Antonowicz, *The effect of microwaves on dc current in an Al-Carbon-Al sandwich*, Physica Status Solidi A **28**, 497 (1975).

[80] N Agrait, J Rodrigo, S Vieira, *On the transition from tunneling regime to point-contact: graphite*, Ultramicroscopy **42–44, Part 1**, 177 (1992).

[81] R Nandkishore, L S Levitov, A V Chubukov, *Chiral superconductivity from repulsive interactions in doped graphene.*, Nat. Phys. **8**, 158 (2012).

[82] A M Black-Schaffer, S Doniach, *Resonating valence bonds and mean-field d-wave superconductivity in graphite*, Phys. Rev. B **75** 134512 (2007).

[83] N B Kopnin, T T Heikkilä, G E Volovik, *High-temperature surface superconductivity in topological flat-band systems*, Phys. Rev. B **83**, 220503 (2011).

[84] N García, P Esquinazi, *Mean field superconductivity approach in two dimensions*, J. Supercond. Nov. Magn. **22**, 439 (2009).

[85] G Profeta, M Calandra, F Mauri, *Phonon-mediated superconductivity in graphene by lithium deposition*, Nat. Phys. **8**, 131 (2012).

[86] B Uchoa, A H C Neto, *Superconducting states of pure and doped graphene*, Phys. Rev. Lett. **98**, 146801 (2007).

[87] N B Kopnin, E B Sonin, *BCS superconductivity of Dirac electrons in graphene layers*, Phys. Rev. Lett. **100**, 246808 (2008).

[88] N Reyren, S Thiel, A D Caviglia, L F Kourkoutis, G Hammerl, C Richter, C W Schneider, T Kopp, A S Rüetschia, D Jaccard, M Gabay, D A Muller, J M Triscone, J Mannhart, *Superconducting interfaces between insulating oxides*, Science **317**, 1196 (2007).

[89] A Gozar, G Logvenov, L F Kourkoutis, A T Bollinger, L A Giannuzzi, L A Muller, I Bozovic, *High-temperature interface superconductivity between metallic and insulating copper oxides*, Nature **455**, 782 (2008).

[90] F Muntyanua, A Gilewski, K Nenkov, J Warchulska, A Zaleski, *Experimental magnetization evidence for two superconducting phases in Bi bicrystals with large crystallite disorientation angle*, Phys. Rev. B **73**, 132507 (2006).

[91] F Muntyanua, A Gilewski, K Nenkov, A Zaleski, V Chistol, *Superconducting crystallite interfaces with T_c up to 21 K in Bi and Bi-Sb bicrystals of inclination type*, Solid State Commun. **147**, 183 (2008).

[92] Y Kopelevich, P Esquinazi, *Graphene physics in graphite*, Adv. Mater. (Weinheim, Ger.) **19**, 4559 (2007).

[93] W A Muñoz, L Covaci, F Peeters, *Tight-binding description of intrinsic superconducting correlations in multilayer graphene*, Phys. Rev. B **87**, 134509 (2013).

[94] Y Kawashima, *Possible room temperature superconductivity in conductors obtained by bringing alkanes into contact with a graphite surface*, AIP Advances **3**, 052132 (2013).

21

High-speed tunable photonic crystal fiber-based femtosecond soliton source without dispersion pre-compensation

Martín Caldarola,[1*] Víctor A. Bettachini,[2] Andrés A. Rieznik,[2] Pablo G. König,[2] Martín E. Masip,[1] Diego F. Grosz,[2,3] Andrea V. Bragas[1,4]

We present a high-speed wavelength tunable photonic crystal fiber-based source capable of generating tunable femtosecond solitons in the infrared region. Through measurements and numerical simulation, we show that both the pulsewidth and the spectral width of the output pulses remain nearly constant over the entire tuning range from 860 to 1160 nm. This remarkable behavior is observed even when pump pulses are heavily chirped (7400 fs^2), which allows to avoid bulky compensation optics, or the use of another fiber, for dispersion compensation usually required by the tuning device.

I. Introduction

Light sources based on the propagation of solitons in optical fibers have emerged as a compact solution to the need of a benchtop source of ultra-short tunable pulses [1–3]. The soliton formation from femtosecond pulses launched into an optical fiber is explained in terms of the interplay between self-phase modulation (SPM) and group-velocity dispersion (GVD) in the anomalous dispersion regime [4]. The wavelength tunability is a consequence of the Raman-induced frequency shift (RIFS) produced on the pulse when traveling through the fiber [5]. The term soliton self-frequency shift (SSFS) [6] was coined to name this effect widely used to produce tunable femtosecond pulses in different wavelength ranges, e.g., from 850 to 1050 nm [7], from 1050 to 1690 nm [8], and from 1566 to 1775 nm [1]. In most cases, photonic crystal fibers (PCF) are used for building these sources since their GVD can be easily tailored to produce solitons in a desired tuning range [9, 10]. For a given choice of the PCF, full experimental characterization of the pump and output pulses, complemented with theoretical predictions, is necessary to understand how nonlinear effects modify the output soliton.

The wavelength tunability in a PCF-based light source is provided by the modulation of the pump power injected into the fiber [11–14]. It is worth noting that the wavelength choice of the output pulse is done without moving any mechanical part, which is clearly attractive for all the proposed and imaginable applications of these soliton sources. Moreover, the wavelength of the output pulse can be chosen as fast as one can modulate the power of the pump pulse, as introduced in Ref. [15, 16]. By introducing an acousto-optic modulator (AOM) in the path of the pump pulse, the output wave-

*E-mail: caldarola@df.uba.ar

[1] Laboratorio de Electrónica Cuántica, Departamento de Física, Universidad de Buenos Aires, Pabellón I, Ciudad Universitaria, (C1428EHA) Buenos Aires, Argentina.

[2] Instituto Tecnológico de Buenos Aires, Eduardo Madero 399, (C1106ACD) Buenos Aires, Argentina.

[3] Consejo Nacional de Investigaciones Científicas y Técnicas, Argentina.

[4] IFIBA, Consejo Nacional de Investigaciones Científicas y Técnicas, Argentina.

length can be changed at a speed which is ultimately limited only by the laser repetition rate. This kind of experimental setup has been presented in some previous reports [14,17], with stunning applications as the one presented in Ref. [18], where a pseudo-CW wideband source for optical coherent tomography is introduced. However, the need to pre-compress the pump pulse to avoid the chirp produced by the AOM contrives against the compact and mechanically robust design of the light source. In this paper, we demonstrate that the PCF-based source presented here is robust against chirped pump pulses. We present a complete set of measurements showing that the temporal and spectral characteristics of the generated solitons in the PCF remain unaltered even when pump pulses are heavily chirped up to ~ 7400 fs^2. Results are presented for the whole range of tunability (860 nm to 1160 nm). We also present numerical simulations which remarkably fit the experimental data and help to understand the soliton behavior.

This paper is organized as follows: In section II, we describe the experimental setup. The numerical simulations are described in section III. In section IV, we present experimental and numerical results and in section V we further analyze the results with numerical simulations. Finally, in section VI, we present our conclusions.

II. Experimental Setup

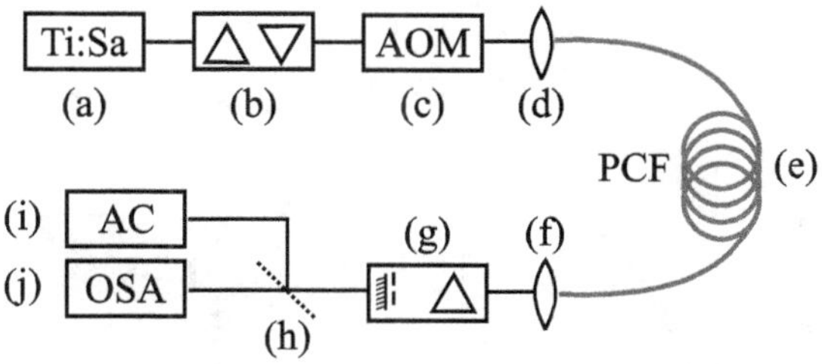

Figure 1: Experimental setup. (a) Titanium-Sapphire (Ti:Sa) laser, (b) Prism compressor, (c) Acousto-optic modulator (AOM), (d) Coupling lens, (e) Photonic crystal fiber (PCF), (f) Collimator objective, (g) Spatial filter, (h) Flipper mirror, (i) Fast-scan interferometric autocorrelator, (j) Optical spectrum analyzer (OSA).

A scheme of the experimental setup is shown in Fig. 1. A Ti:Sa laser (KMLabs) generates ultrashort transform-limited (TL) pulses of $\Delta t = 31$ fs (FWHM-sech^2), $\lambda_{\text{pump}} = 830$ nm, with a spectral width $\Delta\lambda = 23$ nm, and a repetition rate of 94 MHz.

The AOM not only allows high speed (up to MHz) and accurate control of the soliton wavelength, as previously discussed, but also prevents feedback into the Ti:Sa, replacing the optical isolator required in similar setups [19]. As the AOM introduces ~ 56 mm of SF8 glass path, pump pulses gain a positive chirp of about ~ 7400 fs^2, which leads to a time spread by a factor of ~ 3 in them. This can be pre-compensated, for example, by introducing an optical fiber in the anomalous dispersion regime [8, 20] or a prism compressor in the well-known configuration presented in [21]. In this work, the chirp was compensated by a pair of SF18 prisms with an apex separation of 78 cm. Additionally, the prism compressor allowed us to up-chirp pump pulses in a controlled fashion from TL to ~ 1400 fs^2 by introducing an extra glass path at the second prism of the arrangement [22]. This full or partial compensation of the phase distortion introduced by the AOM allowed us to study the role of different chirp figures in the temporal and spectral characteristics of the solitons generated in the PCF.

Pump pulses were coupled into a non-polarization-maintaining microstructured fiber commercially used for supercontinuum generation (Thorlabs, NL-2.3-790-02). Its main parameters are listed in Table 1 and the dispersion curve and SEM image are shown in Fig. 2.[1] Upon propagation down the fiber, the spectrum is highly broadened so a spatial band-pass filter made of a prism and razor blades, similar to the one presented in [23], allowed to filter the spectral region of the solitonic branch (see Fig. 3) without adding any extra chirp to the solitons.

Once the spectral selection was achieved, a flipper mirror directed the filtered beam for analysis either by the optical spectrum analyzer (OSA) or by the interferometric autocorrelator. A fast-scan system [24] allows to perform fast interferometric auto-correlations. Briefly, a platform with a hollow retroreflector is moved sinusoidally back and forth, with a stepper motor at 11 Hz, to produce and op-

[1] Datasheet available in http://www.thorlabs.com.

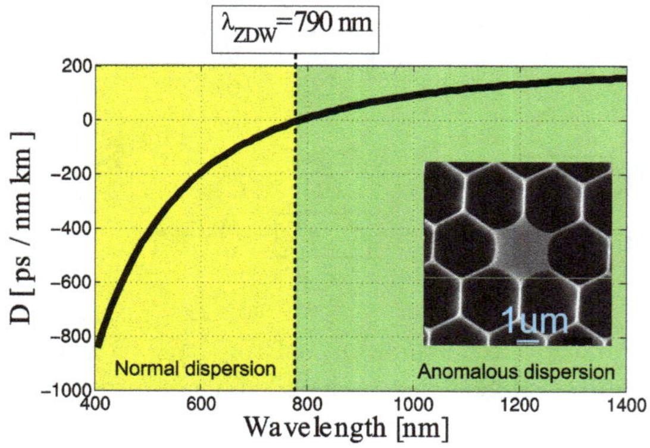

Figure 2: Dispersion curve of the PCF, showing the zero dispersion wavelength (ZDW) at 790 nm. The inset is the scanning electron microscope image of the PCF core. The curve and image were provided by the manufacturer.

L	75 cm
ZDW	790 nm
β_2	-12.4 ps^{2}km^{-1}
β_3	0.07 ps^{3}km^{-1}
$\gamma(\omega)$	$\gamma_0(\omega_0) + (\omega - \omega_0)\gamma_1$
$\gamma_0(\omega_0)$	78 W^{-1}km^{-1}
γ_1	γ_0/ω_0
ω_0	2271 THz

Table 1: PCF parameters relevant to the simulation. Further details can be found in Ref. [14].

tical delay in one of the arms of a Michelson interferometer. The autocorrelation signal is recorded by a PMT and averaged with an oscilloscope.

III. Numerical Simulations

In order to further validate experimental results, we simulated the propagation of femtosecond pulses in the PCF by numerically solving the generalized nonlinear Schrödinger equation (GNLSE) including dispersive, Kerr, instantaneous and delayed Raman response, and self-steepening effects [25], with a conservation quantity error (CQE) adaptive step-size algorithm [26].

The GNLSE reads

$$\begin{aligned} &\frac{\partial A}{\partial z} + \beta_1 \frac{\partial A}{\partial t} + \mathrm{i}\,\beta_2 \frac{\partial^2 A}{\partial t^2} \\ &- \beta_3 \frac{\partial^3 A}{\partial t^3} + \ldots = \mathrm{i}\,\gamma(\omega) \left(1 + \frac{\mathrm{i}}{\omega_0}\frac{\partial}{\partial t}\right) \\ &\times \left(A(z,t) \int_{-\infty}^{\infty} R(t')|A(z,t-t')|^2 dt'\right), \end{aligned} \tag{1}$$

$$\begin{aligned} R(t) &= (1 - f_R)\,\delta(t) + f_R h_R(t), \\ h_R(t) &= (f_a + f_c)\,h_a(t) + f_b h_b(t), \\ h_a(t) &= \tau_1 \left(\tau_1^{-2} + \tau_2^{-2}\right) \mathrm{e}^{-t/\tau_2} \sin(t/\tau_1), \\ h_b(t) &= \left[(2\tau_b - t)/\tau_b^2\right] \mathrm{e}^{-t/\tau_b}, \end{aligned}$$

where $A(z,t)$ is the complex envelope of the electric field, β_n are the expansion terms for the propagation constant around the carrier frequency ω_0 and γ is the nonlinear coefficient. $f_R(t)$ represents the fractional contribution of the delayed Raman effect h_R. Note that Eq. (1) adopts a more accurate description of this effect than the one usually used [4]. In our simulation, we adopted $\tau_1 = 12.2$ fs, $\tau_2 = 32$ fs, $\tau_b = 96$ fs, $f_a = 0.75$, $f_b = 0.21$, $f_c = 0.04$, and $f_R = 0.24$ [27]. The dependence of the fiber non-linear parameter γ with the frequency was modeled as a linear function (see Table 1).

IV. RESULTS

i. Transform-limited pump pulses

First, we present the full characterization of the soliton source seeded by TL pump pulses, in an extended wavelength range if compared with the results presented in our previous paper [14].

In order to investigate the dependence of the output spectrum with the coupled power, managed by the AOM, we skipped spectral filtering at first. Fig. 3 shows the measured spectrum at the PCF output as a function of the coupled power. The infrared solitonic branch appears at ~ 10 mW and undergoes red-shift with increasing power. The maximum wavelength attained is 1130 nm at 55 mW. Spectra in Fig. 3 also shows show that some of the input energy is converted to visible non-solitonic radiation.

The pulsewidth of the filtered soliton as a function of its wavelength, λ_s, is shown in Fig. 4. The

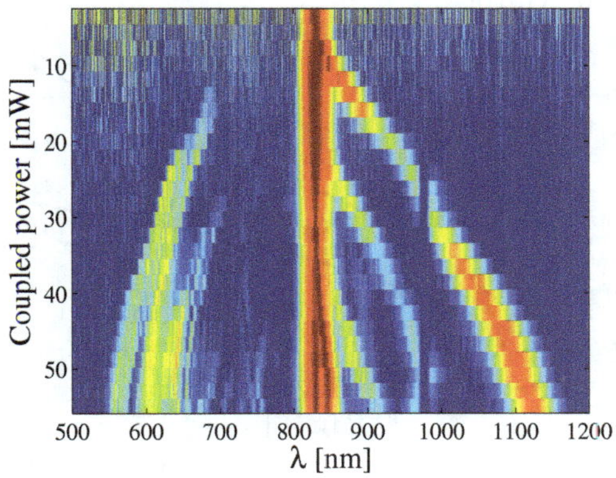

Figure 3: Experimental spectra vs coupled power to the PCF with transform limited (TL) pump pulses. The color map shows spectral intensity. The maximum achieved soliton shift, $\lambda_s \simeq 1130$ nm, was reached at 55 W.

pulsewidth remains constant at $\sim$ 45 fs, for the entire tunability range. Numerical simulations are also plotted in the same figure, showing an excellent agreement with experimental measurements.

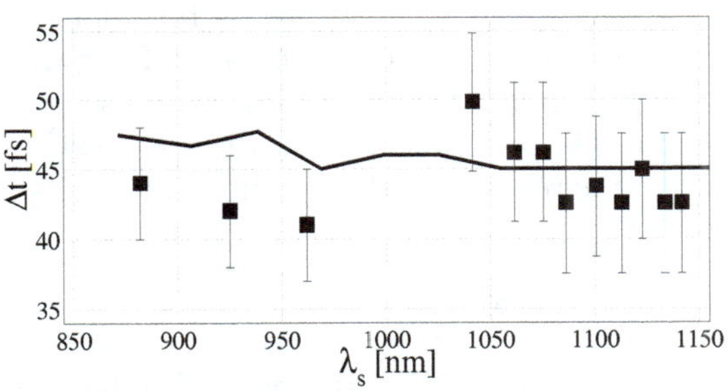

Figure 4: Experimental pulsewidth of the soliton as a function of its wavelength, pumping the PCF with TL pulses. The results for the three lower wavelengths were already present in Ref. [14]. Full line: numerical simulations.

ii. Chirped pump pulses

The effect over the soliton produced by the chirp of pump pulses was studied systematically by introducing a known amount of extra glass path on the second prism of the compressor. This scheme allowed to change the GVD of pump pulses from 0 to 1400 fs^2. Further chirping was achieved by the complete removal of the prism compressor, leading to a total amount of positive chirp $\sim$ 7400 fs^2.

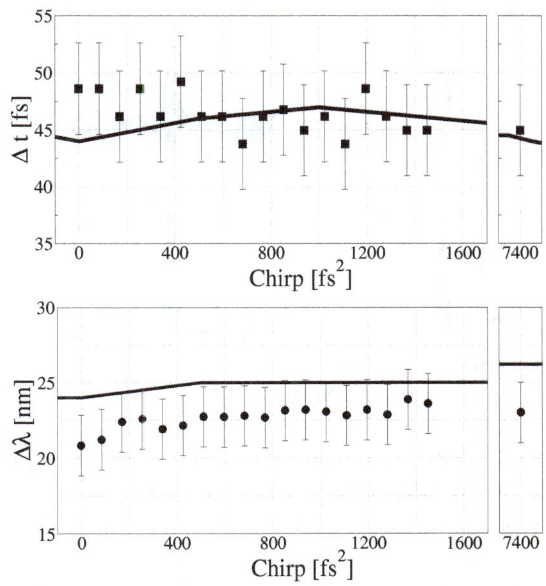

Figure 5: Soliton temporal (a) and spectral width (b) vs. chirp of input pump pulses . Full line: numerical results. The soliton wavelength is $\lambda_s =$ 1075 nm.

Figure 5 (a) shows the pulsewidth of solitons with wavelength $\lambda_s = 1075$ nm upon variation of pump pulses chirp. Even for a $\sim$ 7400 fs^2 chirp, the soliton output pulsewidth remained around 45 fs. Numerical simulations show very good agreement with these observations, as they predict nearly constant pulsewidth regardless of the input chirp (full line in Fig. 5). Measurements and numerical simulations in the spectral domain also indicate that the bandwidth of the output solitons is almost unaffected by the pump pulses chirp [see Fig. 5 (b)]. The product $\Delta t \Delta \nu$ was found to be near 0.315, as it is expected for transform-limited sech^2 pulses.

The effect of this heavy chirping was evident in the auto-correlation traces of pump pulses, as can be seen by comparing Fig. 6 (a) and (c). However, there is not a clear difference between traces of the output solitons for the TL (b) and the highly chirped ($\sim$ 7400 fs^2) case (d).

A color map of the spectra as function of the coupled power, for a highly chirped pump pulse ($\sim$ 7400 fs^2), is shown in Fig. 7. As in the TL case, we observe that a solitonic branch is red shifted

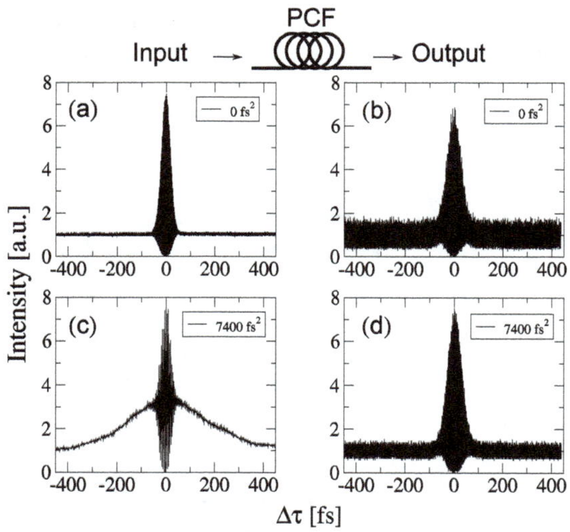

Figure 6: Interferometric auto-correlation traces of TL (a) and heavily chirped, $\sim 7400\ \mathrm{fs}^2$, Ti:Sa pump pulses (c). Interferometric auto-correlation traces of the output solitons are similar in both cases, unchirped (b) and heavily chirped (d). Soliton wavelength is $\lambda_s \simeq 1075$ nm.

by increasing the coupled power. However, in this case, 80 mW of coupled power is required to produce a 1160 nm soliton which represents an increment of about $\sim 45\%$ in comparison to the TL case.

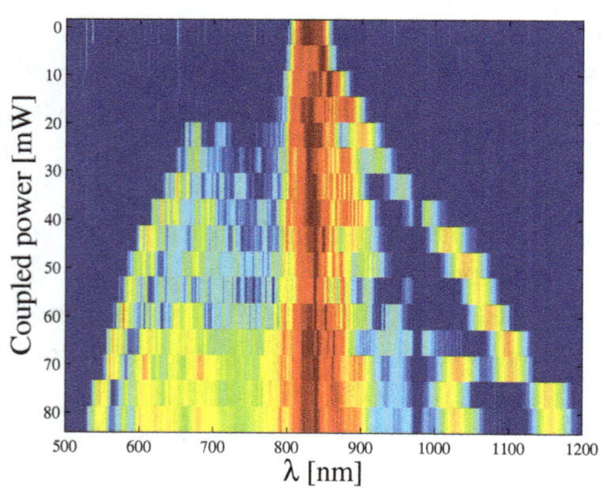

Figure 7: Spectra vs coupled power to the PCF with highly chirped ($\sim 7400\ \mathrm{fs}^2$) input pulses.

A comparison of the soliton red-shift between TL and chirped pump pulse cases is presented Fig. 8. The figure shows that more power is always required to attain the same shift when pump pulses are heavily chirped.

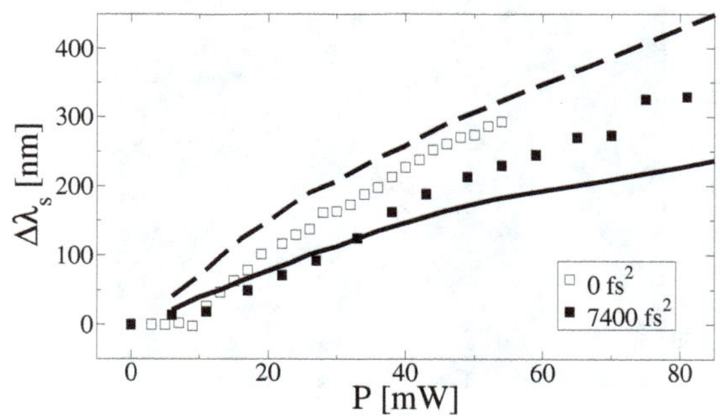

Figure 8: Soliton wavelength shift for chirped (full squares) and TL pump pulses (empty squares) vs pump pulse power. Dashed and full lines correspond to numerical results for TL and highly chirped pump pulses, respectively.

Figure 9 (a) shows the soliton pulsewidth as a function of its wavelength, λ_s, when the pump pulse is heavily chirped ($\sim 7400\ \mathrm{fs}^2$). We observe an approximately constant output pulsewidth (~ 45 fs) in the entire tuning range. Furthermore, the $\Delta t \Delta \nu$ product, shown in Fig. 9 (b), indicates that the generated pulses can be identified as fundamental solitons (sech^2-like), as in the case of TL pump pulses [14]. Numerical simulations were also performed for this case (full lines in Fig. 9) showing an excellent agreement with experimental results.

V. DISCUSSION

i. Fiber soliton self-frequency shift effective length

In order to further analyze soliton formation, we studied the pulse evolution along the fiber by performing numerical simulations. The spectrum evolution along the fiber, for a given coupled power, in the TL and the chirped cases are shown in Fig. 10 (a) and (b), respectively. These simulations show that in the case of chirped pump pulses ($\sim 7400\ \mathrm{fs}^2$), the spectrum broadening and the soliton formation take place farther down into the fiber (see Fig. 10), as compared to the TL case.

The delay in the formation of the soliton can be explained by an interplay of opposite chirping effects: the positive chirp acquired by traversing the AOM is compensated as the pulse advances into the

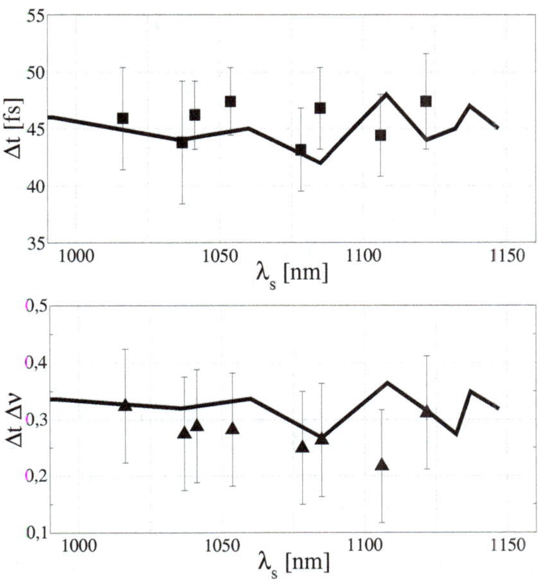

Figure 9: (a) Soliton pulsewidth and (b) the product $\Delta t \Delta \nu$ vs wavelength in the case of highly chirped pump pulses (~7400 fs^2).

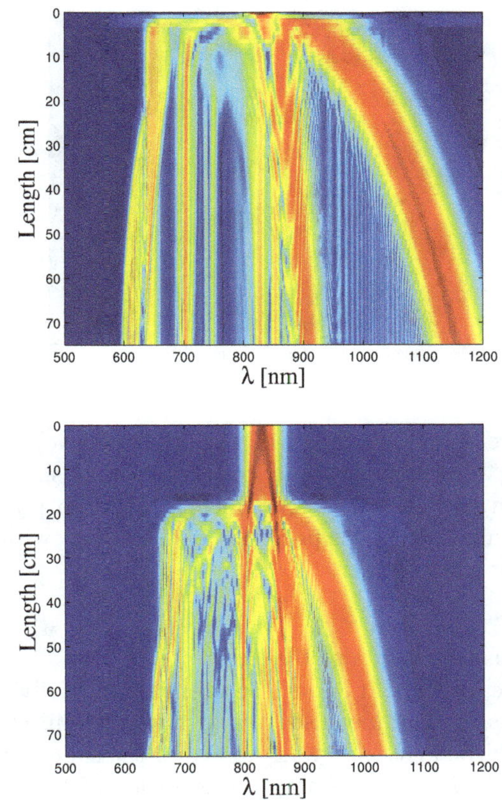

Figure 10: Simulated spectral evolution along the fiber length. Pump pulses with identical peak power produce more soliton shifting with unchirped (a) than with heavily chirped (~7400 fs^2) (b) pump pulses.

PCF, in anomalous propagation, leading to pulse compression. The PCF itself provides pulse compression in the first stretch of the fiber previously to the branching of a soliton. Therefore, the SSFS effective length, i.e., the fiber path where nonlinearity broadens the spectrum, is longer in the TL case. If the chirp is overcompensated and a negatively chirped pulse is fed into the fiber, these pulses would also be compressed within the first stretch of the fiber due to SPM [28] leading to the same behavior than in the positively chirped case, resulting in a narrower tunability range.

Once the soliton is formed and the peak power is high enough, intrapulse Raman scattering redshifts the soliton as it propagates through the remaining of the fiber. This spectral shift increases with both fiber length and soliton peak power [4]. So the fact that the soliton is formed at different lengths explains the red shifts observed for the same coupled power.

However, as a larger wavelength shift can be achieved with a higher input power, this shortening in the effective length in the chirped case could be compensated by coupling more power into the PCF [1]. Another possibility for compensating this effect on the SSFS is using a longer PCF.

ii. Fiber power conversion efficiency

Figure 11 shows a simulation where the same shift wavelength obtained for TL pump pulses is achieved by increasing the coupled power in the shorter effective length fiber (7400 fs^2 chirp). Fission of more than one soliton branch is visible in this case, as compared to the case of TL pump pulses, for which only a single soliton branch appears (Fig. 10). Each soliton branch carries a fundamental soliton ($N = 1$) with a peak power P_0 given by [4]

$$N^2 = 1 = \frac{\gamma P_0 T_0^2}{|\beta_2|}, \tag{2}$$

As γ, β_2 and T_0 (see Figs. 4 and 5) are the same in both, the TL and chirped cases, the peak power

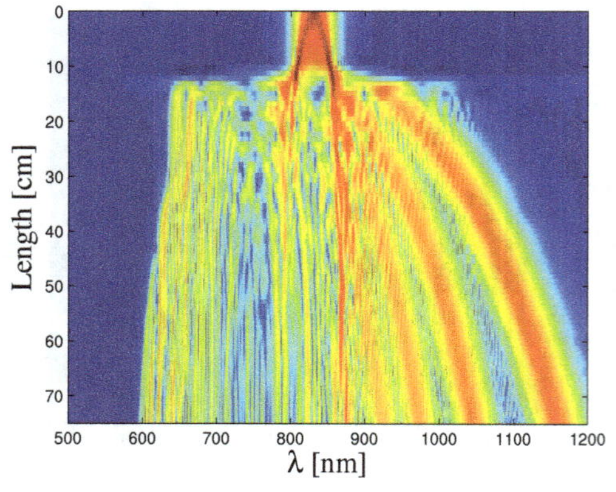

Figure 11: Simulation of the spectral evolution along the fiber showing that the same shifting shown in Fig. 10 (a) for unchirped pump pulses is attained with chirped ones ($\sim$ 7400 fs^2) with a higher pump power.

of the solitons is also the same.

The arising of new soliton branches partially accounts for the increased pump power required in the chirped case (Fig. 11) to attain the same shift. Indeed, the soliton-pump power ratio is 0.2 in the chirped case and 0.44 in the TL case. This result reveals that the use of the PCF as a compressor decreases its power conversion efficiency.

On the other hand, it is possible to achieve the same soliton shift as in the TL case by increasing the fiber length, and keeping the same pump power. In this case, the power conversion efficiency is even lower, 0.17, as predicted by simulations.

VI. Conclusions

We have presented a high-speed tunable soliton infrared source capable of generating $\sim$ 45 fs transform-limited pulses in the range from 860 to 1160 nm. Both the pulsewidth and the spectral width were shown to remain constant over the entire tuning range, even when pump pulses were heavily chirped up to 7400 fs^2. Insensitivity to the chirp of pump pulses points out to the feasibility of avoiding bulky compensation optics prior to the PCF, opening up the possibility to build reliable and compact high-speed tunable femtosecond sources in the near infrared region. A minor drawback of this source is that either more power needs to be coupled or a longer PCF needs to be used in order to achieve the same tuning range obtained with transform-limited pump pulses.

Acknowledgements - This work was supported by ANPCyT PICT 2006-1594, ANPCyT PICT 2006-497 and UBA Programación Científica 2008-2010, Proyecto N X022.

[1] N Nishizawa, T Goto, *Compact system of wavelength-tunable femtosecond soliton pulse generation using optical fibers,* IEEE Photon. Technol. Lett. **11**, 325 (1999).

[2] K Abedin, F Kubota, *Wavelength tunable high-repetition-rate picosecond and femtosecond pulse sources based on highly nonlinear photonic crystal fiber,* IEEE J. Sel. Topics Quantum Electron. **10**, 1203 (2004).

[3] J H Lee, J van Howe, C Xu, X. Liu, *Soliton self-frequency shift: Experimental demonstrations and applications,* IEEE J. Sel. Topics Quantum Electron. **14**, 713 (2008).

[4] G P Agrawal, *Nonlinear fiber optics*, Academic Press, San Diego (2007).

[5] F M Mitschke, L F Mollenauer, *Discovery of the soliton self-frequency shift,* Opt. Lett. **11**, 659 (1986).

[6] J P Gordon, *Theory of the soliton self-frequency shift,* Opt. Lett. **11**, 662 (1986).

[7] B Washburn, S Ralph, P Lacourt, J Dudley, *Tunable near-infrared femtosecond soliton generation in photonic crystal fibers,* Electronics Lett. **37**, 1510 (2001).

[8] J Takayanagi, T Sugiura, M Yoshida, N Nishizawa, *1.0-1.7 μm wavelength-tunable ultrashort-pulse generation using femtosecond yb-doped fiber laser and photonic crystal fiber,* IEEE Photon. Technol. Lett. **18**, 659 (2006).

[9] P Russell, *Photonic crystal fibers,* Science **299**, 358 (2003).

[10] D V Skryabin, F Luan, J C Knight, P St J Russell, *Soliton self-frequency shift cancellation in photonic crystal fibers,* Science **31**, 1705 (2003).

[11] N Nishizawa, Y Ito, T Goto, *0.78-0.90 wavelength-tunable femtosecond soliton pulse generation using photonic crystal fiber,* IEEE Photon. Technol. Lett. **14**, 986 (2002).

[12] K S Abedin, F Kubota, *Widely tunable femtosecond soliton pulse generation at a 10-ghz repetition rate by use of the soliton self-frequency shift in photonic crystal fiber,* Opt. Lett. **28**, 1760 (2003).

[13] N Ishii, C Y Teisset, E E Serebryannikov, T Fuji, T Metzger, F Krausz, A M Zheltikov, *Widely tunable soliton frequency shifting of few-cycle laser pulses,* Phys. Rev. E **74**, 036617 (2006).

[14] M E Masip, A A Rieznik, P G König, D F Grosz, A V Bragas, O E Martínez, *Femtosecond soliton source with fast and broad spectral tunability,* Opt. Lett. **34**, 842 (2009).

[15] S Sanders, *Wavelength-agile fiber laser using group-velocity dispersion of pulsed super-continua and application to broadband absorption spectroscopy,* Appl. Phys. B Lasers Opt. **75**, 799 (2002).

[16] J Walewski, M Borden, S Sanders, *Wavelength-agile laser system based on soliton self-shift and its application for broadband spectroscopy,* Appl. Phys. B Lasers Opt. **79**, 937 (2004).

[17] K Sumimura, T Ohta, N Nishizawa, *Quasi-super-continuum generation using ultrahigh-speed wavelength-tunable soliton pulses,* Opt. Lett.**33**, 2892 (2008).

[18] K Sumimura, Y Genda, T Ohta, K Itoh, N Nishizawa, *Quasi-supercontinuum generation using 1.06 μm ultrashort-pulse laser system for ultrahigh-resolution optical-coherence tomography.* Opt. Lett. **35**, 3631 (2010).

[19] M-C Chan, S-H Chia, T-M Liu, T-H Tsai, M-C Ho, A Ivanov, A Zheltikov, J-Y Liu, H-L Liu, C-K Sun, *1.2- to 2.2- m tunable raman soliton source based on a cr:forsterite laser and a photonic-crystal fiber,* IEEE Photon. Technol. Lett. **20**, 900 (2008).

[20] J Nicholson, A Yablon, P Westbrook, K Feder, M Yan, *High power, single mode, all-fiber source of femtosecond pulses at 1550 nm and its use in supercontinuum generation,* Opt. Express **12**, 3025 (2004).

[21] R L Fork, O E Martínez, J P Gordon, *Negative dispersion using pairs of prisms,* Opt. Lett. **9**, 150 (1984).

[22] R L Fork, C H B Cruz, P C Becker, C V Shank, *Compression of optical pulses to six femtoseconds by using cubic phase compensation,* Opt. Lett. **12**, 483 (1987).

[23] J L A Chilla, O E Martinez, *Direct determination of the amplitude and the phase of femtosecond light pulses,* Opt. Lett. **16**, 39 (1991).

[24] S Costantino, A R Libertun, P D Campo, J R Torga, O E Martínez, *Fast scanner with position monitor for large optical delays,* Opt. Comm. **198**, 287 (2001).

[25] J Dudley, G Genty, S Coen, *Supercontinuum generation in photonic crystal fibers,* Rev. Mod. Phys. **78**, 1135 (2006).

[26] A Heidt, *Efficient adaptive step size method for the simulation of supercontinuum generation in optical fibers,* J. Lightwave Technol. **27**, 3984 (2009).

[27] Q Lin, G Agrawal, *Raman response function for silica fibers,* Opt. Lett. **31**, 3086 (2006).

[28] B R Washburn, J A Buck, S E Ralph, *Transform-limited spectral compression due to self-phase modulation in fibers,* Opt. Lett. **25**, 445 (2000).

22

Master curves for the stress tensor invariants in stationary states of static granular beds. Implications for the thermodynamic phase space

Luis A. Pugnaloni,[1*] José Damas,[2†] Iker Zuriguel,[2‡] Diego Maza[2§]

We prepare static granular beds under gravity in different stationary states by tapping the system with pulsed excitations of controlled amplitude and duration. The macroscopic state—defined by the ensemble of static configurations explored by the system tap after tap—for a given tap intensity and duration is studied in terms of volume, V, and force moment tensor, Σ. In a previous paper [Pugnaloni et al., Phys. Rev. E 82, 050301(R) (2010)], we reported evidence supporting that such macroscopic states cannot be fully described by using only V or Σ, apart from the number of particles N. In this work, we present an analysis of the fluctuations of these variables that indicates that V and Σ may be sufficient to define the macroscopic states. Moreover, we show that only one of the invariants of Σ is necessary, since each component of Σ falls onto a master curve when plotted as a function of $\mathrm{Tr}(\Sigma)$. This implies that these granular assemblies have a common shape for the stress tensor, even though it does not correspond to the hydrostatic type. Although most results are obtained by molecular dynamics simulations, we present supporting experimental results.

I. Introduction

The study of static granular systems is of fundamental importance to the industry to improve the storage of such materials in bulk, as well as to optimize the packaging design. However, far from yielding such benefits of practical interest, physicists have found a fascinating challenge on their way that has stuck them, to a large extent, in the area of static granular matter. Such challenge is finding an appropriate description of the simplest state in which matter can be found: *equilibrium* [1]. At equilibrium, a sample will explore different microscopic configurations over time, in such a way that macroscopic averages over large periods will be well defined, with no aging. Moreover, if not only equilibrium but ergodicity is present, averages over a large number of replicas at a given time should give equivalent results to time averages [2]. Finally, one expects that all macroscopic properties of such equilibrium states can be put in terms of a few independent macroscopic variables. Then, a thermodynamic description and, hopefully, a statistical mechanics approach can be attempted. Theoretical formalisms based on these assumptions have been used to analyze data from experiments and numerical models. However, some of the foundations are still supported by little evidence.

*E-mail: luis@iflysib.unlp.edu.ar
†E-mail: jdelacruz@alumni.unav.es
‡E-mail: iker@unav.es
§E-mail: dmaza@unav.es

[1] Instituto de Física de Líquidos y Sistemas Biológicos, (CONICET La Plata, UNLP), Calle 59 No 789, 1900 La Plata, Argentina.

[2] Departamento de Física y Matemática Aplicada, Facultad de Ciencias, Universidad de Navarra, Pamplona, Spain.

The use of careful protocols to make a granular sample explore microscopic configurations within a seemingly equilibrium macroscopic state has given us the first standpoint [3–6]. In such protocols, the sample is subjected to external excitations of the form of pulses. Well defined, reproducible time averages are found after a transient if the same pulse shape and pulse intensity are applied. However, for low intensity pulses, a previous annealing might be in order since equilibrium is hard to reach —as in supercooled liquids. We will use the expressions *equilibrium state, steady state* or simply *state* to refer to the collection of all configurations generated by using a given external excitation after any transient has disappeared.

More than twenty years ago, Edwards and Oakshott [7] put forward the idea that the number of grains N and the volume V are the basic state variables that suffice to characterize a static sample of hard grains in equilibrium. The NV granular ensemble was then introduced as a collection of microstates, where the sample is in mechanical equilibrium, compatible with N and V. However, newer theoretical works [8–13] suggest that the force moment tensor, Σ, ($\Sigma = V\sigma$, where σ is the stress tensor) must be added to the set of extensive macroscopic variables (i.e., an $NV\Sigma$ ensemble) to adequately describe a packing of real grains.

In the rest of this paper, we show experimental and simulation evidence that the equilibrium states of static granular packings cannot be only described by V (or equivalently the packing fraction, ϕ, defined as the fraction of the space covered by the grains) nor by Σ. We do this by generating states of equal V but different Σ and states of equal Σ but different ϕ. We also show that states of equal V may present different volume fluctuations. Moreover, we show that states of equal V and Σ display the same fluctuations of these variables, suggesting that no other extensive parameter might be required to characterize the state (apart from N). Finally, but of major significance, we show that the shape of the force moment tensor is universal, in the sense that different states that present the same trace of the tensor actually have the same value in all the components of Σ.

II. Simulation

We use soft-particle 2D molecular dynamics [14,15]. Particle–particle interactions are controlled by the particle–particle overlap $\xi = d - |\mathbf{r}_{ij}|$ and the velocities $\dot{\mathbf{r}}_{ij}$, ω_i and ω_j. Here, $\mathbf{r}_{ij}$ represents the center-to-center vector between particles i and j, d is the particle diameter and ω is the particle angular velocity. These forces are introduced in the Newton's translational and rotational equations of motion and then numerically integrated by a velocity Verlet algorithm [16]. The interaction of the particles with the flat surfaces of the container is calculated as the interaction with a disk of infinite radius.

The contact interactions involve a normal force F_n and a tangential force F_t.

$$F_\mathrm{n} = k_\mathrm{n}\xi - \gamma_\mathrm{n} v^\mathrm{n}_{i,j} \tag{1}$$

$$F_\mathrm{t} = -\min\left(\mu|F_\mathrm{n}|, |F_\mathrm{s}|\right) \cdot \mathrm{sign}\left(\zeta\right) \tag{2}$$

where

$$F_\mathrm{s} = -k_\mathrm{s}\zeta - \gamma_\mathrm{s} v^\mathrm{t}_{i,j} \tag{3}$$

$$\zeta\left(t\right) = \int_{t_0}^{t} v^\mathrm{t}_{i,j}\left(t'\right) dt' \tag{4}$$

$$v^\mathrm{t}_{i,j} = \dot{\mathbf{r}}_{ij} \cdot \mathbf{s} + \frac{1}{2} d\left(\omega_i + \omega_j\right) \tag{5}$$

The first term in Eq. (1) corresponds to a restoring force proportional to the superposition ξ of the interacting disks and the stiffness constant k_n. The second term accounts for the dissipation of energy during the contact and is proportional to the normal component $v^\mathrm{n}_{i,j}$ of the relative velocity $\dot{\mathbf{r}}_{ij}$ of the disks.

Equation (2) provides the magnitude of the force in the tangential direction. It implements the Coulomb's criterion with an effective friction following a rule that selects between static or dynamic friction. Notice that Eq. (2) implies that the maximum static friction force $|F_\mathrm{s}|$ used corresponds to $\mu|F_\mathrm{n}|$, which effectively sets $\mu_\mathrm{dynamic} = \mu_\mathrm{static} = \mu$. The static friction force F_s [see Eq. (3)] has an elastic term proportional to the relative shear displacement ζ and a dissipative term proportional to the tangential component $v^\mathrm{t}_{i,j}$ of the relative velocity. In Eq. (5), $\mathbf{s}$ is a unit vector normal to $\mathbf{r}_{ij}$. The elastic and dissipative contributions are characterized by k_s and γ_s, respectively. The shear displacement ζ is calculated through Eq. (4) by integrating $v^{t}_{i,j}$ from the beginning of the contact (i.e., $t = t_0$). The tangential interaction behaves like a damped spring which is formed whenever two grains come into contact and is removed when the contact finishes [17].

The particular set of parameters used for the simulation is (unless otherwise stated): $\mu = 0.5$, $k_n = 10^5(mg/d)$, $\gamma_n = 300(m\sqrt{g/d})$, $k_s = \frac{2}{7}k_n$ and $\gamma_s = 200(m\sqrt{g/d})$. In some cases, we have varied μ and γ_n in order to control the friction and restitution coefficient. The integration time step is set to $\delta = 10^{-4}\sqrt{d/g}$. The confining box (13.39d-wide and infinitely high) contains $N = 512$ monosized disks. Units are reduced with the diameter of the disks, d, the disk mass, m, and the acceleration of gravity, g.

Tapping is simulated by moving the confining box in the vertical direction following a half sine wave trajectory $[A\sin(2\pi\nu t)(1-\Theta(2\pi\nu t-\pi))]$. The excitation can be controlled through the amplitude, A, and the frequency, ν, of the sinusoidal trajectory. We implement a robust criterion based on the stability of particle contacts to decide when the system has reached mechanical equilibrium [14] before a new tap is applied to the sample. Averages were taken over 100 taps in the steady state and over 20 independent simulations for each value of A and ν.

The volume, V, of the system after each tap can be obtained from the packing fraction, ϕ, as $V = N\pi(d/2)^2/\phi$. We measure ϕ in a rectangular window centered in the center of mass of the packing. The measuring region covers 90% of the height of the granular bed (which is of about 40d) and avoids the area close to the walls by 1.5d. We have observed that ϕ is sensitive to the chosen window. However, none of the conclusions drawn in this paper are affected by this choice.

The stress tensor, σ, is calculated from the particle–particle contact forces as

$$\sigma^{\alpha\beta} = \frac{1}{V}\sum_{c_{ij}} r^{\alpha}_{ij} f^{\beta}_{ij}, \tag{6}$$

where the sum runs over all contacts.

The force moment tensor, Σ, is defined as $\Sigma \equiv V\sigma$. During the course of a tap Σ is non-symmetric, however, once mechanical equilibrium is reached in accordance with our criterion, Σ becomes symmetric within a very small error compared with the fluctuations of Σ. Although Σ may depend on depth, we have measured the force moment tensor by simply summing over particle–particle contacts in the entire system.

The fluctuations of ϕ ($\Delta\phi$) and Σ ($\Delta\Sigma$) are calculated as the standard deviation in the 100 taps obtained in each steady state. We average ϕ, Σ, and their fluctuations over 20 independent runs for each steady state and estimate error bars as the standard deviation over these 20 runs.

III. Experimental method

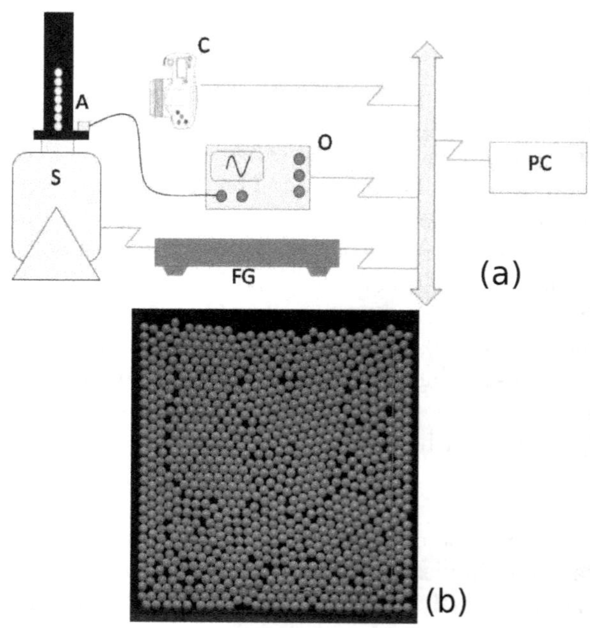

Figure 1: (a) Schematic diagram of the experimental set-up. A: accelerometer, S: shaker, C: camera, O: oscilloscope, FG: function generator, PC: computer. (b) Example of an image of the packing. The region of measurement is indicated by a red square.

The experimental set-up is sketched in Fig. 1(a). A quasi 2D Plexiglass cell (28 mm wide and 150 mm high) is filled with 900 alumina oxide beads of diameter $d = 1 \pm 0.005$ mm. The separation between the front and rear plates was made 15% larger than the bead diameter. The cell is tapped by an electromagnetic shaker (Tiravib 52100) with a trend of half sine wave pulses separated five seconds between them. The tapping amplitude was controlled adjusting the intensity, A, and frequency, ν, of the pulse and was measured by an accelerometer attached to the base of the cell. Averages are taken over 500 taps after equilibration.

High resolution images (more than 10 MPix) are taken after each tap. To calculate the packing fraction, we only consider a rectangular zone at the center of the packing whose limits are at 4 mm from the borders [see Fig. 1(b)]. We determine the centroid of each particle by means of a numerical algorithm with subpixel resolution. Then, we calculate the packing fraction by assuming that the 2D projection of each bead corresponds to a disk of diameter $d = 1$ mm. We estimate the packing fraction with a resolution of ± 0.001. As the separation between plates is larger than the particle diameter, small overlaps between the spheres are present in the 2D projections of the images. Therefore, the calculated packing fraction in some dense configurations might result slightly higher than the hexagonal disk packing limit ($\pi\sqrt{3}/6$).

IV. Tapping characterization and asymptotic equilibrium states

It is often debated [18, 19] what is the appropriate parameter to characterize the external excitation used to drive a granular sample. Dijksman et al. [18] proposed a parameter related to the lift-off velocity of the granular bed. Ludewing et al. [19] presented an energy based parameter. Pugnaloni et al. [15, 20] suggested that the factor of expansion induced on the sample would be a suitable measure [21]. The usual perspective to define a pulse parameter, ϵ, is to achieve a collapse of the ϕ–ϵ curves as different details of the pulse shape are changed (such as amplitude and duration). Parameters defined in all these previous works fail to make such curves collapse for the data presented in this paper. The main reason for this is that our ϕ–ϵ curves are non-monotonic (see for example Fig. 6) presenting a minimum whose depth depends on the details of the pulse shape. Therefore, a simple rescale of the horizontal axis does not suffice to collapse the curves.

Since we are interested in macroscopic states, the actual pulse used to drive the system is merely a control parameter but not a macroscopic variable that describes the state. Therefore, the external pulse does not need to be described with a simplified quantity. The complete functional form of the pulse can be given instead. In our case, we use a sine pulse and both, the pulse amplitude and frequency, are needed to fully describe the excitation. We will employ the usual nondimensional peak acceleration $\Gamma \equiv a_{peak}/g = A(2\pi\nu)^2/g$ (where g is the acceleration of gravity) and the frequency ν to precisely define the external excitation. A detailed study of the dynamical response of a granular bed to a pulse of controlled intensity and duration can be found in Damas et al. [22].

One major issue in studying equilibrium states is the evidence that one can have, indicating if the system is actually at equilibrium [5]. Since the definition of equilibrium is circular [23], we can simply do our best to check if different properties of the system have well defined means (as well as higher order moments of the distributions) which should not depend on the history of the processes applied to the sample.

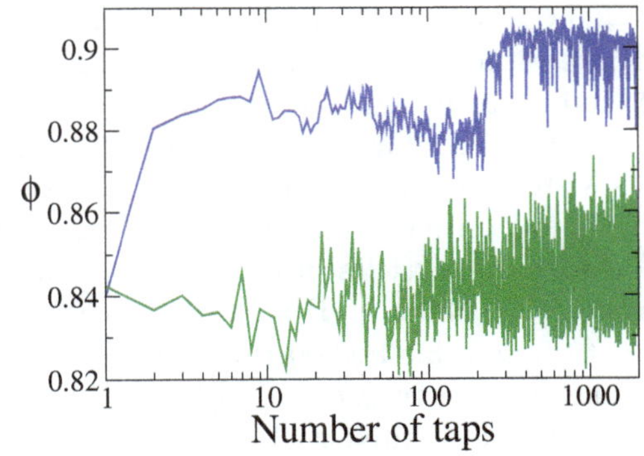

Figure 2: Evolution of ϕ towards the steady state starting from a disordered configuration initially obtained by strong taps (experimental results). Results for two tapping intensities are shown: $\Gamma = 4.8$ (blue), and $\Gamma = 17$ (green). In both experiments, the frequency of the pulse is $\nu = 30$ Hz.

We have generated configurations corresponding to a particular pulse (of a given shape and intensity) by repeatedly applying such pulse to the system and allowing enough time for any transient to fade. To prove that our samples are at equilibrium, we approach a particular pulse through different paths —and starting from different initial conditions (ordered and disordered configurations)— and confirm that the steady states obtained present equivalent mean values and second moments of the distributions of the variables of interest.

In Ref. [26], we showed that the steady states corresponding to excitations of high intensity are reached in a few taps, even if the initial configuration corresponds to a very ordered structure. In Fig. 2, we consider the equilibration process from a highly disordered initial configuration. For low tapping intensities (blue line in Fig. 2), the packing fraction evolves to the steady state in two stages. Just after switching the tapping amplitude to the reference value, the system rapidly evolves to values of ϕ close to the final steady state. Beyond this initial convergence, a slower compaction phase takes the system to the final steady state. For high tapping intensities, the evolution to the steady state is very rapid. The steady state is reached after about a hundred taps [green line in Fig. 2)]. Therefore, we apply a sequence of at least 1000 taps in all our experiments before taking averages to warrant that the steady state has been reached. In our simulations, 400 taps of equilibration were enough.

An interesting example of equilibration is presented in Fig. 3. In this case, we present the values of ϕ and of $\mathrm{Tr}(\Sigma)$ during a very special sequence of pulses obtained in our simulations. The system is initially deposited from a dilute random configuration with all particles in the air. First, we tap the system 200 times at $\Gamma = 4.9$, then 800 times at $\Gamma = 61.5$ and finally, 200 times at $\Gamma = 4.9$. In the whole run, we keep $\nu = 0.5\sqrt{g/d}$. These two values of the pulse intensity have been chosen because they are known to produce packings with the same mean ϕ in the steady state [26]. However, the mean Σ is clearly different. Notice, however, that the same values of ϕ, Σ and its fluctuations, $\Delta\phi$ and $\Delta\Sigma$, are observed for the steady state of low Γ obtained before and after the 800 pulses of high Γ.

Unless otherwise stated, all the results we present in what follows correspond to steady states. We have tested this by obtaining the same states through two different preparation protocols consisting of: (i) application of a large number of identical pulses starting from a disordered configuration, and (ii) application of a reduced number of identical pulses after annealing the system from much higher pulse intensities. In a few cases, the results (mean values and/or fluctuations) from both protocols did not match. This was an indication that a steady state, if existed, was not reached by one of the protocols (or by both protocols). This happens especially for low intensity taps, which require longer equilibration times. Such cases have been removed from the analysis.

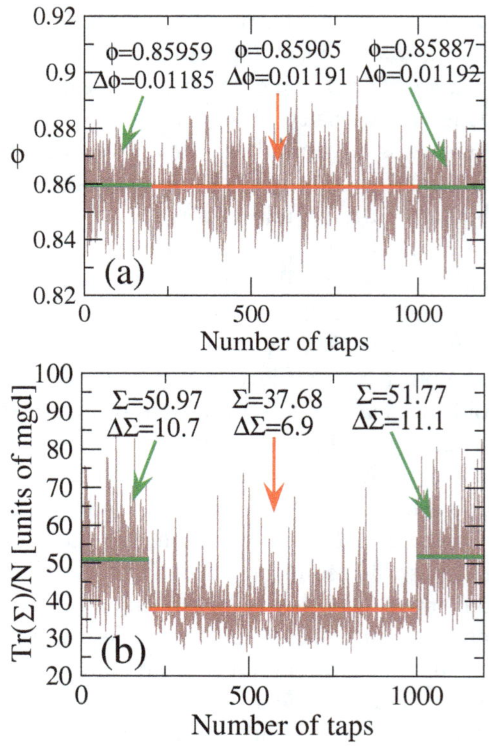

Figure 3: Evolution of ϕ (a) and $\mathrm{Tr}(\Sigma)$ (b) as the pulse intensity is suddenly changed between two values that produce steady states with the same ϕ but different Σ in our simulations. The initial 200 taps correspond to $\Gamma = 4.9$, the middle 800 pulses to $\Gamma = 61.5$ and the final 200 pulses, again, to $\Gamma = 4.9$. In all cases, $\nu = 0.5\sqrt{g/d}$. The mean values and standard deviations in each section are indicated by arrows.

V. Volume and volume fluctuations

In Fig. 4(a), we plot ϕ in the steady state as a function of Γ from our experiments for different pulse frequencies [24]. As we can see, there exists a minimum ϕ at relatively high Γ. A similar experiment on a three-dimensional cell also yielded analogous results [26]. This behavior has also been reported for various models [15, 20, 25]. An explanation for this, based on the formation of arches, has been given in [15]. The position of the minimum in ϕ shifts to larger Γ if the frequency of the pulse is increased (i.e., if the pulse duration is reduced).

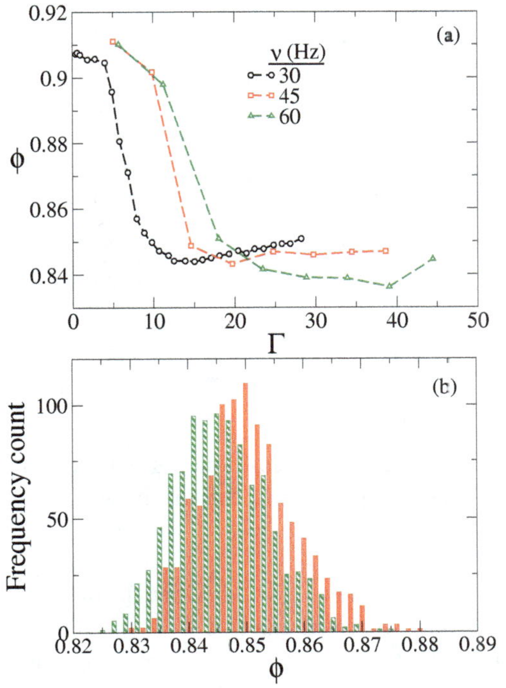

Figure 4: (a) Experimental results for the steady state packing fraction, ϕ, as a function of the reduced peak acceleration, Γ, for different frequencies, ν, of the tap pulse. (b) Histogram of the configurations visited by the system for $\nu = 30$ Hz: $\Gamma = 15$ (green) and $\Gamma = 28$ (red).

Although the increment in the packing fraction beyond the minimum is rather small, it is important to remark that this difference is not an artifact introduced by our experimental resolution. In Fig. 4(b), we show the histogram for the sequence of packings obtained for $\Gamma \simeq 15$ (the minimum packing fraction for $\nu = 30$ Hz) and for $\Gamma = 28$ (the largest excitation explored for $\nu = 30$ Hz). Both steady states are statistically comparable; however, it is possible to distinguish different mean values.

Since steady states of equal ϕ obtained at both

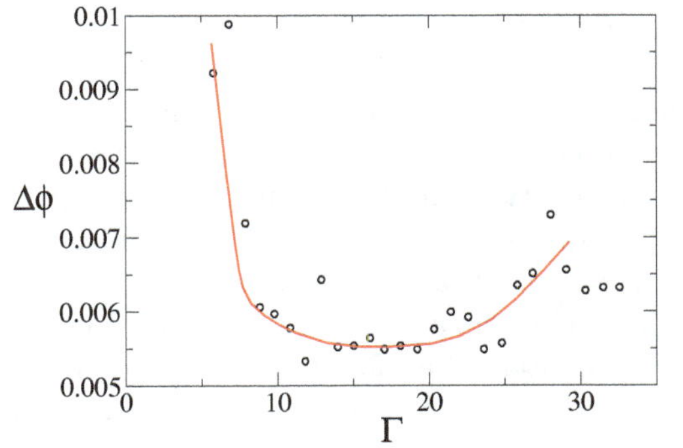

Figure 5: The steady state volume fraction fluctuations, $\Delta\phi$, as a function of Γ from our experiments with $\nu = 30$ Hz. The solid line is only a guide to the eyes.

sides of the ϕ minimum are generated via very different tap intensities, it is worth assessing if such states are, in fact, equivalent. This can be done by comparing the volume fluctuations of such states. A similar analysis was done in Ref. [27] for states generated with different pulse amplitude and duration in liquid fluidized beds.

The fluctuations of ϕ in the steady state, as measured by the standard deviation $\Delta\phi$, are presented in Fig. 5. As we can see, a minimum in the fluctuations is just apparent. The position of such minimum in the fluctuations coincides with the minimum in ϕ. Unfortunately, the resolution of ϕ in our experiments is around the size of the fluctuations. However, the results from the simulations are not limited in this respect and we address to those for a more reliable assessment of the fluctuations.

In Fig. 6(a), we plot ϕ in the steady state as a function of Γ for our simulations. Although the fluctuations of ϕ are large [see Fig. 3], its mean value is well defined with a small confidence interval (see error bars). For low excitations, ϕ decreases as Γ is increased. However, beyond a certain value $\Gamma_{\min}$, the packing fraction grows. The same trend is observed if the tap frequency ν is changed. However, the minimum is deeper for lower ν and its position $\Gamma_{\min}$ shifts to larger values of Γ as ν is increased in coincidence with our experiments (see Fig. 4). Due to the change in the depth of the ϕ minimum in the ϕ–Γ curves, a simple rescaling of Γ is unable to collapse the data for different frequencies. However, rescaling ϕ and Γ with ϕ_{min} and

Γ_{min}, respectively, yields very good collapse, even between simulated and experimental data [26].

In Fig. 6(b), the volume fraction fluctuations, $\Delta\phi$, are plotted as a function of Γ. As we can see, these fluctuations are non-monotonic, as suggested by our experiments (Fig. 5). Non-monotonic volume fluctuations have also been reported in Ref. [6]. For $\Delta\phi$, we obtain a minimum and a maximum. We have also observed a maximum in $\Delta\phi$ for values of Γ below the ones reported here (see for example Refs. [26] and [25]). However, we do not report such low values of Γ in this work and focus on tapping intensities that warrant the steady state with a modest number of pulses.

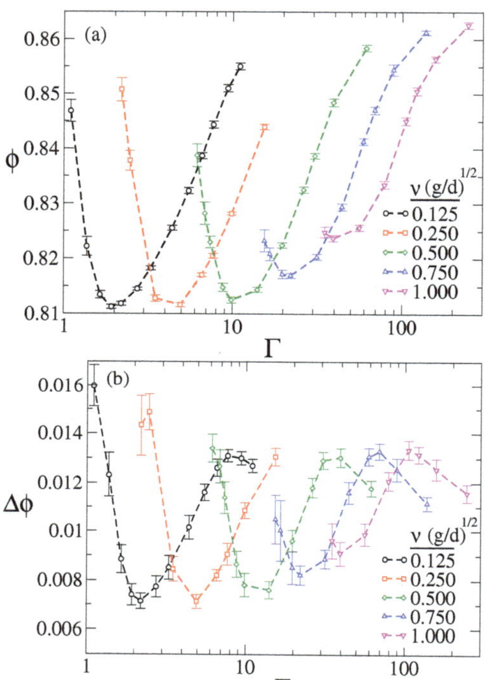

Figure 6: (a) Simulation results for the volume fraction, ϕ, as a function of the reduced peak acceleration, Γ, for different frequencies, ν, of the tap pulse. (b) The corresponding volume fraction fluctuation, $\Delta\phi$, as a function of Γ.

The value of Γ, at which the fluctuations display a minimum, coincides with Γ_{min}, the value at which the minimum packing fraction, ϕ_{min}, is obtained. The maximum coincides with the inflection point in the ϕ–Γ curve at higher Γ. Since one expects to find few mechanically stable configurations compatible with a large volume (low ϕ), it seems reasonable that fluctuations reach a minimum if ϕ does so. Similarly, there should be few low-volume, mechanical stable configurations which implies that at high ϕ fluctuations should diminish. Hence, a maximum in $\Delta\phi$ should be present at intermediate packing fractions. This is more clearly seen in Fig. 7 where we plot the fluctuations as a function of the average value of ϕ.

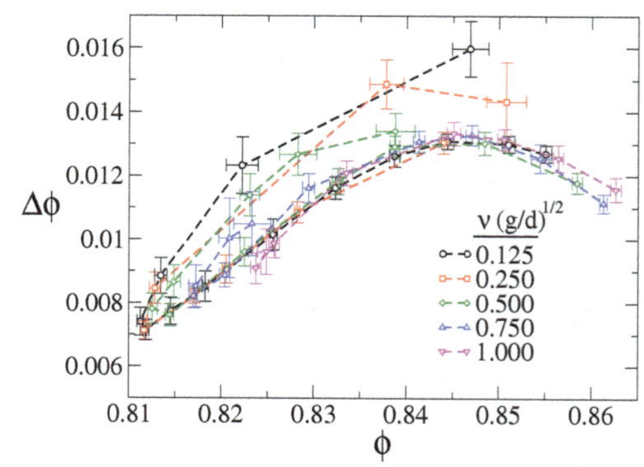

Figure 7: Density fluctuations as a function of ϕ for different frequencies of the tap pulse in the simulations.

Figure 7 presents two distinct branches: the lower branch corresponds to $\Gamma > \Gamma_{min}$ and the upper branch to $\Gamma < \Gamma_{min}$. For $\Gamma > \Gamma_{min}$, the fluctuations corresponding to different tap durations collapse, suggesting that such equal-ϕ, equal-$\Delta\phi$ states might correspond to unique states (below, we will find that this is not the case). For $\Gamma < \Gamma_{min}$, we obtain states of same ϕ as some states in the lower branch but presenting larger fluctuations. This is clear evidence that the equal-ϕ states of the upper and lower branch are indeed distinct and that other macroscopic variables must be used to distinguish one from the other.

We have assessed a number of other structural descriptors (coordination number, bond order parameter, radial distribution function). In all cases, equal–ϕ states from the upper and lower branch of the ϕ–$\Delta\phi$ curve (Fig. 7) present similar values of the structural descriptors with only subtle discrepancies. Although this indicates that the states are not equivalent, it also suggests that such descriptors are not good candidates to form a set of macroscopic variables, along with V, to uniquely identify a given steady state.

In Ref. [26], we have assessed the force moment tensor Σ as a good candidate to complete V and N in describing a stationary state. This has been suggested by some theoretical speculations [12]. However, some authors prefer to directly replace V by Σ. Below, we will show that both, V and Σ, are required for the equilibrium states generated in our simulations. Moreover, we will show that only one of the invariants of Σ is necessary (at least in our 2D systems) and that fluctuations of these variables suggest that no other extra macroscopic parameter may be required.

VI. Stress tensor

Before we focus on the force moment tensor, we will consider the stress tensor, σ, in order to understand the phenomenology of the force distribution in our tapped granular beds. We recall that Σ and σ are simply related through $\Sigma \equiv V\sigma$. However, we have to bear in mind that V is not a simple constant since the volume of the system depends, in a nontrivial way, on the shape and intensity of the excitation.

In Fig. 8, we show the components of σ as a function of Γ for different ν. As a reference, we show results of our simulations for a frictionless system. In a frictionless system, the shear vanishes and σ_{yy} is only determined by the weight of the sample since the Janssen's effect is not present. As we can see, the frictionless sample presents a constant value of σ_{yy} for all Γ. For low Γ, the frictional samples display values of σ_{yy} below the frictionless reference. This is a consequence of the Janssen's effect since part of the weight of the sample is supported by the wall friction. Consequently, in this region, σ_{xy} is also positive [see Fig. 8(b)]. However, for each ν, there is a critical value of Γ, $\Gamma_{shear=0}$. Beyond it, the sample presents an apparent weight above the weight of the packing. In correspondence with this, σ_{xy} changes sign and becomes negative. This indicates that, for $\Gamma > \Gamma_{shear=0}$, the frictional walls are not supporting any weight. Rather, they prevent the packing from expansion by a downward frictional force. As Γ is increased beyond $\Gamma_{shear=0}$, the packing tends to store most of its stress in the horizontal direction (σ_{xx}) while σ_{yy} eventually saturates. For very intense pulses, the sample expands and lifts off significantly during the tap. When the bed falls back, it creates a very compressed structure with most of the stress transmitted in the lateral directions and the wall friction sustaining the system downwards. It is worth mentioning that $\Gamma_{shear=0}$ is always higher than Γ_{min} (see Fig. 6).

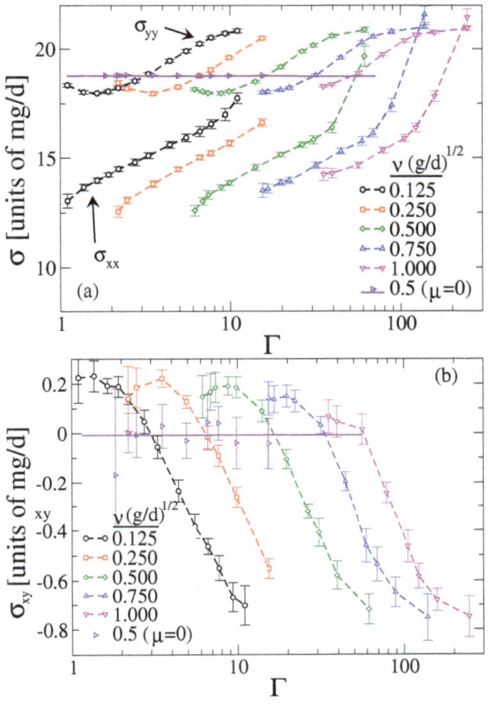

Figure 8: (a) Diagonal components of the stress tensor, σ, as a function of Γ for different frequencies ν of the tap pulse (the upper set of curves corresponds to σ_{yy} and the lower set to σ_{xx}). (b) Off diagonal component of the stress tensor, σ_{xy}. The horizontal line corresponds to σ_{yy} [in panel (a)] and to σ_{xy} [in panel (b)] from simulations of a packing of frictionless disks.

VII. The force moment tensor master curve

Since the stress is not an extensive parameter, the force moment tensor is generally used to characterize the macroscopic state [12,13]. Therefore, we will use Σ in the rest of the paper. Let us simply remark that since V presents a non-monotonic response to Γ, the curves in Fig. 8 present a somewhat different shape if Σ is plotted instead of σ. Particularly, Σ_{yy} does not display a minimum at low Γ, as the one observed for σ_{yy} in frictional disks but a monotonic increase. In Fig. 9, we show the trace of Σ as a function of Γ. There is a clear monotonic increase of $\mathrm{Tr}(\Sigma)$ as Γ is increased. Moreover, for a given Γ, if the frequency of the excitation pulse is increased, a significant reduction in the force moment tensor is observed.

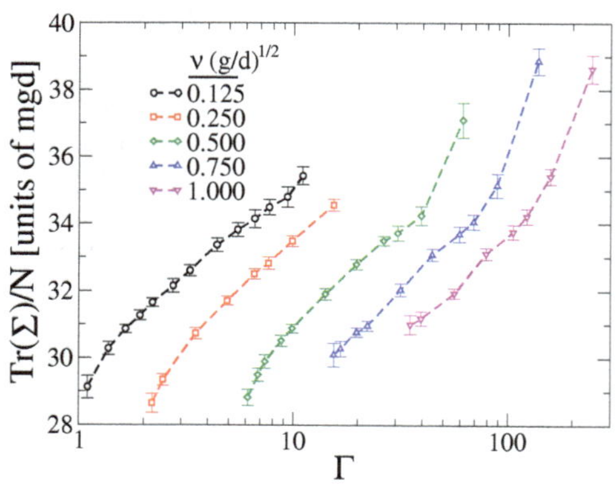

Figure 9: Trace, $\mathrm{Tr}(\Sigma)$, of the force moment tensor as a function of Γ for different frequencies ν of the tap pulse.

In Fig. 10, we plot the components of Σ as a function of its trace for all the steady states generated. We can see that all data for different Γ and ν collapse into three master curves. This indicates that if two equilibrium states present the same $\mathrm{Tr}(\Sigma)$, all the components of Σ are also equal. Here, we point out a relevant piece of information that will be discussed in the next section. Two states may present equal force moment tensor but differ in volume. This means that many points collapsing in Fig. 10 correspond to states of different ϕ. Therefore, at equilibrium, irrespective of the structure of the sample, two states with the same trace in Σ will present equal Σ.

In a liquid at equilibrium, the stress tensor is diagonal and all elements along the diagonal are equal. This hydrostatic property allows us to know the full stress tensor if we only know the hydrostatic pressure (i.e., if we only know the trace of the tensor). In our granular samples, the force moment tensor can also be known if the trace is known. However, the shape of the tensor in static packings under gravity is defined by the three master curves of Fig. 10.

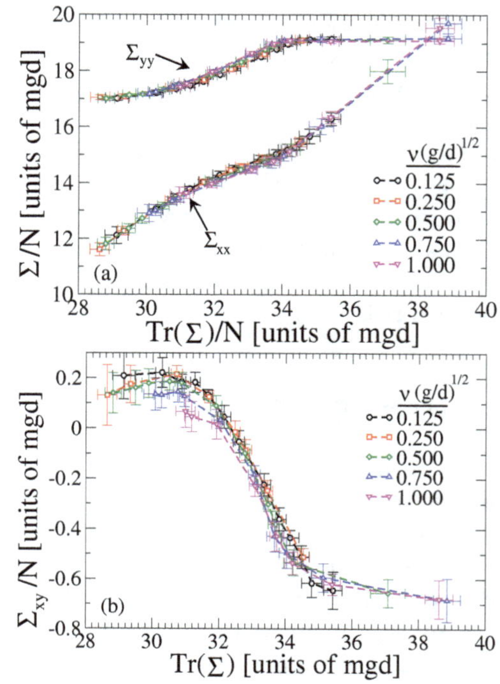

Figure 10: (a) Diagonal components of the force moment tensor, Σ, as a function of $\mathrm{Tr}(\Sigma)$ for different frequencies of the tap pulse (the upper set of curves corresponds to Σ_{yy} and the lower to Σ_{xx}). (b) Off diagonal component of the force moment tensor, Σ_{xy}.

To our knowledge, there is no previous speculation that this property must hold for static granular packings. A more detailed study on the extent of this commonality of the shape of the force moment tensor will be pursued in a future paper [28]. However, we show some suggestive preliminary results below.

In Fig. 11, we show the components of Σ as a function of $\mathrm{Tr}(\Sigma)$ for a range of samples of different materials, for different tapping intensities, and for different tapping frequencies. As we can see, there is reasonable collapse of the data onto the same three master curves shown in Fig. 10. This is an indication that these master curves may be universal and enclose a rather fundamental underlying property (inaccessible to us at this point) of static granular beds.

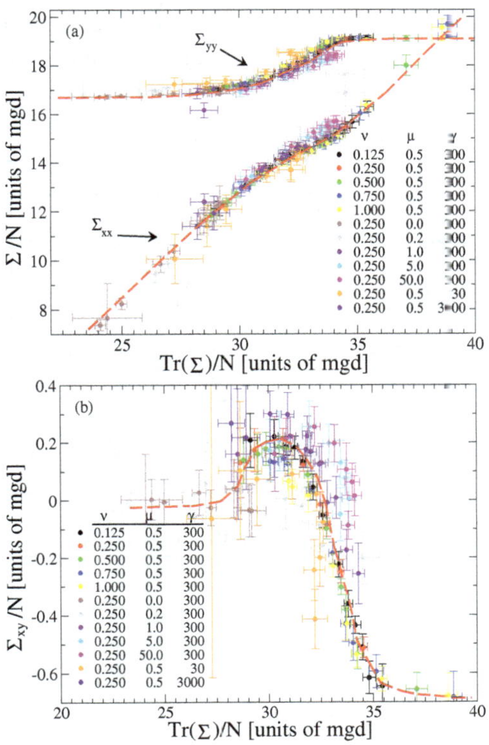

Figure 11: (a) Diagonal components of force moment tensor, Σ, as a function of $\mathrm{Tr}(\Sigma)$ for different frequencies of the tap pulse, different friction coefficient and different restitution (the upper set of curves corresponds to Σ_{yy} and the lower to Σ_{xx}). (b) Off diagonal component of the force moment tensor, Σ_{xy}. The dashed lines are only a guide to the eyes.

VIII. Force moment tensor fluctuations

Since we have shown in the previous section that only the trace of Σ suffices (along with the master curves) to describe the full force moment tensor, we will now focus on this invariant and its fluctuations. In Fig. 12, the fluctuations of $\mathrm{Tr}(\Sigma)$ are plotted as a function of Γ. We obtain a single minimum, in contrast with the minimum and maximum observed in $\Delta\phi$. Interestingly, the states with minimum $\Delta\mathrm{Tr}(\Sigma)$ correspond to the states where the minimums ϕ and $\Delta\phi$ are reached for each ν. However, unlike $\Delta\phi$, the depth of the minimum in $\Delta\mathrm{Tr}(\Sigma)$ is fairly independent of ν. It is unclear why the force moment fluctuations should present a minimum. Provided that the minimum of $\Delta\mathrm{Tr}(\Sigma)$ coincides with the minimum of $\Delta\phi$, it can be speculated that a reduced number of geometric configurations can accommodate a limited number of force configurations. We have seen that all individual components of Σ present the same minimum in their fluctuations, however, the actual values in $\Delta\mathrm{Tr}(\Sigma)$ are dominated by $\Delta\Sigma_{xx}$ which takes values five times larger than $\Delta\Sigma_{yy}$.

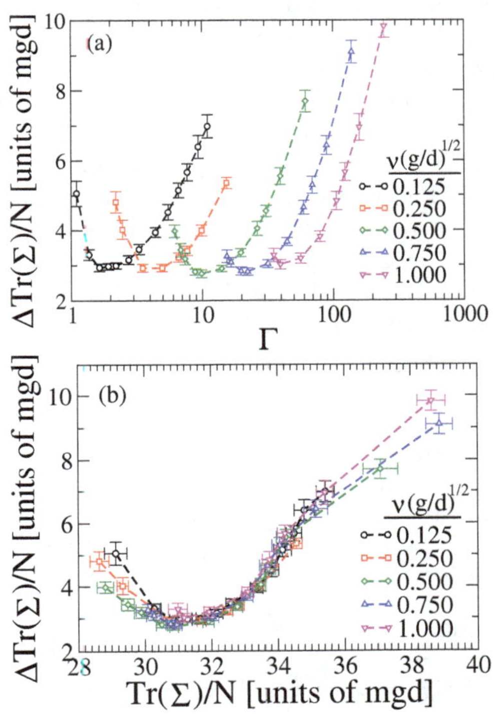

Figure 12: (a) Fluctuations of the trace of the force moment tensor as a function of Γ for different frequencies of the tap pulse. (b) Fluctuations of the trace of the force moment tensor as a function of $\mathrm{Tr}(\Sigma)$.

If we plot $\Delta\mathrm{Tr}(\Sigma)$ in terms of the average value $\mathrm{Tr}(\Sigma)$ [see Fig. 12(b)], we can see that the curves collapse on top of each other over a wide range of $\mathrm{Tr}(\Sigma)$. However, some deviations are apparent at very low and very high forces. Although the fair collapse of the curves suggests that states of equal-Σ may correspond to the same equilibrium states, we will see in the next section that many of these equilibrium states are distinguishable through the volume. The inflection observed at high Tr(Σ) corresponds to the change of regime observed in Fig. 8(a) where the vertical stress saturates and most of the contact forces are directed in the x–direction.

IX. The thermodynamic phase space

As we have suggested, the mean volume of static granular samples is not sufficient to describe the equilibrium state since states of equal-ϕ may present distinct fluctuations. On the other hand, the force moment tensor seems to be able to serve as a standalone descriptor since states of equal Σ do generally present the same Σ fluctuations. However, states of equal-Σ may present different volumes. In Fig. 13, we plot the loci of the equilibrium states generated in our simulations in a hypothetical ϕ–Σ thermodynamic phase space. As we can see, states of equal V but different Σ are obtained as well as states of equal Σ and different V.

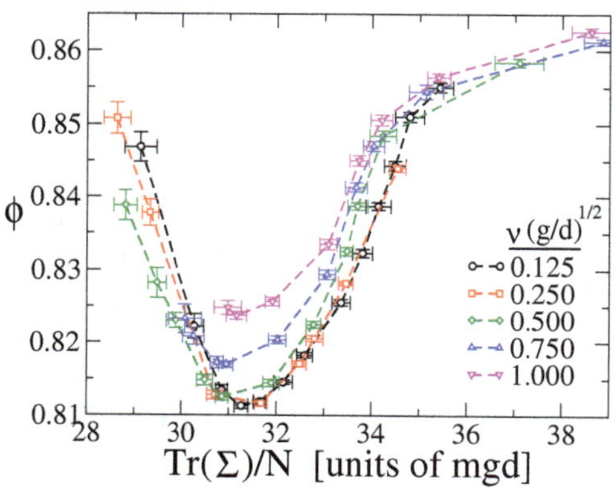

Figure 13: Phase space ϕ–Tr(Σ). Loci visited in the simulations for different frequencies of the tap pulse.

We can ask now if these two state variables suffice to fully describe the equilibrium states. A hint that this may be the case is given by the fact that states generated with different Γ and ν but that correspond to the same state in the ϕ–Σ plot display the same fluctuations of these variables. In Fig. 14, we have highlighted some pairs of neighboring states. We can see that such states do also present similar fluctuations [Fig. 14(b)]. In contrast, states which are distant in the ϕ–Σ plot present distinct fluctuations even if they correspond to an equivalent mean volume or an equivalent mean force moment tensor (see states joined by solid lines in Fig. 14).

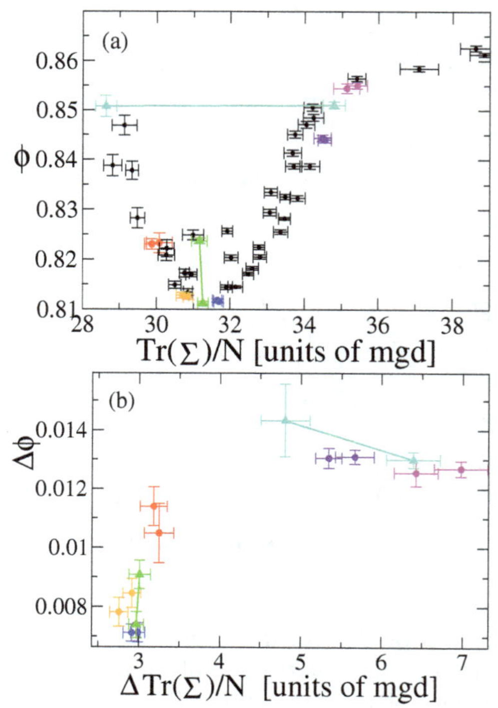

Figure 14: (a) Phase space ϕ–Tr(Σ). (b) Fluctuations of the state variables. Selected neighboring states are colored in pairs for comparison. Distant states of equal V or Tr(Σ) are joined by thick lines.

Let us point out here, that the sole coincidence of fluctuations in the macroscopic variables is not a rigorous proof that the set of chosen variables is a full complete set of thermodynamic parameters. Future explorations of these systems may confirm or disprove that N, V and Tr(Σ) are a full set of macroscopic, extensible variables able to describe all equilibrium states. Meanwhile, it is clear that for moderate tapping intensities, around which the minimum in ϕ is observed, the approximation of a simple NV or $N\Sigma$ ensemble is not warranted in view of the large discrepancies between the curves generated with different pulse frequencies in Fig. 13.

X. Concluding remarks

We have studied steady states of mechanically stable granular samples driven by tap like excitations. We have varied the external excitation by changing both, the pulse amplitude and the pulse duration. We have considered the macroscopic extensive variables V (volume) and Σ (force moment tensor), and their fluctuations. From the results, we can draw the following conclusions:

- There seems to be a rather robust set of master curves for $\Sigma_{\alpha\beta}$ which implies that the knowledge of $\mathrm{Tr}(\Sigma)$ suffices to infer the other components of the force moment tensor.
- The equilibrium states cannot be only described by V or Σ, apart from the number of particles N.
- The equilibrium states seem to be well described by the set $NV\mathrm{Tr}(\Sigma)$.

There exists a number of points to be considered in view of these findings. Here, we mention a few that may serve as starting points for future directions of research:

- What is the extent to which the Σ master curves are applicable? Is this dependent on the dimensionality of the system, the excitation procedure, the chosen contact force law, etc?
- What is the dynamics during a single pulse that leads to the appearance of the ϕ minimum? Is this minimum present in states generated with other types of pulses like fluidization or shear? The ubiquity of this minimum in simulation models [15, 20] suggests that it might be found in numerous conditions.
- Are the fluctuations shown in Figs. 7 and 12 the definitive phenomenological equation of states? Other authors have so far found monotonic density fluctuations [27] or concave up density fluctuations [6].
- How much of the ϕ–$\mathrm{Tr}(\Sigma)$ plane can be explored by changing material properties?
- Are there other excitation protocols (such as shearing) that may give rise to steady states that are thermodynamically equivalent to the ones obtained by tapping?
- Is it possible to construct a phenomenological entropy function from the equations of states (Figs. 7 and 12) by simple integration of a Gibbs–Duhem-like equation? Let us bear in mind that Fig. 7 is multiply valued.

It is worth stressing that if two ensembles generated by arbitrary excitation protocols —such as tapping or shearing— happen to present the same mean values (and fluctuation) for all macroscopic variables, then such macroscopic states should be considered thermodynamically identical. However, it may be the case that a given protocol produces a narrow range of macroscopic states that can be eventually described with a reduced set of macroscopic variables.

Acknowledgements - LAP acknowledges discussions with Massimo Pica Ciamarra. JD acknowledges a scholarship of the FPI program from Ministerio de Ciencia e Innovacin (Spain). This work has been financially supported by CONICET (Argentina), ANPCyT (Argentina), Project No. FIS2008-06034-C02-01 (Spain) and PIUNA (Univ. Navarra).

---

[1] H B Callen, *Thermodynamics and an Introduction to Thermostatistics*, 2nd ed., Wiley-VCH, New York (1985).

[2] R K Pathria, *Statistical Mechanics*, 2nd ed., Butterworth-Heinemann, Oxford (1996).

[3] E R Nowak, J B Knight, M L Povinelli, H M Jeager, S R Nagel. *Reversibility and irreversibility in the packing of vibrated granular material*, Powder Tech. **94**, 79 (1997).

[4] P Richard, M Nicodemi, R Delannay, P Ribire, D Bideau, *Slow relaxation and compaction of granular systems*, Nature Mater. **4**, 121 (2005).

[5] Ph Ribière, P Richard, P Philippe, D Bideau, R Delannay, *On the existence of stationary states during granular compaction*, Eur. Phys. J. E **22**, 249 (2007).

[6] M Schröter, D I Goldman, H L Swinney, *Stationary state volume fluctuations in a granular medium*, Phys. Rev. E **71**, 030301(R) (2005).

[7] S F Edwards, R B S Oakeshott, *Theory of powders*, Physica A **157**, 1080 (1989).

[8] J H Snoeijer, T J H Vlugt, W G Ellenbroek, M van Hecke, J M J van Leeuwen, *Ensemble theory for force networks in hyperstatic granular matter*, Phys. Rev. E. **70**, 061306 (2004).

[9] S F Edwards, *The full canonical ensemble of a granular system*, Physica A **353**, 114 (2005).

[10] S Henkes, C S OHern, B Chakraborty, *Entropy and temperature of a static granular assembly: An ab initio approach*, Phys. Rev. Lett. **99**, 038002 (2007).

[11] S Henkes, B Chakraborty, *Stress correlations in granular materials: An entropic formulation*, Phys. Rev. E. **79**, 061301 (2009).

[12] R Blumenfeld, S F Edwards, *On granular stress statistics: Compactivity, angoricity, and some open issues*, J. Phys. Chem. B **113**, 3981 (2009).

[13] B P Tighe, A R T van Eerd, T J H Vlugt, *Entropy maximization in the force network ensemble for granular solids*, Phys. Rev. Lett. **100**, 238001 (2008).

[14] R Arévalo, D Maza, L A Pugnaloni, *Identication of arches in two-dimensional granular packings*, Phys. Rev. E **74**, 021303 (2006).

[15] L A Pugnaloni, M Mizrahi, C M Carlevaro, F Vericat, *Nonmonotonic reversible branch in four model granular beds subjected to vertical vibration*, Phys. Rev. E **78**, 051305 (2008).

[16] J Schäfer, S Dippel, D E Wolf, *Force schemes in simulations of granular materials*, J. Phys. I (France) **6**, 5 (1996).

[17] H Hinrichsen, D Wolf, *The physics of granular media*, Willey-VCH, Weinheim (2004).

[18] J A Dijksman, M van Hecke, *The role of tap duration for the steady-state density of vibrated granular media*, Eur. Phys. Lett. **88**, 44001 (2009).

[19] F Ludewig, S Dorbolo, T Gilet, N Vandewalle, *Energetic approach for the characterization of taps in granular compaction*, Eur. Phys. Lett. **84**, 44001 (2008).

[20] P A Gago, N E Bueno, L A Pugnaloni, *High intensity tapping regime in a frustrated lattice gas model of granular compaction*, Granular Matter **11**, 365 (2009).

[21] Notice however, that this expansion parameter is not based on the properties of the mechanical excitation itself but on the sample response to it.

[22] J Damas et al., (unpublished)

[23] According to Callen [1], a system is at equilibrium if Thermodynamics holds for such state.

[24] Notice, that in Fig. 1(c) of our previous work [26], the maximum value of Γ reported was half of the value shown here. This discrepancy is due to a deficient filtering of the accelerometers used in Ref. [26]. The accelerations reported here have been corrected and checked against measurements done with a high speed camera.

[25] C M Carlevaro, L A Pugnaloni, *Steady state of tapped granular polygons*, J. Stat. Mech. P01007 (2011).

[26] L A Pugnaloni, I Sánchez, P A Gago, J Damas, I Zuriguel, D Maza, *Towards a relevant set of state variables to describe static granular packings*, Phys. Rev. E **82**, 050301(R) (2010).

[27] M Pica Ciamarra, A Coniglio, M Nicodemi, *Thermodynamics and statistical mechanics of dense granular media*, Phys. Rev. Lett. **97**, 158001 (2006).

[28] L. A. Pugnaloni, et al., (unpublished).

Permissions

All chapters in this book were first published in PIP, by Papers in Physics; hereby published with permission under the Creative Commons Attribution License or equivalent. Every chapter published in this book has been scrutinized by our experts. Their significance has been extensively debated. The topics covered herein carry significant findings which will fuel the growth of the discipline. They may even be implemented as practical applications or may be referred to as a beginning point for another development.

The contributors of this book come from diverse backgrounds, making this book a truly international effort. This book will bring forth new frontiers with its revolutionizing research information and detailed analysis of the nascent developments around the world.

We would like to thank all the contributing authors for lending their expertise to make the book truly unique. They have played a crucial role in the development of this book. Without their invaluable contributions this book wouldn't have been possible. They have made vital efforts to compile up to date information on the varied aspects of this subject to make this book a valuable addition to the collection of many professionals and students.

This book was conceptualized with the vision of imparting up-to-date information and advanced data in this field. To ensure the same, a matchless editorial board was set up. Every individual on the board went through rigorous rounds of assessment to prove their worth. After which they invested a large part of their time researching and compiling the most relevant data for our readers.

The editorial board has been involved in producing this book since its inception. They have spent rigorous hours researching and exploring the diverse topics which have resulted in the successful publishing of this book. They have passed on their knowledge of decades through this book. To expedite this challenging task, the publisher supported the team at every step. A small team of assistant editors was also appointed to further simplify the editing procedure and attain best results for the readers.

Apart from the editorial board, the designing team has also invested a significant amount of their time in understanding the subject and creating the most relevant covers. They scrutinized every image to scout for the most suitable representation of the subject and create an appropriate cover for the book.

The publishing team has been an ardent support to the editorial, designing and production team. Their endless efforts to recruit the best for this project, has resulted in the accomplishment of this book. They are a veteran in the field of academics and their pool of knowledge is as vast as their experience in printing. Their expertise and guidance has proved useful at every step. Their uncompromising quality standards have made this book an exceptional effort. Their encouragement from time to time has been an inspiration for everyone.

The publisher and the editorial board hope that this book will prove to be a valuable piece of knowledge for researchers, students, practitioners and scholars across the globe.

List of Contributors

Pablo M. Jais
Instituto de Física de Buenos Aires, CONICET, Argentina
Laboratorio de Electr_onica Cuántica, Departamento de Física, Facultad de Ciencias Exactas y Naturales, Universidad de Buenos Aires, Intendente Güiraldes 2160, Pabellón I - Ciudad Universitaria, Buenos Aires, C1428EHA, Argentina

Catalina von Bilderling
Instituto de Física de Buenos Aires, CONICET, Argentina
Centro de Microscopías Avanzadas, Facultad de Ciencias Exactas y Naturales, Universidad de Buenos Aires, Argentina

Andrea V. Bragas
Instituto de Física de Buenos Aires, CONICET, Argentina
Laboratorio de Electrónica Cuántica, Departamento de Física, Facultad de Ciencias Exactas y Naturales, Universidad de Buenos Aires, Intendente Güiraldes 2160, Pabellón I - Ciudad Universitaria, Buenos Aires, C1428EHA, Argentina

E. V. Bonzi
Facultad de Matemática, Astronomía y Física, Universidad Nacional de Córdoba. Ciudad Universitaria. 5000 Córdoba, Argentina
Instituto de Física Enrique Gaviola (CONICET), 5000 Córdoba, Argentina

G. B. Grad
Facultad de Matemática, Astronomía y Física, Universidad Nacional de Córdoba, Ciudad Universitaria, 5000 Córdoba, Argentina

R. A. Barrea
Physics Department, DePaul University, Chicago, IL 60614, USA

Rahul Banerjee and S. Kanjilal
St. Paul's Cathedral Mission College, 33/1, Raja Ram-mohan Roy, Sarani, 700 009 Kolkata, India

Iván M. Lengyel and Luis G. Morelli
Departamento de Física, FCEyN UBA and IFIBA, CONICET, Pabell_on 1, Ciudad Universitaria, 1428 Buenos Aires, Argentina

Daniele Soroldoni and Andrew C. Oates
MRC-National Institute for Medical Research, The Ridgeway, Mill Hill, NW7 1AA London, UK
Department of Cell and Developmental Biology, University College London, Gower Street, WC1E 6BT London, UK

T. J. Wilson
Department of Physics & Astronomy, University of Penn-sylvania, Philadelphia, PA 19104-6396, USA
Department of Physics, Illinois Wesleyan University, Bloomington, IL 61702-2900, USA

C. R. Pfeifer
Department of Physics & Astronomy, University of Penn-sylvania, Philadelphia, PA 19104-6396, USA
Department of Physics & Astronomy, Carleton College, Northfield, MN 55057, USA

N. Meysingier
Department of Physics & Astronomy, University of Penn-sylvania, Philadelphia, PA 19104-6396, USA
Strath Haven High School, Wallingford, PA 19086, USA

D. J. Durian
Department of Physics & Astronomy, University of Penn-sylvania, Philadelphia, PA 19104-6396, USA

Ezequiel Goldberg
Universidad Tecnológica Nacional - FRBA, UDB Física, Mozart 2300, C1407IVT Buenos Aires, Argentina

C. Manuel Carlevaro
Universidad Tecnológica Nacional - FRBA, UDB Física, Mozart 2300, C1407IVT Buenos Aires, Argentina
Consejo Nacional de Investigaciones Cientícas y Téecnicas, Argentina

Luis A. Pugnaloni
Consejo Nacional de Investigaciones Cientícas y Téecnicas, Argentina
Dpto. Ingeniería Mecáanica, Facultad Regional La Plata, Universidad Tecnológica Nacional, Av. 60 Esq. 124, 1900 La Plata, Argentina

Paula A. Gago
Dpto. Ingeniería Mecáanica, Facultad Regional La Plata,Universidad Tecnológica Nacional, Av. 60 Esq. 124, 1900 La Plata, Argentina
Department of Earth Science and Engineering, Imperial College London, South Kensington Campus, London SW7 2AZ, UK

Diego Maza
Departamento de Física y Matemtica Aplicada, Facultad de Ciencias, Universidad de Navarra, Navarra, Spain

Luis A. Pugnaloni
Dpto. Ingeniería Mecáanica, Facultad Regional La Plata, Universidad Tecnológica Nacional, Av. 60 Esq. 124, 1900 La Plata, Argentina Consejo Nacional de Investigaciones Cientícasy Técnicas, Argentina

Diego J. R. Sevilla
Departamento de Física y Química, Escuela de For-mación Básica. Facultad de Ciencias Exactas, Ingeniería y Agrimensura. Universidad Nacional de Rosario, Av. Pellegrini 250, S2000BTP Rosario, Argentina

Franco T. Lisandrini and Ariel O. Dobry
Instituto de Física Rosario (CONICET), Bv 27 de Febrero 210 bis, S2000EZP Rosario, Santa Fe, Argentina
Facultad de Ciencias Exactas, Ingeniería y Agrimensura, Universidad Nacional de Rosario, Argentina

Alejandro M. Lobos and Claudio J. Gazza
Instituto de Física Rosario (CONICET), Bv 27 de Febrero 210 bis, S2000EZP Rosario, Santa Fe, Argentina

Dayasindhu Dey, Debasmita Maiti and Manoranjan Kumar
S. N. Bose National Centre for Basic Sciences, Block JD,Sector III, Salt Lake, Kolkata 700106, India

D. R. Parisi
Instituto Tecnol_ogico de Buenos Aires, Lavard_en 389, C1437FBG Ciudad Aut_onoma de Buenos Aires, Argentina
Consejo Nacional de Investigaciones Científicas y Técnicas, Godoy Cruz 2290, 1425 Ciudad Autónoma de Buenos Aires, Argentina

G. A. Patterson
Instituto Tecnológico de Buenos Aires, Lavardén 389, C1437FBG Ciudad Autónoma de Buenos Aires, Argentina

Edgar A. Gómez
Universidad del Quindío, A.A. 2639, 630004 Armenia, Colombia

J. D. Hernández-Rivero
Departamento de Física, Universidade Federal de Minas Gerais, A. A. 486, 31270-901 Pampulha, Belo Horizonte, Minas Gerais, Brazil

Herbert Vinck-Posada
Universidad Nacional de Colombia, A. A. 055051, 111321 Bogotá, Colombia

D. H. Zanette
Centro Atómico Bariloche and Instituto Balseiro (Comisión Nacional de Energía At_omica, Universidad Nacional de Cuyo), Consejo Nacional de Investigaciones Cientícas y Técnicas, Avda. E. Bustillo 9500, 8400 San Carlos de Bariloche, Río Negro, Argentina

L. Martini, D. I. R. Boll and O. A. Fojón
Laboratorio de Colisiones At_omicas, Instituto de Física Rosario (CONICET-UNR), Blvd. 27 de Febrero 210 bis, Rosario 2000, Argentina

V. M. San Martin and A. Figliola
National Scientific and Technical Research Council (CONICET), Argentina

César Llamas and Jesús Vegas
Departamento de Informática, Universidad de Valladolid, 47011 Valladolid, Spain

Miguel Á. González
Departamento de Física de la Materia Condensada, Universidad de Valladolid, 47011 Valladolid, Spain

Manuel Á. González
Departamento de Física Aplicada, Universidad de Valladolid, 47011 Valladolid, Spain

G. C. Dima and G. B. Mindlin
Departamento de Física, Facultad de Ciencias Exactas y Naturales, Universidad de Buenos Aires and IFIBA (CONICET), Pabell_on 1, Cuidad Universitaria, 1428 Buenos Aires, Argentina

M. A. Goldin
Unit de Neurosciences Information et Complexit, Centre National de la Recherche Scientifique UPR-3293, 1 Avenue de la Terrasse, 91190 Gif-sur-Yvette, France

C. Dauphin
Institut Villebon-Georges Charpak, UniversitParis-Sud (Bât. 490) - Rue Hector Berlioz { 91400 Orsay, France
Département de Physique, University Paris-Sud, Universite Paris-Saclay, 91405 Orsay cedex, France

F. Bouquet
Laboratoire de Physique des Solides, CNRS, University Paris-Sud, Université Paris-Saclay, 91405 Orsay cedex, France

Ariel Zylberberg
Department of Vision and Cognition, Netherlands Institute for Neuroscience, an institute of the Royal Netherlands Academy of Arts and Sciences, Meibergdreef 47, 1105 BA Amsterdam, The Netherlands

Luciano Paz
These authors contributed equally to this work
Laboratory of Integrative Neuroscience, Physics Department, FCEyN UBA and IFIBA, CONICET; Pabellón 1, Ciudad Universitaria, 1428 Buenos Aires, Argentina

Pieter R. Roelfsema
Department of Vision and Cognition, Netherlands Institute for Neuroscience, an institute of the Royal Netherlands Academy of Arts and Sciences, Meibergdreef 47, 1105 BA Amsterdam, The Netherlands
Psychiatry Department, Academic Medical Center, Amsterdam, The Netherlands

Stanislas Dehaene
Collge de France, F-75005 Paris, France
University Paris-Sud, Cognitive Neuroimaging Unit, F-91191 Gif/Yvette, France

Mariano Sigman
Laboratory of Integrative Neuroscience, Physics Department, FCEyN UBA and IFIBA, CONICET; Pabell_on 1, Ciudad Universitaria, 1428 Buenos Aires, Argentina

P. Esquinazi
Division of Superconductivity and Magnetism, Institute for Experimental Physics II, Fakultät für Physik und Geowissenschaften, Universität Leipzig, Linnéstrasse 5, D-04103 Leipzig, Germany

Martín Caldarola and Martín E. Masip
Laboratorio de Electrónica Cuántica, Departamento de Física, Universidad de Buenos Aires, Pabellón I, Ciudad Universitaria, (C1428EHA) Buenos Aires, Argentina

Víctor A. Bettachini, Andrés A. Rieznik and Pablo G. König
Instituto Tecnológico de Buenos Aires, Eduardo Madero 399, (C1106ACD) Buenos Aires, Argentina

Diego F. Grosz
Instituto Tecnológico de Buenos Aires, Eduardo Madero 399, (C1106ACD) Buenos Aires, Argentina
Consejo Nacional de Investigaciones Cientficas y Técnicas, Argentina

Andrea V. Bragas
Laboratorio de Electrónica Cuántica, Departamento de Física, Universidad de Buenos Aires, Pabellón I, Ciudad Universitaria, (C1428EHA) Buenos Aires, Argentina
IFIBA, Consejo Nacional de Investigaciones Científicas y Técnicas, Argentina

Luis A. Pugnaloni
Instituto de Física de Líquidos y Sistemas Biológicos, (CONICET La Plata, UNLP), Calle 59 No 789, 1900 La Plata, Argentina

José Damas, Iker Zuriguel and Diego Maza
Departamento de Física y Matemática Aplicada, Facultad de Ciencias, Universidad de Navarra, Pamplona, Spain

Index

www.ingramcontent.com/pod-product-compliance
Lightning Source LLC
LaVergne TN
LVHW080654210926
848485LV00003B/376
9781647283568